Finite Element Method to Model Electromagnetic Systems in Low Frequency

Series Editor
Piotr Breitkopf

Finite Element Method to Model Electromagnetic Systems in Low Frequency

Francis Piriou
Stéphane Clénet

iSTE

WILEY

First published 2023 in Great Britain and the United States by ISTE Ltd and John Wiley & Sons, Inc.

ISTE Ltd
27-37 St George's Road
London SW19 4EU
UK

www.iste.co.uk

John Wiley & Sons, Inc.
111 River Street
Hoboken, NJ 07030
USA

www.wiley.com

Library of Congress Control Number: 2023943094

British Library Cataloguing-in-Publication Data
A CIP record for this book is available from the British Library
ISBN 978-1-78630-811-5

Contents

Introduction

As calculation tools are reaching increasingly high performances, numerical modeling has developed significantly in all sectors of society. It can be used to predict the evolution of a given structure or device starting from an initial state, study physical phenomena by accessing quantities that are not measurable or develop virtual prototypes in order to improve a design process. Applied physics, and in particular low-frequency electromagnetism, which is the subject of this book, are no exceptions. Nowadays, high-performance simulation software is available for students, engineers and researchers. A prerequisite for making the best use of a tool, even in the field of computation, is obviously a good knowledge of its foundations and principles. In this context, it seemed interesting to propose a book that may grasp, under the best conditions, the path leading to building these numerical models.

The modeling of electromagnetic phenomena relies on two partial differential equations, known as Maxwell's equations:

$$\mathbf{curl\ E} = -\,\partial \mathbf{B} / \partial t$$

$$\mathbf{curl\ H} = \mathbf{J} + \partial \mathbf{D} / \partial t$$

These two equations should be completed by behavior laws that describe the reaction of media to electromagnetic fields, which are associated with physical phenomena such as dielectric polarization, electric conduction and ferromagnetism. Finally, for a proper formulation of the problem, boundary conditions should be added, for either a finite or infinite studied domain. Although it may appear simple, this problem, composed of several equations, has no analytical solution, except for the case of elementary geometry, with linear behavior laws.

As the exact solution to the problem is not available, there are two possibilities for reaching an approximation of this solution:

– Formulate hypotheses on the geometry, the behavior laws of the materials and the spatial distribution of electromagnetic fields. The objective is to make the analytical solution possible. This approach requires the "model builder" to have very deep, expert knowledge on the studied system, to be able to formulate the "right hypotheses". If the latter are not valid, there is a very high risk of reaching a low-quality solution, which is very far from the exact solution of the initial problem. Moreover, this approach is not always possible if complex phenomena, such as nonlinearities, are predominant.

– Or, reformulate the initial problem in discrete form, leading to a system of differential algebraic equations. An approximation of the exact solution is then obtained at the cost of a significant amount of computation, which can be readily processed by the computers that are available nowadays. This reformulation, requiring few or almost no hypotheses, is obtained by implementing numerical methods. In the field of electromagnetism, the most widespread such method is the finite element method.

This book focuses on the second approach, often referred to as "computational electromagnetics", providing a detailed description of the implementation of the finite element method in low-frequency electromagnetism. Our purpose is to explain the process starting from equations verified by electromagnetic fields in the continuous domain, in order to arrive at a system of equations that will be solved using a computer. This process, often called "discretization", will be conducted with a permanent concern for maintaining a link between physics, i.e. the properties of electromagnetic fields, and numerical analysis, through the finite element method.

Furthermore, this book is mainly addressed to students, engineers and researchers in the field of electrical engineering. They will be able to better understand the intricate details of (open-source or commercial) software that models the behavior of electromagnetic fields. They will thus have the possibility of better using these tools and therefore have a good knowledge of their limits. This book is also addressed to students, engineers and researchers in the field of numerical analysis who are interested in better understanding the links between numerical methods and physics in the field of electromagnetism.

Even though this book offers few pieces of information on numerical implementation, it provides all the elements required for understanding the theoretical foundations. It also allows us to conceive the link between physics and numerical methods and therefore between the applications and the software used.

The above-stated Maxwell's equations allow for the study of all electromagnetic phenomena. For certain low-frequency applications it is, however, possible to derive them in a "static" or "quasi-static" state. Under certain hypotheses, these simpler problems lead to solutions that are equal or very close to those that would have been obtained using the full Maxwell equation system. After discretization using a numerical method, they can be used to obtain smaller size systems of equations that are easier to solve due to their mathematical properties.

Approximations by problems under a static or quasi-static state are widely used in many domains such as power grids, electrical machines, power electronics and non-destructive testing. This book focuses in particular on three static problems, namely electrostatics, electrokinetics (when electric charges travel at constant speed, the fields do not depend on time) and magnetostatics. In the quasi-static state, Maxwell's equations can be written in the magnetoquasistatic form (more often referred to as "magnetodynamics") or in the electroquasistatic form. In this quasi-static case, our focus will be on magnetodynamics. On the contrary, electroquasistatic problems will not be considered, but the developments remain similar to those used in the case of magnetodynamics.

This book has four chapters, each corresponding to a stage of the process leading to the discretization of Maxwell's equations.

The objective of Chapter 1 is to formulate various problems in the static state and the magnetodynamic state, and then to solve them. For each problem, the equilibrium equations are written, as well as the behavior laws and the boundary conditions on the electromagnetic fields. A review of the properties of these fields also highlights their behavior at the interface between two media, and the nature of their integral forms. A key point of this chapter is the definition of electric and magnetic quantities, referred to as "source terms", which are at the origin of the creation of electromagnetic fields. These terms can be located inside the studied domain (electric charges, inductors, permanent magnets) or imposed on the boundary of the domain (electromotive or magnetomotive forces, current density or magnetic flux).

Chapter 2 is dedicated to the introduction of functional spaces associated with vector operators: gradient, curl and divergence. As these operators are used when writing the equations of static and quasi-static problems, they can be used to define the functional spaces to which various electromagnetic fields belong. An analysis is conducted on the properties of functional spaces and in particular on the images and kernels of the vector operators in relation to the topology of the studied domain. These properties lead quite naturally to the notion of scalar and vector potentials, widely used as intermediary for solving static and quasi-static problems, which will be introduced in Chapter 3. The notion of gauge is also presented, which imposes the uniqueness of a field when defining by a single vector operator. Gauge conditions will therefore be very useful to impose the uniqueness of potentials so that the problem is properly posed. They are also used in the construction of source terms.

Chapter 3 focuses on the potential-based formulations for static and magnetodynamic problems. In the case of static problems, the introduction of these potentials allows for the reduction of the number of unknowns, passing from two unknown fields to only one unknown potential. This potential can be a scalar or a vector quantity. For each problem, two formulations in terms of potential referred to as "scalar" or "vector" are obtained. In magnetodynamics, two potentials are used in close relation to those introduced for static problems. Two formulations known as "electric" and "magnetic" are then deduced.

These potentials are not necessarily introduced in a direct manner, requiring instead a reformulation of the source terms of the initial problem, located either inside or on the boundary of the domain. The first part of this chapter is dedicated to this reformulation. The number of sources is often limited, facilitating a focus on the essential, which is a systematic method for imposing source terms. However, as shown by the examples presented, the methodology is readily applicable to problems with a greater number of sources, using the superposition theorem (even though the behavior laws are not linear).

Chapter 4 is dedicated to the discretization of formulations of static and magnetodynamic problems. Successful completion of this discretization requires first of all finding the proper spaces of approximation within which the approximate solutions will be sought. These spaces must have a finite dimension for implementation on a computer. In the case of the finite element method, the spaces of approximation are defined from a mesh, which is obtained by splitting the studied domain into elements of simple shapes (tetrahedron, hexahedron, prism, etc.). A field is then perfectly defined by a vector, whose entries are the coefficients of the basis of the approximation space. The entries of this vector are then the degrees of freedom to be determined. These spaces of finite dimension must be included in the functional spaces to which electromagnetic fields belong. This means they must meet the properties introduced in Chapter 2. This condition leads to physically acceptable field approximations in the sense that they verify the continuity conditions. Whitney finite elements are currently the most commonly used, and they generate spaces that allow for a real physical interpretation of the degrees of freedom, which are then fluxes, circulations and densities. Moreover, imposing gauge conditions is natural, as well as the calculation of source terms. It is important to note that, in this book, our developments are limited to first-order finite elements for a tetrahedron-based mesh. Very similar approaches can be applied with higher-order functions and elements of other shapes (hexahedron, prism, pyramid, etc.). The introduction of discrete forms of fields in the potential-based formulations of static and magnetodynamic problems does not allow us to directly build a system of equations. Our objective is to use the weighted residual method, allowing for the transformation of the initial problem, based on local equations, into a problem based on integral equations. For each of the potential-based formulations developed in Chapter 3, the weighted residual method is used in association with Whitney finite elements in order to build a system of equations to be solved, which is then the numerical model we seek.

1

Equations of Electromagnetism

1.1. Maxwell's equations

Maxwell's equations can be written in a general form as follows (Stratton 1941; Ida 2020):

$$\mathbf{curlE} = -\frac{\partial \mathbf{B}}{\partial t} \quad [1.1]$$

$$\mathbf{curlH} = \mathbf{J} + \frac{\partial \mathbf{D}}{\partial t} \quad [1.2]$$

$$\text{div}\mathbf{B} = 0 \quad [1.3]$$

$$\text{div}\mathbf{D} = \rho \quad [1.4]$$

The vector fields denoted by **E**, **B**, **H** and **D** represent, respectively, the electric field, the magnetic flux density, the magnetic field and the electric displacement field. Electric current density **J** and electric charge density ρ can be considered source terms.

Finally, it is common to adopt the quasi-static approximation for electromagnetic devices operating at industrial frequencies. In this case, the term $\partial \mathbf{D}/\partial t$ in equation [1.2] can be considered negligible. Equation [1.2] can then be written as:

$$\mathbf{curlH} = \mathbf{J} \quad [1.5]$$

If the divergence operator is now applied to equation [1.5], considering the properties of vector operators ($\text{div}\,\mathbf{curl}\,\mathbf{H} = 0$), the following property can be deduced for the electric current density:

$$\text{div}\mathbf{J} = 0 \tag{1.6}$$

1.2. Behavior laws of materials

Maxwell's equations, as presented above, are independent of the media. But electric fields (**E**, **J**, **D**) and magnetic fields (**H**, **B**) are related by behavior laws.

1.2.1. *General case*

It can be noted that in vacuum these behavior laws are linear and have the following form:

$$\mathbf{D} = \varepsilon_0 \mathbf{E} \tag{1.7}$$

$$\mathbf{B} = \mu_0 \mathbf{H} \tag{1.8}$$

where ε_0 represents the electric permittivity constant ($\varepsilon_0 = 10^{-9}/36\pi$ F/m) and μ_0 is the magnetic permeability constant ($\mu_0 = 4\pi 10^{-7}$ H/m). They are linked by the classical relation: $\varepsilon_0 \mu_0 c^2 = 1$, where c represents the speed of light in vacuum.

On the contrary, electromagnetic fields in media interact with their environment. These interactions also depend on fields of different physical natures, such as temperature T or mechanical stress $\boldsymbol{\sigma}_m$. In this case, the behavior laws become significantly more complex, and the following relations can be written:

$$\mathbf{D} = f(\mathbf{E}, \mathrm{T}, \boldsymbol{\sigma}_m, \ldots) \tag{1.9}$$

$$\mathbf{B} = g(\mathbf{H}, \mathrm{T}, \boldsymbol{\sigma}_m, \ldots) \tag{1.10}$$

$$\mathbf{J} = h(\mathbf{E}, \mathrm{T}, \boldsymbol{\sigma}_m, \ldots) \tag{1.11}$$

These various functions may depend on the history of the material. As an example, in the case of ferromagnetic materials, the value of the field **B** at the time t depends not only on the value of the field **H** at time t, but also on its previous values in the time interval [0, t]. This phenomenon is known as magnetic hysteresis. Some materials, used for their electric properties linking fields **D** and **E**, also exhibit this hysteresis phenomenon. They are referred to as ferroelectric materials.

Behavior laws are often at the origin of the links between various physics domains. This is the case with coupling the equations of electromagnetism with the equations of thermodynamics and mechanics. For example, electric current density **J**, for a given value of the electric field **E**, decreases as a function of temperature due to thermal energy, which tends to reduce the conductivity of the material. As a first approximation, conductivity σ(T) can be written as a function of temperature:

$$\sigma(\mathrm{T}) = \frac{\sigma_{\mathrm{ref}}}{1+\alpha\Delta \mathrm{T}} \quad [1.12]$$

In this expression, $\Delta\mathrm{T} = (\mathrm{T} - \mathrm{T}_{\mathrm{ref}})$ with $\mathrm{T} > \mathrm{T}_{\mathrm{ref}}$, σ_{ref} is the conductivity at temperature $\mathrm{T}_{\mathrm{ref}}$ and α is a temperature coefficient of the considered material also dependent on $\mathrm{T}_{\mathrm{ref}}$. However, in many applications, multiphysics couplings can be considered negligible in a first approach. In this case, the behavior laws can be written in a simplified form:

$$\mathbf{D} = \mathbf{f}(\mathbf{E}) \quad [1.13]$$

$$\mathbf{B} = \mathbf{g}(\mathbf{H}) \quad [1.14]$$

$$\mathbf{J} = \mathbf{h}(\mathbf{E}) \quad [1.15]$$

1.2.2. *Simplified forms*

Even in the form presented in equations [1.13], [1.14] and [1.15], the behavior laws in material media may be relatively complex (anisotropy, hysteresis). It is nevertheless often possible to simplify them without affecting the precision of the results, which will be shown in the following section.

1.2.2.1. *Dielectric materials*

Consider the behavior law of dielectric materials written using relation [1.13]. Introducing the electric polarization vector $\mathcal{P}_e$, which depends on the electric field **E**, the following relation can be written (Ida 2020):

$$\mathbf{D} = \varepsilon_0\mathbf{E} + \mathcal{P}_e(\mathbf{E}) \quad [1.16]$$

In the case of ferroelectricity, polarization $\mathcal{P}_e$ follows a hysteresis loop when the electric field varies as a function of time. However, the hypothesis of isotropy and linearity is acceptable for many dielectric materials. Polarization can therefore be considered to be directly proportional to the electric field strength. It can then be written as follows:

$$\mathcal{P}_e = \varepsilon_0 \chi_e \mathbf{E} \tag{1.17}$$

where χ_e represents the electric susceptibility of the material. Grouping equations [1.16] and [1.17] leads to:

$$\mathbf{D} = \varepsilon_0 (1 + \chi_e)\mathbf{E} = \varepsilon_0 \varepsilon_r \mathbf{E} \tag{1.18}$$

This expression involves ε_r, a dimensionless number, which represents the relative permittivity of the considered material. Introducing permittivity $\varepsilon = \varepsilon_0\varepsilon_r$, the commonly used behavior law of dielectric materials is obtained:

$$\mathbf{D} = \varepsilon \mathbf{E} \tag{1.19}$$

1.2.2.2. *Conductive materials*

For conductive materials, assuming thermal effects are negligible, electric current density is proportional to the electric field. The electrical behavior law is then expressed as:

$$\mathbf{J} = \sigma \mathbf{E} \tag{1.20}$$

where σ is the electrical conductivity.

1.2.2.3. *Magnetic properties of materials*

The magnetic properties of materials can be expressed using relation [1.14]. Similar to the dielectric materials, a magnetic polarization vector, denoted by $\mathcal{P}_m$, can be introduced (Bozorth 1993; Benabou 2002). Using this vector, it is possible to express the magnetic flux density in the following form:

$$\mathbf{B} = \mu_0 \mathbf{H} + \mathcal{P}_m(\mathbf{H}) \tag{1.21}$$

In this expression, the magnetization vector **M** can also be introduced, posing $\mathcal{P}_m = \mu_0 \mathbf{M}(\mathbf{H})$. Equation [1.21] can then be written as:

$$\mathbf{B} = \mu_0 (\mathbf{H} + \mathbf{M}(\mathbf{H})) \tag{1.22}$$

Equations [1.21] and [1.22] contain a first term corresponding to the magnetic flux density created in vacuum ($\mu_0\mathbf{H}$) and a second term, respectively, $\mathcal{P}_m(\mathbf{H})$ or $\mu_0\mathbf{M}(\mathbf{H})$, corresponding to the response of the material medium to the external magnetic field.

There are various behaviors leading to the following classification:

– diamagnetic materials that, when subjected to a magnetic field, create a magnetization that opposes the external field. In this case, equation [1.22] can be written as:

$$\mathbf{B}=\mu_0(\mathbf{H}+\chi_m\mathbf{H})=\mu_0(1+\chi_m)\mathbf{H} \qquad [1.23]$$

where χ_m represents the magnetic susceptibility with $\chi_m < 0$ and is of the order of -10^{-5}. It can be noted that this reaction magnetization is very weak for most materials used in electrical engineering. As an example, the magnetic susceptibility of copper is equal to $-1.18\ 10^{-5}$;

– Paramagnetic materials that, when subjected to a magnetic field, create a very weak magnetization in the same direction as the external field. In this case, the expression [1.23] is unchanged, but $\chi_m>0$ may range between 10^{-3} and 10^{-5}. As an example, magnetic susceptibility of molybdenum is equal to $1.05\ 10^{-4}$.

– Ferromagnetic materials that, when subjected to a magnetic field of several hundred amperes per meter, may create a magnetization $\mathbf{M}(\mathbf{H})$ of the order of 10^6 A/m. Moreover, in the absence of an external field, they may present a remanent magnetization. Considering their exceptional magnetic properties, these materials are understandably used in the field of conversion of electromagnetic energy.

For industrial frequencies, with the exception of ferromagnetic materials, it is commonly accepted that the vacuum behavior law (see relation [1.8]) is perfectly suited for the modeling of the magnetic behavior law of material media.

1.2.2.4. *Ferromagnetic materials*

As shown in Figure 1.1, for ferromagnetic materials, magnetization describes a hysteresis loop when the magnetic field varies alternatively. In this figure, M_r represents the remanent magnetization, M_s represents the saturation magnetization and H_c represents the coercive field. It can be readily shown that the energy dissipated as losses, during a loop, is equal to the loop area. It can also be noted that magnetization $\mathbf{M}(\mathbf{H})$ varies with temperature, and beyond the Curie temperature the material exhibits paramagnetic behavior. Finally, ferromagnetic materials often have anisotropic magnetic properties, which means that their behavior varies according to the direction of the applied magnetic field.

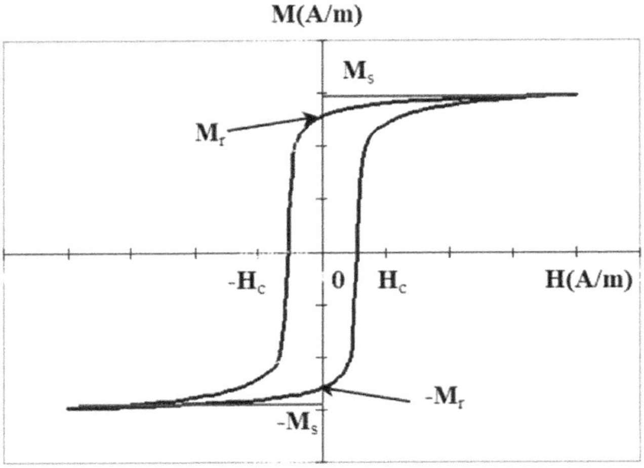

Figure 1.1. *M(H) characteristic of a ferromagnetic material*

Depending on the value of the coercive field intensity H_c, two families of magnetic materials can be identified:

– Soft magnetic materials, for which H_c is below several hundred amperes per meter. They are mainly used for concentrating and driving the circulation of the magnetic field. The material then behaves as a "good magnetic conductor".

– Hard magnetic materials, for which the coercive field intensity H_c is very high. These materials are used as permanent magnets.

1.2.2.4.1. Soft magnetic materials

In order to represent soft ferromagnetic materials, it is possible, for certain applications, to ignore the anisotropy and also the hysteresis phenomenon. This new behavior law is then deduced from the **M(H)** characteristic, shown in Figure 1.1, by considering the anhysteretic curve (one-to-one curve experimentally measured according to a standard). In this case, magnetic flux density can be written as:

$$\mathbf{B} = \mu_0(\mathbf{H} + \mathbf{M}(\mathbf{H})) = \mu_0(\mathbf{H} + \chi_m(\mathbf{H})\mathbf{H}) = \mu_0\mathbf{H}(1 + \chi_m(\mathbf{H})) \qquad [1.24]$$

or by introducing a nonlinear magnetic permeability $\mu(\mathbf{H}) = \mu_0(1 + \chi_m(\mathbf{H}))$:

$$\mathbf{B} = \mu(\mathbf{H})\mathbf{H} \qquad [1.25]$$

Finally, it is also possible to take into account only the linear part of the magnetic characteristic. This yields:

$$\mathbf{B}=\mu\mathbf{H} \quad [1.26]$$

1.2.2.4.2. Permanent magnets

As for the hard magnetic materials, used as permanent magnets, they are of various types. For example, we can mention iron–nickel–aluminum–cobalt (Alnico) alloys, rare-earth-based alloys (samarium-cobalt (SmCo) and neodymium–iron–boron (Nd–Fe–B)) and ferrites.

Figure 1.2(a) shows the useful characteristics (part of the hysteresis loop that is exploited when the material "operates" as a permanent magnet) of the materials commonly used as permanent magnets.

This figure shows that except for the case of Alnico-type alloys, the useful characteristic of permanent magnets can be represented by a linear characteristic, as shown in Figure 1.2(b). This is:

$$\mathbf{B}=\mu_A\mathbf{H}+\mathbf{B}_r \quad [1.27]$$

where μ_A represents the magnetic permeability of the permanent magnet (close to μ_0) and $\mathbf{B}_r$ the remanent flux density.

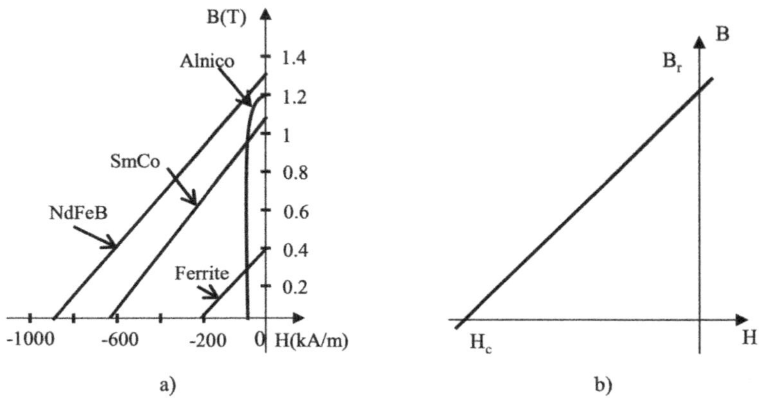

Figure 1.2. *a) Characteristics of the most common hard ferromagnetic materials used as permanent magnets; b) simplified representation*

1.3. Interface between two media and boundary conditions

Before studying the behavior of materials at the interface between two media with different properties, it is important to recall the definition of a vector field based on its normal and tangential components at a point on a surface. Considering a vector, denoted by **u** (see Figure 1.3), at a point M of a surface Γ, and **n** the normal vector to the surface at this point, it can be decomposed into its normal component $\mathbf{u}_n$ and tangential component $\mathbf{u}_t$ as follows:

$$\mathbf{u} = \mathbf{u}_n + \mathbf{u}_t \qquad [1.28]$$

where $\mathbf{u}_n = (\mathbf{n}.\mathbf{u})\mathbf{n}$ is the normal component, and $\mathbf{u}_t = \mathbf{n} \wedge (\mathbf{u} \wedge \mathbf{n})$ is the tangential component (the operator "." denotes the scalar product and "∧" denotes the vector product).

NOTE.– For the sake of simplicity, when conditions are imposed on the tangential component, it is preferable to use the term $\mathbf{u} \wedge \mathbf{n}$. Indeed, it can be verified that if $\mathbf{u} \wedge \mathbf{n} = 0$, then $\mathbf{u}_t = 0$. Similarly, if two vectors $\mathbf{u}_1$ and $\mathbf{u}_2$ have equal tangential components, this is equivalent to having $\mathbf{u}_1 \wedge \mathbf{n} = \mathbf{u}_2 \wedge \mathbf{n}$.

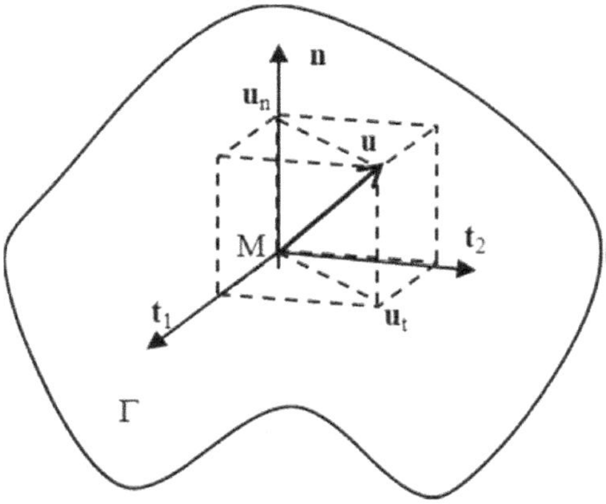

Figure 1.3. *Definition of a vector field based on its normal and tangential components*

Having set these definitions, the next section describes how they are used to define the continuity conditions between two media and the boundary conditions.

1.3.1. *Continuity conditions between two media*

The five already defined vector fields, **E**, **H**, **B**, **D** and **J**, have certain properties when passing from one medium to another (Ida 2020). These properties are derived from equations [1.1], [1.3], [1.4], [1.5] and [1.6].

1.3.1.1. *Electric and magnetic fields*

It can be shown that if the curl of a field is defined, then the tangential component of this field is continuous at the interface between two media that may have different characteristics. Since the curl of the electric field is defined by equation [1.1], it can be deduced that:

$$\mathbf{E}_{1t} = \mathbf{E}_{2t} \qquad [1.29]$$

where $\mathbf{E}_{kt}$ ($k \in \{1,2\}$) represents the component of the electric field tangential to the interface. This result shows that at the interface between two media with different properties, the tangential component of the electric field is conserved.

For the magnetic field, as defined by equation [1.5], similar to the case of the electric field, the following can be written:

$$\mathbf{H}_{1t} = \mathbf{H}_{2t} \qquad [1.30]$$

This relation shows that at the interface between two media the tangential component of the magnetic field is conserved.

1.3.1.2. *Electric displacement field, magnetic flux density and current density*

Likewise, it can be shown that if the divergence of a field is defined, then its normal component is continuous at the interface between two media with different physical properties.

The divergence of the fields **D**, **B** and **J** is defined, and this property is then applied. For the electric displacement field, based on equation [1.4] and in the absence of surface charge density, the following expression can be written:

$$\mathbf{D}_{1n} = \mathbf{D}_{2n} \qquad [1.31]$$

where $\mathbf{D}_{kn}$ ($k \in \{1,2\}$) represents the normal component of the electric displacement field on the interface.

Concerning the magnetic flux density, based on equation [1.3], the following expression can be written:

$$\mathbf{B}_{1n} = \mathbf{B}_{2n} \tag{1.32}$$

where $\mathbf{B}_{kn}$ ($k \in \{1,2\}$) represents the normal component of the magnetic flux density. Therefore, at the interface between two media, the normal component of the magnetic flux density is conserved.

As already noted for the electric displacement field and the magnetic flux density, equation [1.6] leads to the relation:

$$\mathbf{J}_{1n} = \mathbf{J}_{2n} \tag{1.33}$$

therefore at the interface between two media the normal component of the current density is conserved.

1.3.1.3. *Refraction of field lines*

In order to alleviate the developments, this section considers a two-dimensional (2D) case, limited to the pair of fields composed of the magnetic field and the magnetic flux density {**H**,**B**}. The conclusions that will be drawn are, however, valid for the three-dimensional (3D) case and can be extended to the case of the {**E**,**D**} and {**E**,**J**} pairs that verify the same conservation conditions as the {**H**,**B**} pair.

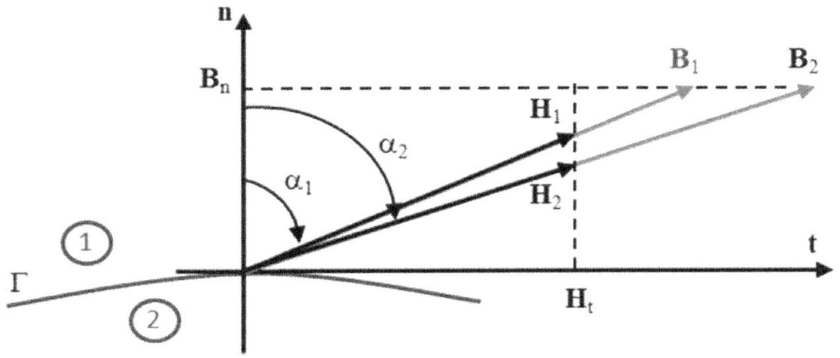

Figure 1.4. *Normal and tangential components of fields B_1, H_1, B_2 and H_2*

As shown in Figure 1.4, let us consider an interface Γ between two magnetic materials denoted by 1 and 2. The behavior law of the materials, assumed to be isotropic and linear, is given by relation [1.26] with permeabilities μ_1 and μ_2. Under these conditions, the magnetic fields $\mathbf{H}_1$ and $\mathbf{H}_2$ are, respectively, collinear with magnetic flux densities $\mathbf{B}_1$ and $\mathbf{B}_2$. Finally (see Figure 1.4), the projection of the fields onto the two axes "**n**" and "**t**", corresponding to the normal and tangential components, meets the properties expressed by [1.30] and [1.32].

α_1 and α_2 are the angles made with the normal **n** directed from medium 2 to medium 1 by the two pairs of fields $\{\mathbf{H}_1,\mathbf{B}_1\}$ and $\{\mathbf{H}_2,\mathbf{B}_2\}$, respectively.

An elementary calculation, based on the continuity properties of the normal and tangential components [1.30] and [1.32] and on the behavior law [1.26], leads to the relation:

$$\mathrm{tg}\alpha_2 = \mathcal{K}_r \mathrm{tg}\alpha_1 \qquad [1.34]$$

where $\mathcal{K}_r$ represents a refractive index such that $\mathcal{K}_r = \mu_2 / \mu_1$.

Let us now consider the following case: if $\mu_1 \rightarrow \infty$, then $\mathcal{K}_r \rightarrow 0$, $\mathrm{tg}\alpha_2 \approx 0$ and $\alpha_2 = 0$. This implies that the pair $\{\mathbf{H}_1,\mathbf{B}_1\}$ is normal to the surface and therefore $\mathbf{H}_t = 0$. On the contrary, if $\mu_1 \rightarrow 0$, then $\mathcal{K}_r \rightarrow \infty$, $\mathrm{tg}\alpha_2 \rightarrow \infty$ and $\alpha_2 = \pi / 2$. In this case, the pair $\{\mathbf{H}_1,\mathbf{B}_1\}$ is tangential to the surface and the component $\mathbf{B}_n = 0$.

When considering the pairs $\{\mathbf{E}_1,\mathbf{D}_1\}$ and $\{\mathbf{E}_2,\mathbf{D}_2\}$ with the behavior law [1.19] or the pairs $\{\mathbf{E}_1,\mathbf{J}_1\}$ and $\{\mathbf{E}_2,\mathbf{J}_2\}$ with the behavior law [1.20], the same conclusions are reached. Under these conditions, the refractive index $\mathcal{K}_r$ is equal to $\varepsilon_2 / \varepsilon_1$ and σ_2 / σ_1, respectively.

As an example, let us consider the case of a conductive material whose conductivity has a finite value σ_2. If it is brought into contact with another conductive material, whose conductivity σ_1 tends to infinity, then $\mathcal{K}_r \rightarrow 0$ and the tangential component of the electric field strength **E** and of the current density **J** at the interface is equal to zero. In this case, the interface can be considered a gate for the current density.

On the contrary, if the conductive material, of conductivity σ_2, is in contact with an insulating material, whose conductivity is $\sigma_1 = 0$, then $\mathcal{K}_r \rightarrow \infty$ and the normal component (of **E** and **J**) to the interface between the two media is equal to zero. This interface will be considered a wall.

These considerations will be very useful in the following section, particularly when boundary conditions imposed at the boundary of a domain are imposed.

1.3.2. *Boundary conditions*

For the study of electromagnetic systems, a well-posed formulation of the problem requires imposing spatial boundary conditions to the fields. For an infinite domain, these conditions are applied to infinity. In the case of numerical simulation, the domain is often limited to a part of the space. In this case, boundary conditions should be imposed at the boundaries of the domain. These boundary conditions may be derived either from symmetry conditions of the problem or from properties of the materials that are in contact with the boundary (see section 1.3.1). For example, if the boundary is in contact with a highly insulating material, then the normal component of the current density is imposed to zero. To have a physical meaning, these conditions always relate to the conservative (normal or tangential) component of the concerned field. Therefore, if a condition applies to the magnetic field, it concerns the tangential component. On the contrary, in the case of magnetic flux density, it relates to the normal component.

However, in the context of problems evolving in time, a generally imposed condition is that the value of fields at the initial instant t = 0 is equal to zero.

Taking into account the notations introduced after equation [1.28] for the normal and tangential components, the boundary conditions on the fields **E**, **H**, **D**, **B** and **J** can be written, for a large number of applications, as follows:

$$\mathbf{n} \wedge \mathbf{E}\big|_{\Gamma_e} = 0 \qquad [1.35]$$

$$\mathbf{n} \wedge \mathbf{H}\big|_{\Gamma_h} = 0 \qquad [1.36]$$

$$\mathbf{B.n}\big|_{\Gamma_b} = 0 \qquad [1.37]$$

$$\mathbf{J.n}\big|_{\Gamma_j} = 0 \qquad [1.38]$$

$$\mathbf{D.n}\big|_{\Gamma_d} = 0 \qquad [1.39]$$

These conditions, known as "homogeneous boundary conditions", can be interpreted as follows:

– Equation [1.35] indicates that the tangential component of the electric field is equal to zero on the boundary Γ_e and therefore the electric field **E** is normal to this surface. Using the expression introduced at the end of section 1.3.1.3, this boundary can be considered a gate for the field **E**. These are gate-type boundary conditions.

– Equation [1.36] leads to the same interpretations for the magnetic field intensity **H** on the boundary Γ_h. It can also be considered a gate for the magnetic field.

– Equation [1.37] indicates that the normal component of the magnetic flux density is equal to zero on the boundary Γ_b. As already seen in section 1.3.1.3, this condition requires the boundary to behave as a wall for the magnetic flux density **B**. These are referred to as wall-type boundary conditions.

– Equation [1.38] leads to the same interpretations for the current density, i.e. Γ_j behaves as a wall for the current density **J**.

– Equation [1.39], similar to equations [1.37] and [1.38], shows that the boundary Γ_d behaves as a wall for the electric displacement field **D**.

It can be shown that relations [1.1] and [1.35] imply equation [1.37]. Similarly, relations [1.5] and [1.36] imply equation [1.38]. On the contrary, the reverse is not true and depends on the topology of the domain (see Chapter 2).

1.4. Integral forms: fundamental theorems

The above-stated Maxwell's equations provide local information on electromagnetic fields. The integral form of these equations leads to general theorems that are commonly used in electromagnetism. These theorems can be used to connect local quantities (vector fields) and global quantities such as the electromotive force "e", the current density flux "I", the magnetic flux "ϕ", the magnetomotive force "f_m" and the total charges "Q".

1.4.1. *Faraday's law*

Given a rigid loop C_S, the boundary of a surface denoted by S (see Figure 1.5), consider equation [1.1] that connects the electric field and the magnetic flux density. Integrating the equation over the surface S yields:

$$\iint_S \mathbf{curlE}.\mathbf{n}\mathrm{dS} = -\iint_S \frac{\partial \mathbf{B}}{\partial t}.\mathbf{n}\,\mathrm{dS} \qquad [1.40]$$

Using the Stokes theorem and inverting the operator differentiated with respect to time with the surface integral (which is possible, as the loop is assumed rigid), the following can be written:

$$\oint_{C_S} \mathbf{E}.\mathbf{dl} = -\frac{\mathrm{d}}{\mathrm{dt}}\iint_S \mathbf{B}.\mathbf{n}\,\mathrm{dS} \qquad [1.41]$$

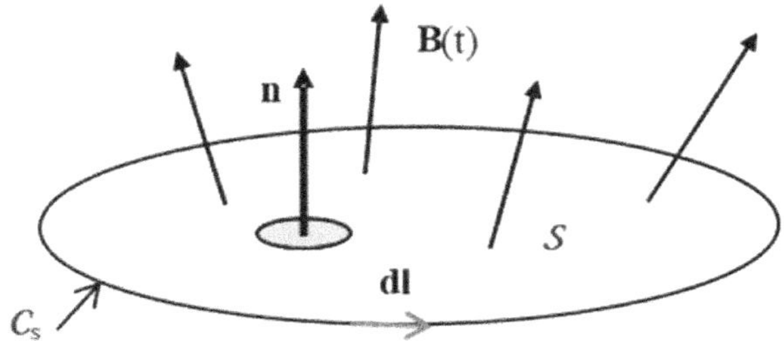

Figure 1.5. *Faraday's law implementation example*

The left-hand side term of this equation corresponds to the electromotive force "e" induced in the loop, and the right-hand side term corresponds to the time derivative of the magnetic flux through the surface S (denoted by ϕ). This leads to Faraday's law:

$$e = -\frac{d\phi}{dt} \quad [1.42]$$

1.4.2. *Ampère's law*

As shown in Figure 1.6, consider a conductor carrying a current density **J** and a surface S bounded by a contour C_S. Based on equation [1.5], using the same approach as for Faraday's law, the following can be written:

$$\iint_S \mathbf{curlH.n}dS = \iint_S \mathbf{J.n}dS \quad [1.43]$$

Using the Stokes theorem, the term on the left-hand side is replaced by the circulation of the magnetic field along the contour C_S. The term on the right-hand side, which represents the flux of **J**, is therefore equal to the value of the electric current (denoted by "I") flowing through the surface S. This relation leads to Ampère's law, namely:

$$\oint_{C_s} \mathbf{H.dl} = I \quad [1.44]$$

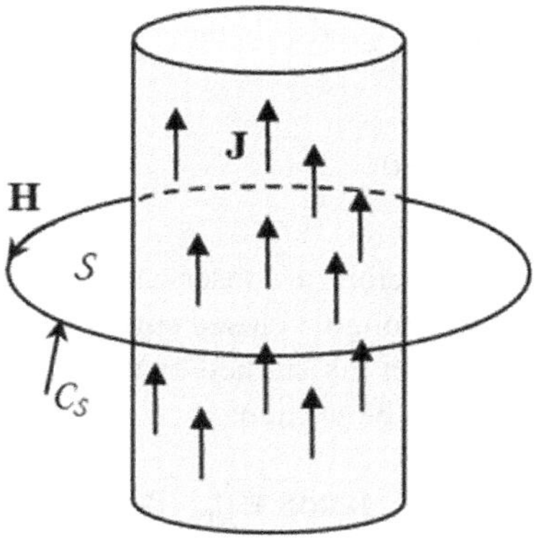

Figure 1.6. *Illustration of Ampère's law: conductor carrying a current*

1.4.3. *Law of conservation of the magnetic flux*

Equation [1.3] provides information related to the behavior of the magnetic flux density, i.e. it is divergence free. In order to analyze this property, consider the case of the domain Ω, of boundary $\Gamma = \Gamma_{h1} \cup \Gamma_b \cup \Gamma_{h2}$, defined in Figure 1.7. A magnetic flux density **B** flows through this domain. The boundary condition on the boundaries Γ_{h1} and Γ_{h2} is [1.36], and on the lateral boundary Γ_b it is [1.37]. This is known as the flux tube.

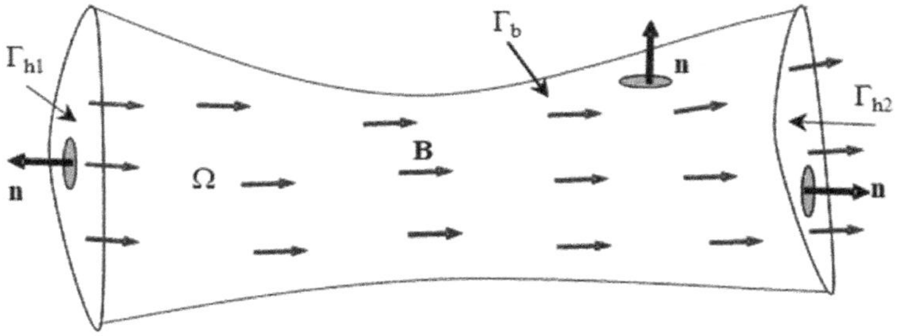

Figure 1.7. *Flux tube: law of conservation of the flux*

Calculating now the volume integral over the domain Ω of equation [1.3], we obtain:

$$\iiint_{\Omega} \operatorname{div} \mathbf{B} \, d\tau = 0 \qquad [1.45]$$

Applying to this equation Ostrogradski's theorem, also known as the "divergence theorem", we have:

$$\iiint_{\Omega} \text{div}\mathbf{B}d\tau = \oiint_{\Gamma} \mathbf{B.n}dS = 0 \qquad [1.46]$$

Magnetic flux density is therefore a conservative flux vector field. This means that the magnetic flux flowing through a closed surface (in this case, the surface Γ of the domain Ω) is equal to zero. In the studied example, decomposing the boundary Γ (Γ_b, Γ_{h1}, Γ_{h2}), the following can be written:

$$\oiint_{\Gamma} \mathbf{B.n}dS = \iint_{\Gamma_b} \mathbf{B.n}dS + \iint_{\Gamma_{h1}} \mathbf{B.n}dS + \iint_{\Gamma_{h2}} \mathbf{B.n}dS = 0 \qquad [1.47]$$

Considering the boundary conditions on the lateral surface Γ_b ($\mathbf{B.n} = 0$), it can be deduced that the incoming flux through Γ_{h1} (see the orientation of the normal vectors in Figure 1.7) is naturally equal to the outgoing flux through Γ_{h2}.

It is important to note that the divergence of the current density is also zero (see equation [1.6]). Under these conditions, it has the same properties as the magnetic flux density, i.e. it is a conservative flux vector field. This reflects the fact that electric current is conserved all along a conductor.

1.4.4. *Gauss' law*

This section focuses on equation [1.4] that links the electric displacement field to the electric charge density ρ. To study the properties of this equation, consider the domain Ω of boundary Γ_e enclosing a charge density ρ (see Figure 1.8).

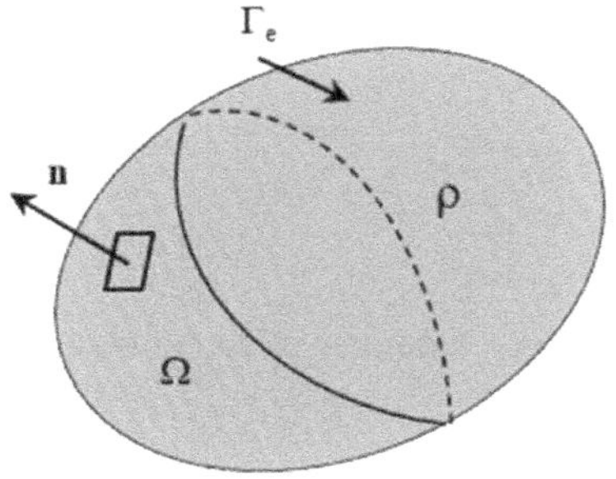

Figure 1.8. *Illustration of Gauss' law*

Let us calculate, for equation [1.4], the volume integral over the domain Ω. This yields:

$$\iiint_{\Omega} \mathrm{div}\mathbf{D}\, d\tau = \iiint_{\Omega} \rho d\tau \qquad [1.48]$$

Applying Ostrogradski's theorem for the divergence operator leads to:

$$\oiint_{\Gamma_e} \mathbf{D}.\mathbf{n}dS = \iiint_{\Omega} \rho d\tau \qquad [1.49]$$

The term on the right-hand side corresponds to the total charges Q inside the domain, therefore:

$$\oiint_{\Gamma_e} \mathbf{D}.\mathbf{n}dS = Q \qquad [1.50]$$

As for the term on the left-hand side, it corresponds to the electric flux ϕ_e through the surface of the domain Ω. This reflects Gauss' law, i.e. the electric flux through a closed surface is equal to the total charges Q enclosed by the volume defined by this surface.

1.5. Various forms of Maxwell's equations

Depending on the given problem, in the context of low-frequency electromagnetism (see section 1.1), it is possible to simplify the initial model defined by equations [1.1], [1.3], [1.4] and [1.5]. Static and quasi-static problems are then identified. Concerning static problems, our focus is on studying the problems of electrostatics, electrokinetics and magnetostatics. As far as quasi-static problems are concerned, this book focuses only on magnetoquasistatics, commonly referred to as "magnetodynamics". The following section studies these various forms and introduces, for each of them, the boundary conditions and the notion of source term.

1.5.1. *Electrostatics*

Electrostatics aims to study, within a given domain Ω, the distribution of the electric field and of the electric displacement field in the presence of static source terms. The study is conducted at electrostatic equilibrium; therefore, the problem to be addressed is stationary in time. As an example, Figure 1.9 shows a domain of permittivity ε_0 inside of which there is a subdomain Ω_1 of permittivity ε_1. On the boundary, there are two types of boundary conditions, Γ_{dk} (see equation [1.39]) with $k \in \{1,2\}$ and Γ_{ek} (see equation [1.35]) with $k \in \{1,2\}$. It is important to recall that

Γ_{dk} represents a wall for the electric displacement field and Γ_{ek} is a gate for the electric field. The two gates, Γ_{e1} and Γ_{e2}, are in contact with electrodes $\mathcal{E}_1$ and $\mathcal{E}_2$.

For this example, the source term can be the circulation f_s of the electric field strength **E** along an arbitrary path γ_{12} (see Figure 1.9) linking the two electrodes:

$$f_s = \int_{\gamma_{12}} \mathbf{E}.\mathbf{dl} \quad [1.51]$$

At the surface of the electrodes, located on the boundary of the domain Ω, the electric displacement field has the following property:

$$\mathbf{D} = \pm\sigma_s \mathbf{n} \quad [1.52]$$

where σ_s is the surface density of charges on the boundary with the electrode and **n** is the outgoing unit normal vector. The expression of the amount of charges Q_σ on each electrode is:

$$\iint_{\Gamma_{ek}} \mathbf{D}.\mathbf{n}dS = \pm Q_\sigma \quad [1.53]$$

In this case, Maxwell's equations (see equations [1.1]–[1.4]) in electrostatics and in the absence of electric charge density within the domain lead to solving two equations:

$$\mathbf{curlE} = 0 \quad [1.54]$$

$$\text{div}\mathbf{D} = 0 \quad [1.55]$$

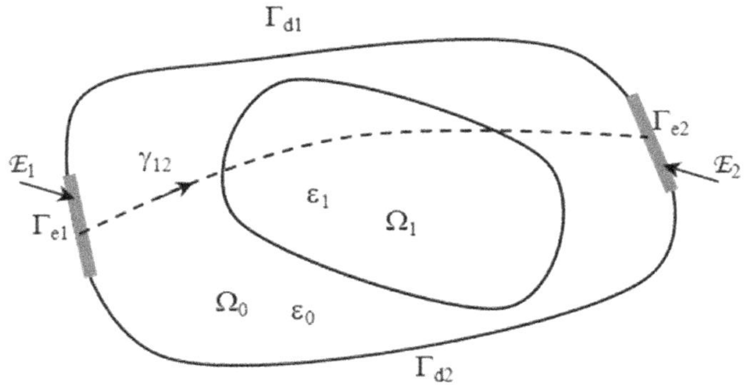

Figure 1.9. *Representation of an electrostatic problem*

The solution to these equations aims to find a curl-free electric field [1.54] that verifies equation [1.55] via the dielectric behavior law [1.19] and also the boundary conditions on the boundary of the domain, defined in Figure 1.9.

1.5.2. *Electrokinetics*

Electrokinetics studies the distribution of the electric field and of the current density in a conductor in the presence of charges in motion, when the speed of these charges is constant.

As an example, consider the set-up represented in Figure 1.10. The conductive domain Ω is composed of a main region of conductivity σ_1 surrounding two subregions of conductivity σ_2 and σ_3. On the boundary of the domain, there are two wall-type boundaries for the current density (Γ_{j1} and Γ_{j2}) and two other gate-type boundaries for the electric field Γ_{e1} and Γ_{e2}. The boundary conditions on these boundaries are defined, respectively, by equations [1.38] and [1.35].

Two types of source terms can be applied on the boundaries Γ_{e1} and Γ_{e2}:

– the first is an electromotive force, denoted by "e", which corresponds to the circulation of the electric field on a path γ_{12} (see Figure 1.10) inside the domain, linking the two surfaces Γ_{e1} and Γ_{e2}, such that:

$$\int_{\gamma_{12}} \mathbf{E.dl} = \mathrm{e} \qquad [1.56]$$

– the second consists of imposing the current density flux, denoted by I, to the surfaces Γ_{e1} and Γ_{e2}. Its expression is:

$$\iint_{\Gamma_{ek}} \mathbf{J.n}\mathrm{dS} = \pm \mathrm{I} \qquad [1.57]$$

where $k \in \{1,2\}$.

In this context, Maxwell's equations can be written as:

$$\mathbf{curlE} = 0 \qquad [1.58]$$

$$\operatorname{div} \mathbf{J} = 0 \qquad [1.59]$$

which can be completed by the electric behavior law [1.20] and the boundary condition [1.35] on the boundaries Γ_{ek} and [1.38] on the boundaries Γ_{jk}.

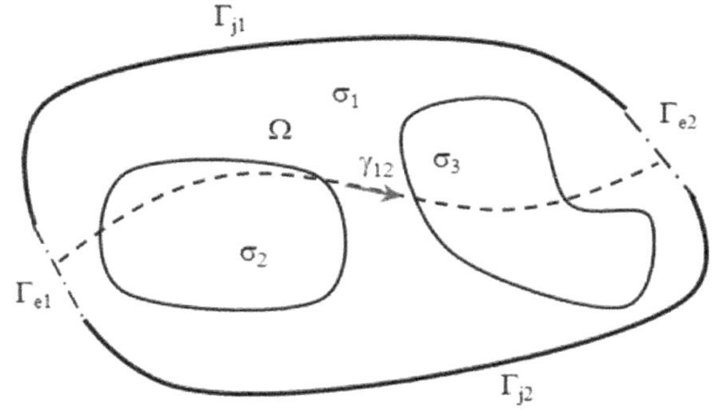

Figure 1.10. *Representation of an electrokinetic problem*

1.5.3. *Magnetostatics*

The magnetostatics problem aims to study the distribution of the magnetic field **H** and of the magnetic flux density **B** for source terms that are time invariant. In this context, the distribution of the current density, denoted by $\mathbf{J}_0$, is assumed to be known, unlike in the case of electrokinetics.

For the study of magnetostatics, the general case is considered, as illustrated in Figure 1.11. Given a domain Ω of boundary Γ, such that: $\Gamma = \Gamma_{b1} \cup \Gamma_{b2} \cup \Gamma_{h1} \cup \Gamma_{h2}$. The boundaries Γ_{bk} ($k \in \{1,2\}$) are of wall type for the magnetic flux density (see equation [1.37]). On the contrary, Γ_{h1} and Γ_{h2} represent a gate for the magnetic flux density (see equation [1.36]).

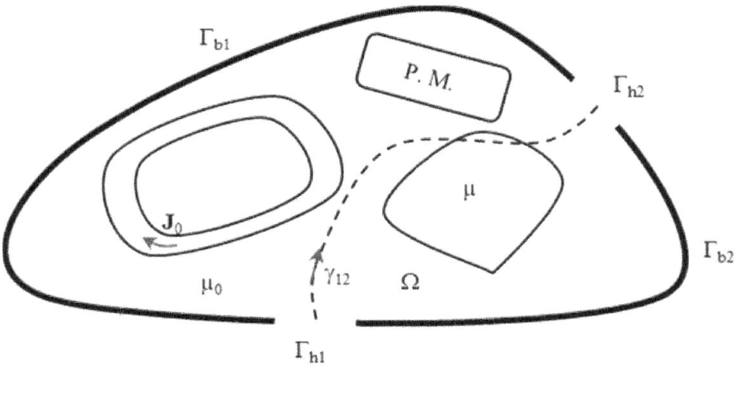

Figure 1.11. *Representation of a magnetostatic problem*

The studied domain, of permeability μ_0, contains a ferromagnetic material of permeability μ [1.26] and source terms, hence a permanent magnet (denoted by "PM" in Figure 1.11) and a conductor carrying a current density $\mathbf{J}_0$ that is also referred to as inductor. Between the two gates, Γ_{h1} and Γ_{h2}, it is possible to impose a magnetomotive force (or a magnetic flux through both of them).

The following section details the various source terms with the associated equations:

– the magnetomotive force f_m that is imposed between the two boundaries Γ_{h1} and Γ_{h2} is defined by:

$$f_m = \int_{\gamma_{12}} \mathbf{H}.\mathbf{dl} \qquad [1.60]$$

where γ_{12} represents a path through the domain Ω linking the boundaries Γ_{h1} to Γ_{h2}, as shown in Figure 1.11;

– the magnetic flux, denoted by ϕ, can be imposed on the two surfaces Γ_{h1} and Γ_{h2} such that:

$$\iint_{\Gamma_{h_k}} \mathbf{B}.\mathbf{n}dS = \pm\phi \qquad [1.61]$$

where $k \in \{1,2\}$. When the two boundaries, Γ_{h1} and Γ_{h2}, are separated by surfaces of type Γ_b, the incoming flux through Γ_{h1} is equal to the outgoing flux through Γ_{h2} (see equation [1.47] related to the law of conservation of the magnetic flux). The two source terms f_m and ϕ are exclusive, in the sense that they cannot be imposed simultaneously;

– an inductor, carrying a current density $\mathbf{J}_0$. In the case of multi-wire winding, by knowing the intensity I of the current through a conductor, the current density $\mathbf{J}_0$ is defined by:

$$\mathbf{J}_0 = \frac{I}{S_c}\mathbf{n} \qquad [1.62]$$

where S_c represents the cross-section of the wire conductors and $\mathbf{n}$ is the unit normal vector of current density whose direction corresponds to the geometrical orientation of the conductors;

– a permanent magnet, characterized by its behavior law. A simplified characteristic is generally used, as shown in Figure 1.2(b), which can be written in the form of equation [1.27]. It can be easily verified that the coercive field $\mathbf{H}_c$ has the following expression:

$$\mathbf{H}_c = -\frac{\mathbf{B}_r}{\mu_A} \qquad [1.63]$$

Based on equations [1.27] and [1.63], the magnetic field strength **H** in the permanent magnet can be written as:

$$\mathbf{H} = \frac{\mathbf{B}}{\mu_A} - \mathbf{H}_c \qquad [1.64]$$

Based on these four source terms, the Maxwell's equations to be solved in this context are:

$$\mathbf{curl H} = \mathbf{J}_0 \qquad [1.65]$$

$$\mathrm{div}\mathbf{B} = 0 \qquad [1.66]$$

complemented by the magnetic behavior law [1.26] in the air and ferromagnetic material and equation [1.64] for the permanent magnet. These are completed by the homogeneous boundary conditions [1.36] and [1.37] for the magnetic field and the magnetic flux density, respectively, and also the source terms [1.60] or [1.61], as applicable.

1.5.4. *Magnetodynamics*

Magnetodynamics studies, in a conductive domain Ω_c, the electromagnetic phenomena at industrial frequencies based on the quasi-static approximation.

As an example of the studied set-up, let us consider the system shown in Figure 1.12. It is a domain Ω (of conductivity $\sigma = 0$) with a boundary Γ_b inside of which there is an inductor, carrying a current density $\mathbf{J}_0$ assumed to be known and varying in time, and a subdomain Ω_c of boundary Γ_j where conductivity is assumed to be non-zero. In this example, the source term is an inductor, but this can be a permanent magnet in motion or any other device. In order to alleviate the developments for magnetodynamics, this section considers only the subdomain Ω_c and its associated equations. Section 3.6 explains the coupling between the equations of magnetodynamics on Ω_c and those of magnetostatics, defined on the subdomain Ω-Ω_c.

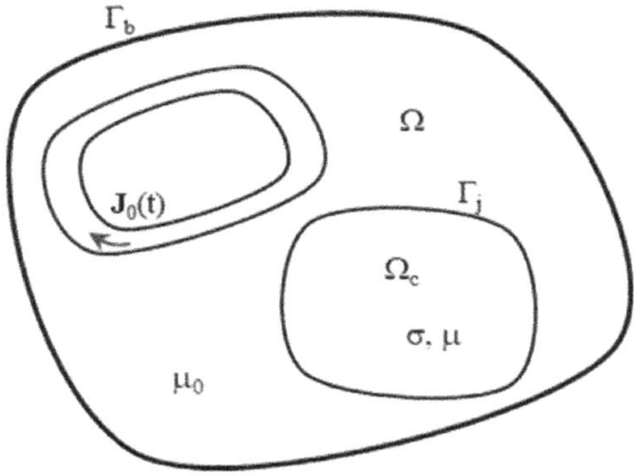

Figure 1.12. *Example of studied domain in magnetodynamics with a source term J_0 and a conductive subdomain Ω_c*

In the domain Ω_c, the equations to be solved, on the basis of Maxwell's equations, are written as follows:

$$\mathbf{curlE} = -\frac{\partial \mathbf{B}}{\partial t} \qquad [1.67]$$

$$\mathbf{curlH} = \mathbf{J} \qquad [1.68]$$

$$\mathrm{div}\mathbf{B} = 0 \qquad [1.69]$$

$$\mathrm{div}\mathbf{J} = 0 \qquad [1.70]$$

These equations are completed by the behavior laws [1.20] and [1.26], and also by the boundary conditions. For the example shown in Figure 1.12, the boundary conditions of wall type impose that, on the boundary of conductor Γ_j (domain Ω_c), the normal component of the current density is equal to 0. The continuity property of the magnetic field **H**, at the boundary Γ_j between the equations of magnetostatics (domain Ω-Ω_c) and magnetodynamics (domain Ω_c), provides the other boundary conditions required for the correct formulation of the problem. If the conductive domain Ω_c is in direct contact with the external boundary, boundary conditions must be imposed on the tangential component of the magnetic field intensity **H** or of the electric field strength **E**.

2

Function Spaces

2.1. Introduction

This chapter presents the function spaces that host the various electromagnetic fields introduced in Chapter 1. This presentation requires a review of some definitions. Next, the focus will be on kernels and images of vector operators (**grad**, **curl**, div) and on the extent to which the latter are strongly related. The dependencies of these relations on the topology of the studied domain will be explained. These properties will be very useful in Chapter 3 for building the potential-based formulations.

2.2. Spaces of differential operators

2.2.1. *Definitions*

Consider an open and bounded domain Ω in $\Re^3$, its boundary being denoted by Γ. Let $L^2(\Omega)$ be the space of square integrable scalar functions over the domain Ω. The scalar product of two functions u and v from $L^2(\Omega)$ is given by:

$$(u, v)_\Omega = \int_\Omega u\, v\, d\tau \quad \forall u, v \in L^2(\Omega) \qquad [2.1]$$

Similarly, consider $\mathbf{L}^2(\Omega)$ the space of square integrable vector functions over the domain Ω. The scalar product of two fields $\mathbf{u}$ and $\mathbf{v}$ from $\mathbf{L}^2(\Omega)$ is written as:

$$(\mathbf{u}, \mathbf{v})_\Omega = \int_\Omega \mathbf{u}.\mathbf{v}\, d\tau \quad \forall \mathbf{u}, \mathbf{v} \in \mathbf{L}^2(\Omega) \qquad [2.2]$$

Two elements u and v $\in$ $L^2(\Omega)$ or **u** and **v** $\in$ $\mathbf{L}^2(\Omega)$ are orthogonal if their scalar product is equal to zero or respectively:

$$(u, v)_\Omega = 0 \qquad [2.3]$$

$$(\mathbf{u}, \mathbf{v})_\Omega = 0 \qquad [2.4]$$

In physics, the spaces $L^2(\Omega)$ and $\mathbf{L}^2(\Omega)$ can be interpreted as finite energy scalar and vector fields.

2.2.2. *Function spaces of grad, curl, div*

As already noted in section 1.1, Maxwell's equations are defined using curl and divergence vector operators and also, as shown in what follows, the gradient operator. The function spaces of these operators, i.e. the set of fields to which the operator can be applied, are subspaces of $L^2(\Omega)$ and $\mathbf{L}^2(\Omega)$, hence:

$$H(\mathbf{grad}, \Omega) = \left\{ u \in L^2(\Omega); \mathbf{grad}\, u \in \mathbf{L}^2(\Omega) \right\} \qquad [2.5]$$

$$H(\mathbf{curl}, \Omega) = \left\{ \mathbf{u} \in \mathbf{L}^2(\Omega); \mathbf{curl}\,\mathbf{u} \in \mathbf{L}^2(\Omega) \right\} \qquad [2.6]$$

$$H(\mathrm{div}, \Omega) = \left\{ \mathbf{u} \in \mathbf{L}^2(\Omega); \mathrm{div}\,\mathbf{u} \in L^2(\Omega) \right\} \qquad [2.7]$$

If homogeneous boundary conditions are introduced on the boundary Γ of the domain, three new subspaces can be defined as follows:

$$H_0(\mathbf{grad}, \Omega) = \left\{ u \in H(\mathbf{grad}, \Omega); u|_\Gamma = 0 \right\} \qquad [2.8]$$

$$H_0(\mathbf{curl}, \Omega) = \left\{ \mathbf{u} \in H(\mathbf{curl}, \Omega); \mathbf{u} \wedge \mathbf{n}|_\Gamma = 0) \right\} \qquad [2.9]$$

$$H_0(\mathrm{div}, \Omega) = \left\{ \mathbf{u} \in H(\mathrm{div}, \Omega); \mathbf{u}.\mathbf{n}|_\Gamma = 0 \right\} \qquad [2.10]$$

It is important to recall that in these expressions **n** represents a unit vector normal to the boundary Γ of Ω, outwardly directed.

NOTE.– It can be noted that, for the space of definition associated with a curl or divergence operator, boundary conditions are applied, respectively, to the tangential or normal component of the field. This is fully consistent with the continuity properties of the fields, related to the curl or divergence operator, described in section 1.3.1. As for the gradient operator, since it is applied to a scalar function, the condition is obviously related to the value of the function.

2.2.3. *Kernel of vector operators*

The subspaces defined in section 2.2.2 can be associated with the kernel of vector operators, such that:

$$\ker(\mathbf{grad}) = \{\ \mathrm{u} \in \mathrm{H}\,(\mathbf{grad}, \Omega)\,; \mathbf{grad}\ \mathrm{u} = 0\} = \mathrm{H}(\mathbf{grad}0, \Omega) \qquad [2.11]$$

$$\ker(\mathbf{curl}) = \{\ \mathbf{u} \in \mathrm{H}\,(\mathbf{curl}, \Omega); \mathbf{curl}\mathbf{u} = 0\} = \mathrm{H}(\mathbf{curl}\,0, \Omega) \qquad [2.12]$$

$$\ker(\mathrm{div}) = \{\ \mathbf{u} \in \mathrm{H}\,(\mathrm{div}, \Omega); \mathrm{div}\mathbf{u} = 0\} = \mathrm{H}(\mathrm{div}0, \Omega) \qquad [2.13]$$

The kernel of an operator (**grad**, **curl** or div) contains the set of functions for which this operator is equal to zero. Considering homogeneous boundary conditions, new subspaces of the kernels defined in [2.11], [2.12] and [2.13] can be defined:

$$\mathrm{H}_0(\mathbf{grad}0, \Omega) = \{\ \mathrm{u} \in \mathrm{H}\,(\mathbf{grad}, \Omega)\,;\ \mathrm{u}|_\Gamma = 0,\ \mathbf{grad}\mathrm{u} = 0\} \qquad [2.14]$$

$$\mathrm{H}_0(\mathbf{curl}\,0, \Omega) = \{\ \mathbf{u} \in \mathrm{H}\,(\mathbf{curl}, \Omega);\ \mathbf{u} \wedge \mathbf{n}|_\Gamma = 0,\ \mathbf{curl}\mathbf{u} = 0\} \qquad [2.15]$$

$$\mathrm{H}_0(\mathrm{div}0, \Omega) = \{\ \mathbf{u} \in \mathrm{H}\,(\mathrm{div}, \Omega);\ \mathbf{u}.\mathbf{n}|_\Gamma = 0,\ \mathrm{div}\mathbf{u} = 0\} \qquad [2.16]$$

2.2.4. *Image spaces of operators*

Furthermore, images of the gradient, curl and divergence can be introduced, denoted, respectively, by $\mathbf{grad}\mathrm{H}(\Omega)$, $\mathbf{curl}\mathrm{H}(\Omega)$, $\mathrm{div}\mathrm{H}(\Omega)$. For example, the subspace $\mathbf{grad}\mathrm{H}(\Omega)$ contains the fields **v** of $\mathbf{L}^2(\Omega)$ such that there is u, belonging to $\mathrm{H}(\mathbf{grad}, \Omega)$, with $\mathbf{v} = \mathbf{grad}\mathrm{u}$. These images can be completed by taking into account homogeneous boundary conditions on the boundary, which are $\mathbf{grad}\mathrm{H}_0(\Omega)$, $\mathbf{curl}\mathrm{H}_0(\Omega)$, $\mathrm{div}\mathrm{H}_0(\Omega)$. For example, $\mathbf{grad}\mathrm{H}_0(\Omega)$ is the space containing the fields **v** of $\mathbf{grad}\mathrm{H}(\Omega)$, such that $\mathrm{u} = 0$ on the boundary Γ of Ω.

It should be noted that there is a fundamental property, on the one hand, between the image of the gradient and the kernel of the curl, and, on the other hand, between the image of the curl and the kernel of the divergence:

$$\mathbf{grad}\mathrm{H}(\Omega) \subset \mathrm{H}(\mathbf{curl}0,\Omega) \quad [2.17]$$

$$\mathbf{curl}\mathrm{H}(\Omega) \subset \mathrm{H}(\mathrm{div}0,\Omega) \quad [2.18]$$

These two properties recall, on the one hand, that the curl of a vector field resulting from a gradient is zero ($\mathbf{curl}(\mathbf{grad}) = 0$) and, on the other hand, that the divergence of a curl is also zero ($\mathrm{div}(\mathbf{curl}) = 0$). However, the reverse is not true, as will be seen when considering the topology of the studied domains in the following sections (see sections 2.3 and 2.4). As an example, there may be a field $\mathbf{w}$ whose curl is zero ($\mathbf{curl}\ \mathbf{w} = 0$), but that does not derive from a gradient, which means there is no field u belonging to $\mathbf{grad}\ \mathrm{H}(\Omega)$, so that $\mathbf{w} = \mathbf{grad}\mathrm{u}$.

If we now consider the boundary conditions, the image of the gradient operator $\mathbf{grad}\mathrm{H}_0(\Omega)$ obviously belongs to $\mathrm{H}_0(\mathbf{curl}0, \Omega)$. The same is true for the image of the curl operator $\mathbf{curl}\mathbf{H}_0(\Omega)$ that belongs to $\mathrm{H}_0(\mathrm{div}0, \Omega)$. This is expressed by the following relations:

$$\mathbf{grad}\mathrm{H}_0(\Omega) \subset \mathrm{H}_0(\mathbf{curl}0,\Omega) \quad [2.19]$$

$$\mathbf{curl}\,\mathrm{H}_0(\Omega) \subset \mathrm{H}_0(\mathrm{div}0,\Omega) \quad [2.20]$$

It is easy to verify these two properties enable the boundary conditions to be taken into account. For example, consider a function $\mathbf{v} \in \mathbf{grad}\mathrm{H}_0(\Omega)$. Then, there is a scalar function u such that $\mathbf{v} = \mathbf{grad}\mathrm{u}$ with the boundary condition on Γ: $\mathrm{u}\,|_\Gamma = 0$. The boundary is therefore an equipotential surface for the function u. The vector $\mathbf{grad}\mathrm{u}$ is then normal to Γ and therefore the tangential component of $\mathbf{grad}\mathrm{u}$ is zero, which means $\mathbf{n} \wedge \mathbf{grad}\mathrm{u} = \mathbf{n} \wedge \mathbf{v} = 0$. The vector $\mathbf{v}$ thus meets the right boundary conditions in order to belong to $\mathrm{H}_0\ (\mathbf{curl}0, \Omega)$.

In this context, restrictive conditions should be introduced on the topology of the domain or complementary spaces, in order to transform the inclusions of equations [2.17], [2.18], [2.19] and [2.20] into equality. This will be addressed in section 2.4.

2.3. Studied topologies

This section focuses on the definition of various domain topologies: connected or disconnected, simply connected and not simply connected, and, finally, contractible and non-contractible.

2.3.1. *Connected and disconnected domain*

A domain is connected if, for any two points of the domain, there is a continuous path connecting them, fully within the domain. As an illustration, the domain in Figure 2.1(a) is connected; indeed, for all arbitrary points P_1 and P_2 of the domain, there is a path γ, belonging to the domain, that connects them. On the other hand, the domain in Figure 2.1(b) is disconnected since, to go from P_1 to P_2, regardless of the path considered, the latter is not fully included in the studied domain.

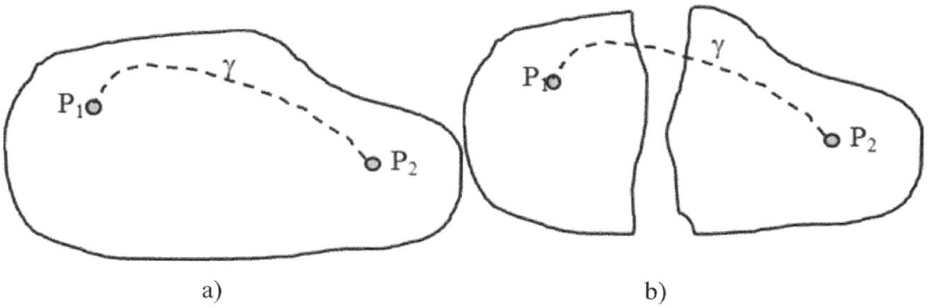

Figure 2.1. *a) Example of a connected domain; b) disconnected domain*

2.3.2. *Simply connected and not simply connected domain*

By definition, a domain is simply connected if a closed path, arbitrarily chosen, can be contracted to a point by continuous transformation. If this is not the case, the domain is not simply connected.

As an illustration, it can be seen that the domain in Figure 2.2(a) is simply connected. Indeed, any closed contour γ can be contracted to a point by successive continuous transformations. On the contrary, the domain in Figure 2.2(b) is not simply connected as, given the presence of a hole, it is not possible to contract by successive deformations a contour γ to a point while remaining inside the domain. A torus is not a simply connected domain. However, a sphere, having a cavity inside, is a simply connected domain. Indeed, any arbitrary contour surrounding the cavity, belonging to the domain, can be contracted to a point by "sliding", if required, on the surface of the cavity.

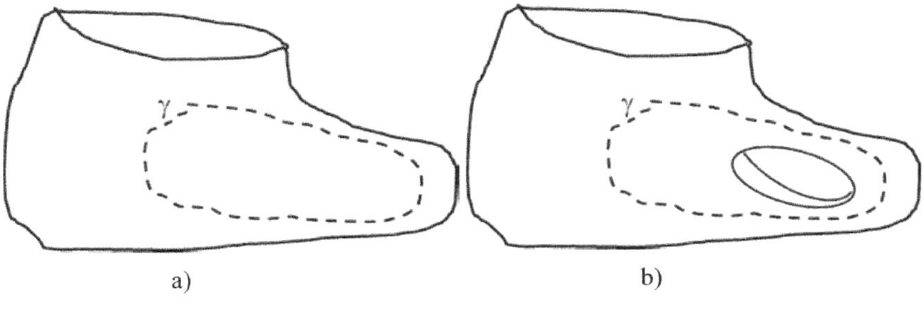

Figure 2.2. *a) Simply connected domain; b) not simply connected domain*

2.3.3. *Contractible and non-contractible domain*

A domain is "contractible" if a contour or an arbitrary closed surface, taken inside the domain, can be contracted to a point by successive transformations. A further definition can be that a domain is contractible if it is simply connected with a connected boundary, in the sense that two arbitrary points of the boundary can be connected by a path belonging to this boundary.

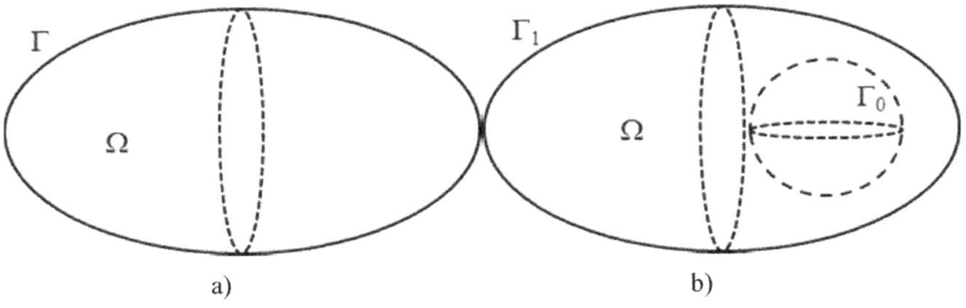

Figure 2.3. *a) Contractible domain; b) non-contractible domain*

The domain Ω represented in Figure 2.3(a) is contractible. However, the presence of the cavity in Figure 2.3(b) makes the domain non-contractible. Indeed, if we consider a closed surface surrounding the cavity, it is not possible, by continuous transformation, to contract it to a point within the domain. The aforementioned surface cannot be reduced more than the surface of the cavity, otherwise it will be out of the domain. It can be noted, on the same Figure 2.3(b), that the domain is simply connected (any closed contour can be contracted to a point by continuous transformation within the domain), but with disconnected boundaries. Indeed, the union of Γ_0 and Γ_1 forms a disconnected surface.

A sphere is typically contractible. On the contrary, a hollow sphere, though simply connected, is not contractible.

2.3.4. *Properties of function spaces*

Based on the topological notions introduced, inclusions [2.17] and [2.18] can be rewritten as follows:

– For a simply connected domain, [2.17] is transformed into equality:

$$\mathbf{grad}H(\Omega) = H(\mathbf{curl}0, \Omega) \quad [2.21]$$

In this case, if the curl of a vector $\mathbf{v}$ is zero, there is a function p such that $\mathbf{v} = \mathbf{grad}p$.

– For a domain with a connected boundary, equation [2.18] then becomes:

$$\mathbf{curl}H(\Omega) = H(\text{div}0, \Omega) \quad [2.22]$$

If a function $\mathbf{v}$ is such that $\text{div}\mathbf{v} = 0$, then there is a function $\mathbf{u}$ such that $\mathbf{curl}\mathbf{u} = \mathbf{v}$.

– If the domain is contractible (simply connected and with a connected boundary), then the two properties [2.21] and [2.22] are simultaneously met.

2.4. Relations between vector subspaces

The focus in what follows is on four images: $\mathbf{grad}H(\Omega)$, $\mathbf{curlH}(\Omega)$, $\mathbf{grad}H_0(\Omega)$ and $\mathbf{curlH}_0(\Omega)$ and also on four subspaces corresponding to the kernel of the curl and divergence: $H(\mathbf{curl}0, \Omega)$, $H(\text{div}0, \Omega)$, $H_0(\mathbf{curl}0, \Omega)$ and $H_0(\text{div}0, \Omega)$. Taking into account the topology of the domain Ω, the properties of these eight function subspaces and their possible interconnections are analyzed.

2.4.1. *Orthogonality of function spaces*

In order to analyze the properties of function spaces, a first step is to study the orthogonality of four image spaces.

Given the image of the gradient, **grad**H(Ω), the aim is to define the space that is orthogonal to it, which is denoted at first by $(\mathbf{grad}H(\Omega))^{\perp}$. Consider $u \in H(\mathbf{grad}, \Omega)$ and $\mathbf{v} \in (\mathbf{grad}H(\Omega))^{\perp}$. Using the formulas on vector operators, the following relation can be written:

$$\int_{\Omega} \mathbf{grad}u.\mathbf{v}\, d\tau = -\int_{\Omega} u\,\mathrm{div}\mathbf{v}\, d\tau + \int_{\Gamma} u\mathbf{v}.\mathbf{n}dS \qquad [2.23]$$

where Γ represents the boundary of the domain Ω. The subspace $(\mathbf{grad}H(\Omega))^{\perp}$ is orthogonal to **grad**H(Ω) if relation [2.23] is equal to zero $\forall u \in H(\mathbf{grad}, \Omega)$. For this, **v** must meet the properties:

$$\mathrm{div}\mathbf{v} = 0 \text{ and } \mathbf{v.n}\big|_{\Gamma} = 0 \qquad [2.24]$$

or, considering [2.16], $\mathbf{v} \in H_0(\mathrm{div}0, \Omega)$. Moreover, it can be shown that the two subspaces **grad**H(Ω) and $H_0(\mathrm{div}0, \Omega)$ are supplementary in $\mathbf{L}^2(\Omega)$ (Bossavit 1988). Then:

$$\mathbf{L}^2(\Omega) = \mathbf{grad}H(\Omega) \oplus H_0(\mathrm{div}0, \Omega) \qquad [2.25]$$

According to this relation, any vector function $\mathbf{w} \in \mathbf{L}^2(\Omega)$ can be decomposed into two functions **v** and **u** belonging, respectively, to $H_0(\mathrm{div}0, \Omega)$ and **grad**H(Ω).

Consider now the image of the gradient considering the boundary conditions: $\mathbf{grad}H_0(\Omega)$. According to the above-mentioned reasoning, its associated orthogonal space is: H(div0, Ω). Therefore, it can be deduced that:

$$\mathbf{L}^2(\Omega) = \mathbf{grad}H_0(\Omega) \oplus H(\mathrm{div}0, \Omega) \qquad [2.26]$$

Let us consider the image of the curl curlH(Ω) and find the subspace orthogonal to it and denoted by $(\mathbf{curl}H(\Omega))^{\perp}$. The approach is similar to the one mentioned above. Consider $\mathbf{u} \in (\mathbf{curl}H(\Omega))^{\perp}$ and $\mathbf{v} \in H(\mathbf{curl}, \Omega)$. In order to be orthogonal, the two subspaces **curl**H(Ω) and $(\mathbf{curl}H(\Omega))^{\perp}$) must meet the following property:

$$\int_{\Omega} \mathbf{u.curlv}\, d\tau = \int_{\Omega} \mathbf{curlu.v}\, d\tau - \int_{\Gamma} (\mathbf{n} \wedge \mathbf{u}).\mathrm{v}dS = 0 \qquad [2.27]$$

u verifies equation [2.27] with the condition $\forall \mathbf{v} \in H(\mathbf{curl}, \Omega)$ by meeting the conditions:

$$\mathbf{curlu} = 0 \text{ and } \mathbf{u} \wedge \mathbf{n}\big|_{\Gamma} = 0 \qquad [2.28]$$

Given [2.15], it can then be shown that $\mathbf{u} \in H_0(\mathbf{curl}0, \Omega)$. The two spaces are also supplementary (Bossavit 1988), therefore:

$$\mathbf{L}^2(\Omega) = \mathbf{curl}H(\Omega) \oplus H_0(\mathbf{curl}0, \Omega) \quad [2.29]$$

Considering now $\mathbf{v} \in \mathbf{curl}(H_0, \Omega)$, then, in order to meet equation [2.27], $\mathbf{u} \in H(\mathbf{curl}0, \Omega)$, the following can be written:

$$\mathbf{L}^2(\Omega) = \mathbf{curl}H_0(\Omega) \oplus H(\mathbf{curl}0, \Omega) \quad [2.30]$$

2.4.2. *Analysis of function subspaces*

Making use of the previous properties, this section describes how the space $\mathbf{L}^2(\Omega)$ of vector functions can be decomposed, by means of the four images: $\mathbf{grad}H(\Omega)$, $\mathbf{curlH}(\Omega)$, $\mathbf{grad}H_0(\Omega)$ and $\mathbf{curlH}_0(\Omega)$ and four subspaces corresponding to the kernel of the curl and divergence: $H(\mathbf{curl}0, \Omega)$, $H(\mathrm{div}0, \Omega)$, $H_0(\mathbf{curl}0, \Omega)$ and $H_0(\mathrm{div}0, \Omega)$. As already seen in section 2.4.1, these eight subspaces are linked by properties [2.25], [2.26], [2.29] and [2.30]. In addition to these properties, there are the inclusion relations [2.17], [2.18], [2.19] and [2.20]. Depending on the topology of the studied domains, inclusions [2.17] and [2.18] can be substituted, respectively, by relations [2.21] and [2.22]. Our analysis will be conducted for domains with various topological properties.

As a first step, consider the decomposition of space $\mathbf{L}^2(\Omega)$ without a priori on the topology of the domain Ω. The properties [2.25], [2.26], [2.29] and [2.30], associated with inclusions [2.17], [2.18], [2.19] and [2.20], allow for the decomposition of space $\mathbf{L}^2(\Omega)$ into five arbitrarily chosen equal segments, as shown in Figure 2.4.

The fact that spaces are supplementary is graphically shown in Figure 2.4, as their association "covers" the domain $\mathbf{L}^2(\Omega)$. As can be seen, the subspaces $\mathbf{grad}H(\Omega)$ and $H_0(\mathrm{div}0, \Omega)$ are a good illustration of equation [2.25]. The same applies to subspaces $\mathbf{grad}H_0(\Omega)$ and $H(\mathrm{div}0, \Omega)$ for equation [2.26]. On the contrary, concerning subspaces $\mathbf{curl}H(\Omega)$ and $H_0(\mathbf{curl}0, \Omega)$, they correspond to equation [2.29], and equation [2.30] is illustrated by the positioning of subspaces $\mathbf{curl}H_0(\Omega)$ and $H(\mathbf{curl}0, \Omega)$. Relations [2.17], [2.18], [2.19] and [2.20] can be readily identified in Figure 2.4.

The following four topologies are considered for the domain Ω: contractible, simply connected but not contractible, not simply connected with a connected boundary and the general case.

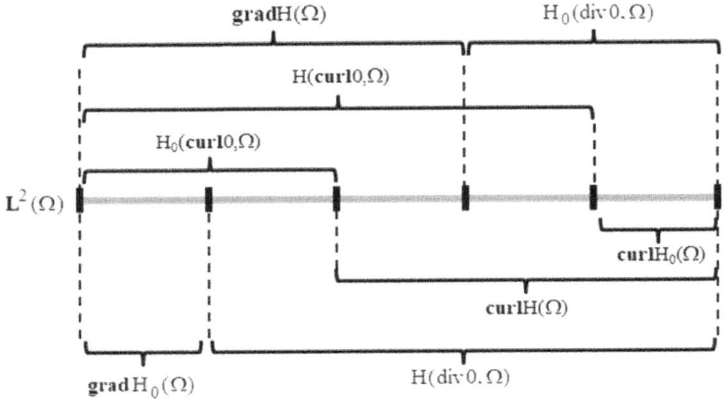

Figure 2.4. *Decomposition of space $L^2(\Omega)$ without a priori on the topology of domain Ω*

2.4.2.1. *Contractible domain*

For a contractible domain, i.e. simply connected and with a connected boundary, the properties of function spaces (see equations [2.25], [2.26], [2.29] and [2.30]) are preserved. On the contrary, the topology of the domain allows for the use of equalities [2.21] and [2.22] and, as shown in Figure 2.5, the space $\mathbf{L}^2(\Omega)$ is divided into three segments instead of the initial five (see Figure 2.4). Using various properties and equalities, the decomposition $\mathbf{L}^2(\Omega)$ can be achieved, as shown in Figure 2.5.

In order to illustrate the diagram in Figure 2.5, consider a field $\mathbf{w} \in H_0(\text{div}0, \Omega)$. The equality $H_0(\text{div}0, \Omega) = \mathbf{curl}H_0(\Omega)$ is applicable, i.e. $\mathbf{w}$ can be expressed from a field of vectors $\mathbf{v}$ such that:

$$\mathbf{w} = \mathbf{curlv} \;\; \forall \;\; \mathbf{v} \in H_0(\mathbf{curl}, \Omega) \qquad [2.31]$$

where $\mathbf{v}$ represents a field known as "vector potential". In Chapter 3, the importance of this notion of potential when applied to Maxwell's equations will be highlighted.

Consider now the field of vectors $\mathbf{f} \in H(\mathbf{curl}0, \Omega)$. Again using the diagram in Figure 2.5 and the definition of the gradient image space, the following can be written:

$$\mathbf{f} = \mathbf{grad}\text{p} \;\; \forall \;\; \text{p} \in H(\mathbf{grad}, \Omega) \qquad [2.32]$$

In this expression, it can be seen that the field $\mathbf{f}$ can be similarly represented by a scalar potential p.

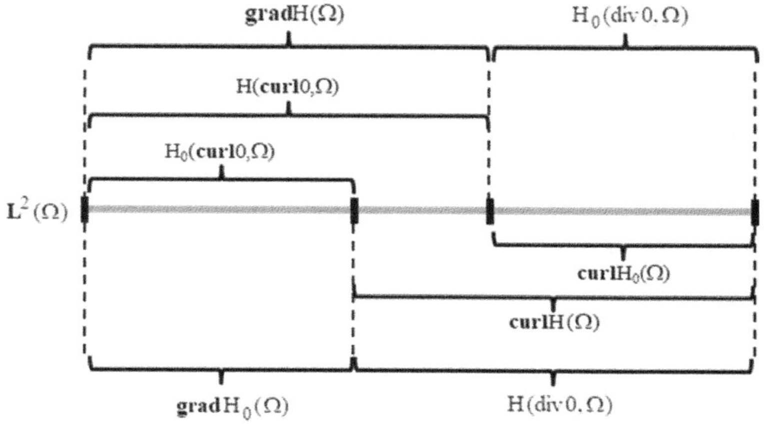

Figure 2.5. *Decomposition of space $L^2(\Omega)$ for a contractible domain Ω (simply connected domain with a connected boundary)*

2.4.2.2. *Not simply connected domain with a connected boundary*

For a not simply connected domain with a connected boundary, the four properties introduced in section 2.4.1 (see equations [2.25], [2.26], [2.29] and [2.30]) can be applied. Given the topological characteristics of the studied domain compared to the case of section 2.4.2.1, relation [2.22] is replaced by inclusion [2.18]. Therefore, the diagram in Figure 2.5 is modified and its form is presented in Figure 2.6.

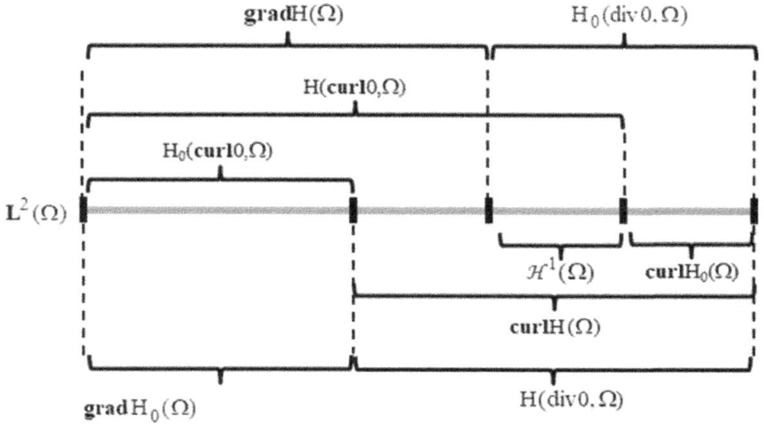

Figure 2.6. *Decomposition of space $L^2(\Omega)$ for a not simply connected domain Ω with connected boundary*

This modification leads to a new subspace, denoted by $\mathcal{H}^1(\Omega)$, with zero curl vector fields that do not derive from a gradient. This subspace is orthogonal to **grad**H(Ω). It is therefore in $H_0(\text{div}0, \Omega)$ and since it is in the kernel of the curl:

$$\mathcal{H}^1(\Omega) = H(\mathbf{curl}0, \Omega) \cap H_0(\text{div}0, \Omega) \qquad [2.33]$$

Based on the definitions of subspaces H(**curl**0, Ω) and $H_0(\text{div}0, \Omega)$, the function space $\mathcal{H}^1(\Omega)$ (Bossavit 1988) can be defined as follows:

$$\mathcal{H}^1 = \left\{ \mathbf{u} \in \mathbf{L}^2(\Omega),\ \mathbf{curl}\mathbf{u} = 0,\ \text{div}\mathbf{u} = 0, \mathbf{u}.\mathbf{n}|_\Gamma = 0 \right\} \qquad [2.34]$$

The dimension of this space is finite and equal to the number "k" of holes within the domain. A possible interpretation is to consider the subspace $\mathcal{H}^1(\Omega)$ as allowing for the introduction of additional functions to the space H(**curl**0, Ω) in order to "make" the domain Ω a simply connected domain, with the relation:

$$H(\mathbf{curl}0, \Omega) = \mathbf{grad}H(\Omega) \oplus \mathcal{H}^1(\Omega) \qquad [2.35]$$

In order to analyze the influence of the domain topology, the two cases presented in section 2.4.2.1 are studied. There is no change for the vector field $\mathbf{w} \in H_0(\text{div}0, \Omega)$, and it can be expressed using the curl of a vector field **v**, as expressed by [2.31].

For a vector field $\mathbf{f} \in H(\mathbf{curl}0, \Omega)$, Figure 2.6 shows that it can be decomposed as follows:

$$\mathbf{f} = \mathbf{grad}p + \sum_{i=1}^{k} K_i \mathbf{h}_i \qquad [2.36]$$

In this expression, K_i is a constant and $\mathbf{h}_i$ are k basis functions of $\mathcal{H}^1(\Omega)$. As for the field p, it is defined as follows:

$$p \in H(\mathbf{grad}, \Omega) \qquad [2.37]$$

where p is a scalar potential.

2.4.2.3. *Simply connected domain with a disconnected boundary*

Consider a simply connected domain with a disconnected boundary. Compared to the case of the contractible domain, instead of relation [2.22], inclusion relation

[2.20] is applicable. As shown in Figure 2.7, this inclusion leads to a new function space denoted by $\mathcal{H}^2(\Omega)$. Then:

$$\mathcal{H}^2(\Omega) = \mathrm{H}_0(\mathbf{curl}0, \Omega) \cap \mathrm{H}(\mathrm{div}0, \Omega) \qquad [2.38]$$

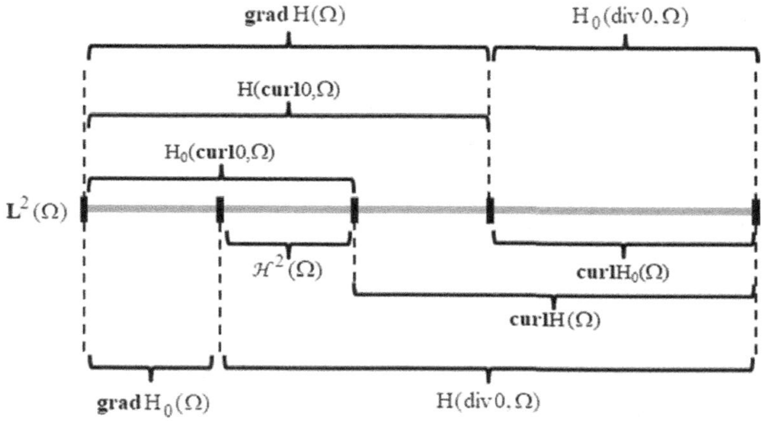

Figure 2.7. *Decomposition of space $L^2(\Omega)$ for a not simply connected domain Ω with disconnected boundary*

The properties of this subspace are deduced (see Figure 2.7) from the intersection of $\mathrm{H}_0(\mathbf{curl}0, \Omega)$ and $\mathrm{H}(\mathrm{div}0, \Omega)$. They are therefore written as (Bossavit 1988):

$$\mathcal{H}^2 = \{\mathbf{u} \in \mathbf{L}^2(\Omega) \;, \mathbf{curlu} = 0, \; \mathbf{divu} = 0, \; \mathbf{u} \wedge \mathbf{n}|_\Gamma = 0\} \qquad [2.39]$$

The dimension of this subspace is finite and equal to the number of cavities of the domain. Similarly to the simply connected case, introducing this space makes it possible to define the following property:

$$\mathrm{H}(\mathrm{div}0, \Omega) = \mathbf{curl}\mathrm{H}(\Omega) \oplus \mathcal{H}^2(\Omega) \qquad [2.40]$$

This relation is similar to equation [2.22], which is applicable to a contractible domain. In other terms, the boundary Γ of the domain is composed of ℓ disconnected closed surfaces Γ_j, with $j = 1, \ldots, \ell$.

Let us consider the two fields introduced in section 2.4.2.1. For the field $\mathbf{w} \in \mathrm{H}(\mathrm{div}0, \Omega)$, Figure 2.7 and equation [2.40] show that it can be decomposed as follows:

$$\mathbf{w} = \mathbf{curl}\mathbf{v} + \sum_{j=1}^{\ell} \mathrm{K}_j \mathbf{h}_j \tag{2.41}$$

where K_j is a constant and $\mathbf{h}_j$ are ℓ basis functions of $\mathcal{H}^2(\Omega)$ and:

$$\mathbf{v} \in \mathrm{H}(\mathbf{curl}, \Omega) \tag{2.42}$$

where $\mathbf{v}$ represents a vector potential. Considering the vector field $\mathbf{f} \in \mathrm{H}(\mathbf{curl}0, \Omega)$, Figure 2.7 clearly shows that it can be written using expression [2.32].

2.4.2.4. *General case: not simply connected domain with a disconnected boundary*

In the general case, space $\mathbf{L}^2(\Omega)$ is decomposed using properties [2.25], [2.26], [2.29] and [2.30]. To take into account the topology of the studied domain, inclusions [2.17], [2.18], [2.19] and [2.20] can be used. Based on Figure 2.4, the function subspaces are decomposed as shown in Figure 2.8.

This general case features the two already defined subspaces $\mathcal{H}^1(\Omega)$ and $\mathcal{H}^2(\Omega)$.

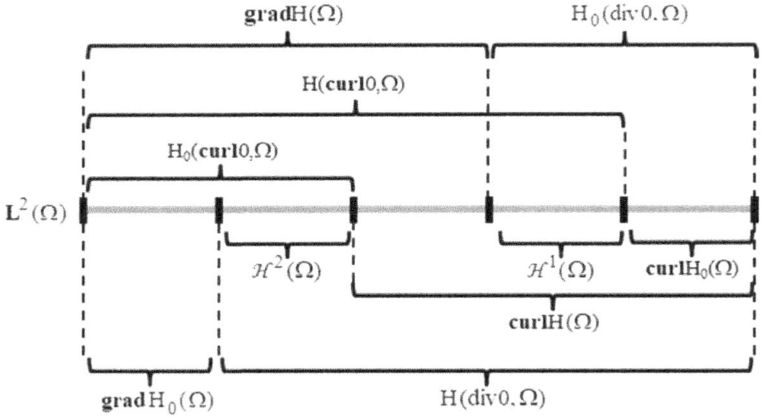

Figure 2.8. *Decomposition of space $L^2(\Omega)$ for a not simply connected domain with disconnected boundary*

Considering as an example the two fields introduced in section 2.4.2.1, the vector field $\mathbf{w} \in H_0(\text{div}0, \Omega)$ is decomposed according to relation [2.41]. Similarly, a vector field $\mathbf{f} \in H(\mathbf{curl}0, \Omega)$ is decomposed by means of relations [2.36].

2.4.3. *Organization of function spaces*

Based on the properties of function spaces described in the previous sections, it is possible to associate them into a sequence using vector operators. Consider the general case of a not simply connected domain with a disconnected boundary. In this context, a four-level graphical representation is built, linking the function spaces introduced in section 2.2 and the **grad**, **curl** and div operators. This representation features the properties mentioned in section 2.2.4, and particularly the following:

$$\begin{aligned} &H(\mathbf{grad},\Omega) \subset L^2, \ \mathbf{grad}H(\Omega) \subset H(\mathbf{curl},\Omega), \\ &\mathbf{curl}\,H(\Omega) \subset H(\text{div}\Omega), \ \text{div}H(\Omega) \subset L^2(\Omega) \end{aligned} \qquad [2.43]$$

Consider the scalar function space $H(\mathbf{grad}, \Omega) \subset L^2(\Omega)$, corresponding to the first line in Figure 2.9. We have highlighted the kernel of the gradient operator $H(\mathbf{grad}0, \Omega)$. If the gradient operator is applied to the first line, the second line, representing $H(\mathbf{curl}, \Omega)$, shows that the image of the gradient is included in the kernel of the curl operator. On the contrary, the vector function space $\mathcal{H}^1(\Omega)$ included in the kernel of the curl operator does not belong to the gradient image subspace, as noted in section 2.4.2.2. Applying the curl operator to the second line in Figure 2.9, the third line corresponding to $H(\text{div}, \Omega)$ shows that its image is included in the kernel of the divergence operator. In this case, this graphical representation shows that the subspace $\mathcal{H}^2(\Omega)$, which is in the kernel of the divergence operator, does not derive from the curl of a function from $H(\mathbf{curl}, \Omega)$.

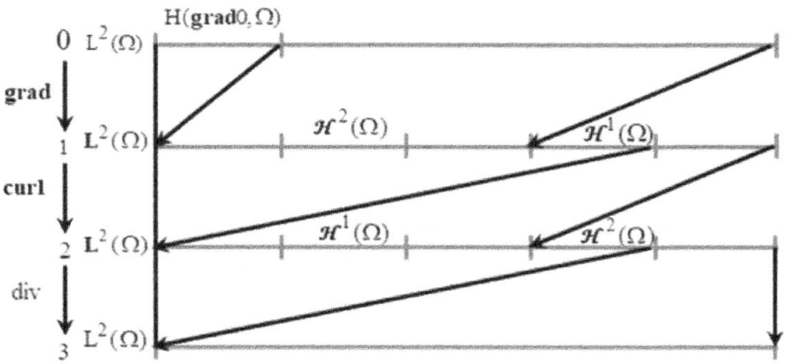

Figure 2.9. *Not simply connected domain with disconnected boundary: graphical representation of function spaces of grad, curl and div operators*

This analysis leads to a series of function spaces, as shown in Figure 2.10.

$$L^2(\Omega) \xrightarrow{\mathbf{grad}} H(\mathbf{curl},\Omega) \xrightarrow{\mathbf{curl}} H(\mathrm{div},\Omega) \xrightarrow{\mathrm{div}} L^2(\Omega)$$

Figure 2.10. *Series of function spaces of grad, curl and div operators*

Consider now a contractible domain (simply connected domain with a connected boundary). Compared to the case of the study presented in Figure 2.10, the representation is simpler as the dimension of subspaces $\mathcal{H}^1$ and $\mathcal{H}^2$ is zero. As Figure 2.11 shows, the first line is unchanged. As for the second line, it can be noted that the image of the gradient operator is equal to the kernel of the curl operator, according to equation [2.21]. Similarly, for the third line, the image of the curl operator is equal to the kernel of the divergence operator (see equation [2.22]).

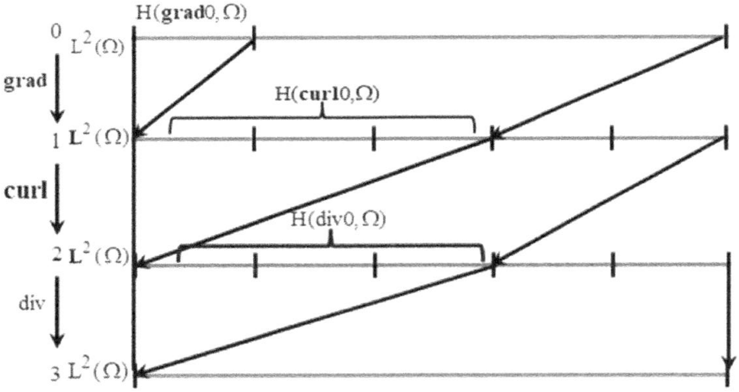

Figure 2.11. *Contractible domain: graphical representation of function spaces of grad, curl and div operators*

2.5. Vector fields defined by a vector operator

It turns out that there are an infinite number of fields that verify a partial differential equation defined by a **grad**, **curl** or div vector operator. As an example, consider a known scalar field f, such that div**u** = f, there are an infinite number of vectors **u** verifying this equation. This section reviews the vector fields defined by a gradient, curl or divergence operator and explores how uniqueness can be imposed by adding the so-called "gauge" condition.

2.5.1. *Infinite number of solutions*

In the case of Maxwell's equations, as noted in section 1.1, vector fields are defined by means of vector operators. However, via behavior laws, the various fields are connected in twos. If the objective is to find a solution that verifies only one equation and the associated boundary conditions, this is referred to as an admissible field. It is a solution provided it verifies a second equation via the behavior law and also the other boundary conditions. For example, this is the case with the electric field and the current density in electrokinetics (see section 1.5.2).

On the contrary, as will be seen in Chapter 3, it is possible to search for a vector field or a scalar uniquely defined by a vector operator. In this case, there are an infinite number of solutions.

For the sake of simplicity, the following study refers to the case of a contractible domain Ω. Consider a field $\mathbf{u}$ belonging to $\mathbf{grad}H(\Omega)$. By definition, there is at least one scalar field, denoted by p_1, which verifies the following relation:

$$\mathbf{u} = \mathbf{grad}p_1 \qquad [2.44]$$

Let us now consider a scalar field p_2 such that $p_2 = p_1 + K$, where K represents a constant. It can be readily verified that $\mathbf{grad}p_2 = \mathbf{u}$. Therefore, there are an infinite number of scalar fields p verifying [2.44]. Similarly, consider a field $\mathbf{w}$ belonging to $\mathbf{curl}H(\Omega)$, there is at least one function $\mathbf{v}_1$ such that:

$$\mathbf{w} = \mathbf{curl}\mathbf{v}_1 \qquad [2.45]$$

Let us now introduce a function $\mathbf{v}_2$ built as the sum of field $\mathbf{v}_1$ and a second field belonging to the space $\mathbf{grad}H(\Omega)$. As shown by equation [2.21], this second field can be expressed as the gradient of a scalar α. Then:

$$\mathbf{v}_2 = \mathbf{v}_1 + \mathbf{grad}\alpha \qquad [2.46]$$

It is easy to show that $\mathbf{v}_1$ and $\mathbf{v}_2$ have the same curl $\mathbf{w}$. Therefore, there are an infinite number of fields $\mathbf{v}$ verifying equation [2.45].

Finally, given a field of vectors $\mathbf{w}_1$ defined using the divergence operator through the relation:

$$q = \mathrm{div}\mathbf{w}_1 \qquad [2.47]$$

It is easy to show that any function $\mathbf{w}_2 \in H(\mathrm{div}, \Omega)$, which is equal to the sum of $\mathbf{w}_1$ and a function deriving from a curl of a function $\mathbf{v}$, such that:

$$\mathbf{w}_2 = \mathbf{w}_1 + \mathbf{curl}\mathbf{v} \qquad [2.48]$$

is also a solution to equation [2.47]. Under these conditions, for equation [2.47], there are also an infinite number of solutions.

As a conclusion, it is important to note that for a function $\mathbf{f}$ (or p) belonging to images $\mathbf{grad}H(\Omega)$, $\mathbf{curl}H(\Omega)$ and $\mathrm{div}H(\Omega)$, there are an infinite number of functions u (or $\mathbf{u}$) that make it possible to write, respectively, $\mathbf{f} = \mathbf{grad}u$, $\mathbf{f} = \mathbf{curl}\mathbf{u}$ and $p = \mathrm{div}\mathbf{u}$. Imposing the uniqueness of the solution requires an additional condition referred to as the "gauge condition".

2.5.2. *Gauge conditions*

As noted in section 2.4.3, depending on the properties of fields, it is possible to introduce a scalar or vector potential. These potentials, defined by a vector operator, are not physical quantities but mathematical entities. Moreover, as noted in section 2.5.1, they are not unique. To have a unique solution it is necessary to impose an additional condition, which is referred to as the "gauge condition".

This section focuses on how to obtain the uniqueness of a field defined by the gradient, curl or divergence.

2.5.2.1. *Gradient operator*

As noted in section 2.5.1, if p_1 is the solution to equation $\mathbf{u} = \mathbf{grad}p_1$ with $\mathbf{u}$ a known field, then $p_2 = p_1 + K$ is also a solution. For this solution to be unique, it is then sufficient to set a value of the solution $p(\mathbf{x})$ at a given point "$\mathbf{x}_0$" of the domain:

$$p(\mathbf{x}_0) = p_0 \qquad [2.49]$$

In this case, as the potentials p_1 and p_2 must verify the constraint $p_1(\mathbf{x}_0) = p_2(\mathbf{x}_0) = p_0$, $K = 0$ is automatically imposed. This reflects the uniqueness of the solution.

2.5.2.2. *Curl operator*

As already noted, there are an infinite number of functions $\mathbf{v}$ such that $\mathbf{curl}\mathbf{v} = \mathbf{w}$, with $\mathbf{w}$ being a known vector field. Indeed, if $\mathbf{v}_1$ is a solution, then any field $\mathbf{v}_1 = \mathbf{v}_2 + \mathbf{grad}\alpha$ is also a solution (see equation [2.46]). The literature proposes several gauge conditions (Stratton 1941; Ida 2020), but one of the most widely known is the Coulomb gauge:

$$\mathrm{div}\mathbf{v} = 0 \quad [2.50]$$

In the numerical simulation, another gauge is often used to set the value of potential α (see equation [2.46]). This gauge consists of imposing (Albanese and Rubinacci 1990):

$$\mathbf{v}.\boldsymbol{\eta} = 0 \quad [2.51]$$

where $\boldsymbol{\eta}$ is an arbitrary vector field whose field lines do not close.

Consider the case of the path γ, defined in Figure 2.12, and let us calculate the circulation of potentials $\mathbf{v}_1$ and $\mathbf{v}_2$ introduced in equation [2.46] that verify [2.51]. Then:

$$\int_P^Q \mathbf{v}_2.\mathbf{dl} = \int_P^Q \mathbf{v}_1.\mathbf{dl} + \alpha_Q - \alpha_P \quad [2.52]$$

Consider the path γ to be a field line of the vector $\boldsymbol{\eta}$, where the term $\mathbf{dl}$ is collinear to $\boldsymbol{\eta}$ along γ. In fact, the scalar products $\mathbf{v}_1.\mathbf{dl}$ and $\mathbf{v}_2.\mathbf{dl}$ are zero (see equation [2.51]). It can therefore be verified, using equation [2.52], that α_Q and α_P are equal.

This property is applicable to any pair of points P and Q of the curve γ and along all the field lines of the vector $\boldsymbol{\eta}$ in the studied domain. This leads to setting the scalar potential α at a constant value. With the gradient of α being zero, by means of equation [2.46], it can be found that $\mathbf{v}_1$ and $\mathbf{v}_2$ are equal, and therefore the vector potential is unique.

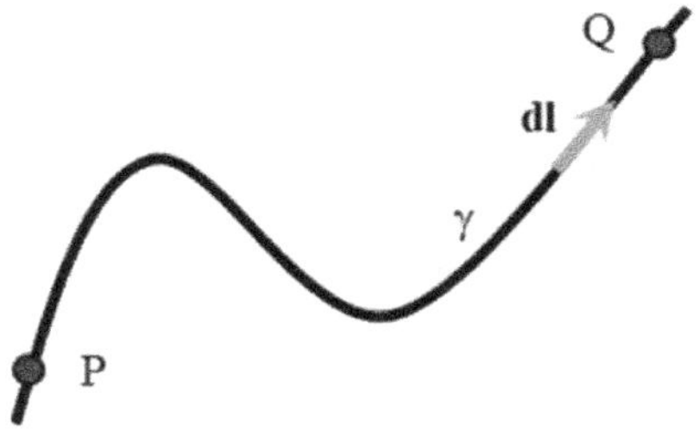

Figure 2.12. *Example of path γ for the gauge $v.\eta = 0$*

2.5.2.3. *Divergence operator*

If a field $\mathbf{w}_1$ is a solution to the equation div$\mathbf{w}$ = q, then the field $\mathbf{w}_2 = \mathbf{w}_1 + \mathbf{curl v}$ is also a solution. A possible gauge condition is to impose the value of the curl of **w**, for example **curlw** = 0. Boundary conditions on the boundary of the domain must also be imposed on **w**.

2.6. Structure of function spaces

This section explores the link of the sequence of function spaces depending on vector operators and boundary conditions. To alleviate the developments, a contractible domain Ω is considered, with a boundary Γ composed of two parts Γ_n and Γ_m with homogeneous boundary conditions.

2.6.1. *Adjoint operators*

Assume that the boundary Γ of the domain Ω is composed of two complementary boundaries Γ_n and Γ_m. Consider the function subspaces of the gradient, curl and divergence operators (see equations [2.8], [2.9] and [2.10]) that are rewritten considering the boundary conditions. The following subspaces can be defined:

$$H_{\Gamma_k}(\mathbf{grad},\Omega) = \left\{u \in H(\mathbf{grad},\Omega); u\big|_{\Gamma_k} = 0\right\} \quad [2.53]$$

$$H_{\Gamma_k}(\mathbf{curl},\Omega) = \left\{\mathbf{u} \in H(\mathbf{curl},\Omega); \mathbf{u} \wedge \mathbf{n}\big|_{\Gamma_k} = 0)\right\} \quad [2.54]$$

$$H_{\Gamma_k}(\mathrm{div},\Omega) = \left\{\mathbf{u} \in H(\mathrm{div},\Omega); \mathbf{u.n}\big|_{\Gamma_k} = 0\right\} \quad [2.55]$$

where k = n or m depending on the considered boundary.

In what follows, we again consider the definitions provided for the scalar product over the domain Ω and the notations introduced in relations [2.1] and [2.2]. Consider a vector operator $\mathcal{L}$ that can be the gradient, curl or divergence. The aim is to find its adjoint operator, denoted by $\mathcal{L}^*$, which by definition satisfies:

$$(\mathcal{L}u, v)_\Omega = (u, \mathcal{L}^* v)_\Omega \quad [2.56]$$

Given a function u belonging to space $\mathcal{D}_\Omega$, let us denote by $\mathcal{D}^*_\Omega$ the space of functions v adjoint to u. For a given vector operator and its associated function space, the objective is to find the adjoint vector operator and also its associated function space.

2.6.1.1. *Adjoint operator of the gradient*

Consider a scalar function $u \in H_{\Gamma n}(\mathbf{grad}, \Omega)$ defined over a domain Ω of boundary Γ with $\Gamma = \Gamma_n \cup \Gamma_m$. The aim is to find, for a vector function $\mathbf{v}$, the operator $\mathcal{L}^*$ and the adjoint function space denoted at first by $(H(\mathbf{grad}, \Omega))^*$. Using equation [2.23], the following can be written:

$$\int_\Omega \mathbf{grad}u.\mathbf{v}\, d\tau + \int_\Omega u \operatorname{div}\mathbf{v}\, d\tau = \int_\Gamma u\mathbf{v}.\mathbf{n} dS \qquad [2.57]$$

It can be noted that this equation is equivalent to [2.56] if the surface integral on the right-hand side is equal to zero. As $u \in H_{\Gamma n}(\mathbf{grad}, \Omega)$ and therefore $u\,|_{\Gamma n} = 0$, this integral over the boundary Γ is zero provided that:

$$\mathbf{v}.\mathbf{n}\big|_{\Gamma_m} = 0 \qquad [2.58]$$

Under these conditions, the following can be written:

$$\int_\Omega \mathbf{grad}u.\mathbf{v}\, d\tau = -\int_\Omega u \operatorname{div}\mathbf{v}\, d\tau \qquad [2.59]$$

which leads to:

$$\mathcal{L}^* = -\operatorname{div} \text{ and } \mathcal{D}^*_\Omega = H_{\Gamma_m}(\operatorname{div}, \Omega) \qquad [2.60]$$

2.6.1.2. *Adjoint operator of the curl*

A similar reasoning as for the gradient is applied, but using relation [2.27] and a vector field $\mathbf{u} \in H_{\Gamma n}(\mathbf{curl}, \Omega)$. The following relations are obtained for the operator $\mathcal{L}^*$ and the associated function space $\mathcal{D}^*$:

$$\mathcal{L}^* = \mathbf{curl} \text{ and } \mathcal{D}^*_\Omega = H_{\Gamma_m}(\mathbf{curl}, \Omega) \qquad [2.61]$$

2.6.1.3. *Adjoint operator of the divergence*

The case of the divergence uses relation [2.23] and a vector field $\mathbf{u} \in H_{\Gamma n}(\operatorname{div}, \Omega)$. For the operator $\mathcal{L}^*$ and the associated function space $\mathcal{D}^*$, the following expressions are found:

$$\mathcal{L}^* = -\mathbf{grad} \text{ and } \mathcal{D}^*_\Omega = H_{\Gamma_m}(\mathbf{grad}, \Omega) \qquad [2.62]$$

2.6.1.4. *Synthesis of adjoint operators and associated function spaces*

Table 2.1 summarizes, for each vector operator, the domain of definition $\mathcal{D}$, and also the associated function space $\mathcal{D}^*$ and the adjoint operator $\mathcal{L}^*$.

Vector operator $\mathcal{L}$	Function space $\mathcal{D}$	Function space $\mathcal{D}^*$	Adjoint operator $\mathcal{L}^*$
grad	$H_{\Gamma_n}(\mathbf{grad},\Omega)$	$H_{\Gamma_m}(\text{div},\Omega)$	$-\text{div}$
curl	$H_{\Gamma_n}(\mathbf{curl},\Omega)$	$H_{\Gamma_m}(\mathbf{curl},\Omega)$	**curl**
div	$H_{\Gamma_n}(\text{div},\Omega)$	$H_{\Gamma_m}(\mathbf{grad},\Omega)$	$-\mathbf{grad}$
grad	$H_{\Gamma_m}(\mathbf{grad},\Omega)$	$H_{\Gamma_n}(\text{div},\Omega)$	$-\text{div}$
curl	$H_{\Gamma_m}(\mathbf{curl},\Omega)$	$H_{\Gamma_n}(\mathbf{curl},\Omega)$	**curl**
div	$H_{\Gamma_m}(\text{div},\Omega)$	$H_{\Gamma_n}(\mathbf{grad},\Omega)$	$-\mathbf{grad}$

Table 2.1. *Synthesis of function spaces of vector operators and adjoint operators*

2.6.2. *Tonti diagram*

Consider the two sequences of function spaces shown in Figure 2.10, but displayed vertically and differing by their boundary conditions (see Figure 2.13). For the sequence on the left, the conditions relate to the boundary Γ_n and for the sequence on the right to the boundary Γ_m. The adjoint operators as well as the corresponding function spaces are also displayed vertically, taking into account the constraints on the boundary conditions. The first structure of the Tonti diagram is thus obtained.

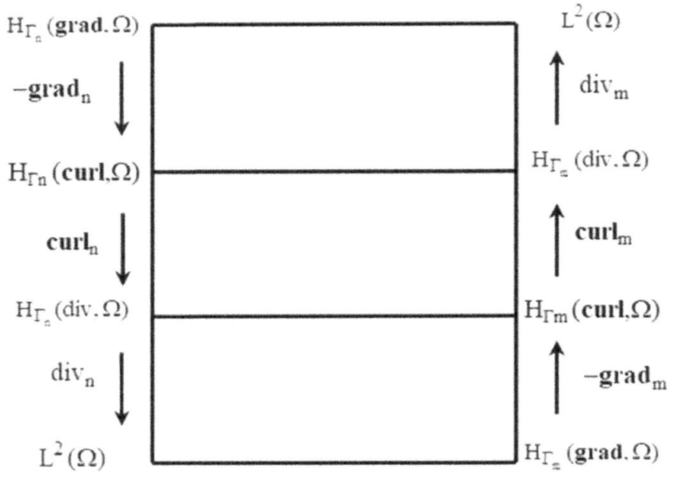

Figure 2.13. *Tonti diagram in the general case*

This diagram contributes to a better understanding of the links between the various fields present in Maxwell's equations, both in the continuous and discrete domains. This diagram will be used in Chapter 3 as it facilitates, on the one hand, the display of the interaction of the electromagnetic fields and their potentials through the vector operator and behavior laws and, on the other hand, easy deduction of the properties.

3

Maxwell's Equations: Potential Formulations

3.1. Introduction

Maxwell's equations, in a static or quasi-static state, are rarely solved by considering the (**E**, **H**, **B**, **J**, **D**) fields as unknowns. They are generally solved using formulations based on potentials instead of fields. The use of potentials, inter alia, makes it possible to simplify the equation system.

In order to facilitate the implementation of these potential formulations, the source terms presented in section 1.5 can be rewritten. This is the first step in this chapter, with the introduction of source fields in the case of partial differential equations. The results will then be applied to the equations of electrostatics, electrokinetics, magnetostatics and magnetodynamics. This approach will show that the use of source fields allows for a quite natural introduction of potential formulations.

3.2. Consideration of source terms

Section 1.5 presented the various source terms generally encountered in low-frequency electromagnetism. These source terms can be classified into two categories:

– global quantities, imposed on the boundaries of the domain, such as the current density flux or the magnetic flux or the circulation of the electric or magnetic field;

– local quantities inside the domain, such as charge density in electrostatics or current density in an inductor or permanent magnet in magnetostatics and magnetodynamics.

In what follows, these source terms will be addressed independently. It should be noted, however, that it is possible to consider a problem involving source terms imposed both on the boundaries (surface terms) and inside the domain (volume terms). The contributions associated with each of the sources must then be calculated, and based on the linearity of differential operators, combined in order to have a complete problem accounting for all the sources. The multisource case is addressed in sections related to electrostatics, electrokinetics and magnetostatics.

Finally, to simplify the presentation, it is assumed that the domain is contractible. Nevertheless, the methodology presented can be transposed to non-contractible domains using the properties introduced in section 2.4.2.

3.2.1. *Global source quantities imposed on the boundaries*

As for the sources, imposed on the boundaries, the link with the local physical quantities is achieved via an integration along a path between two gates (circulation of a field) or from a surface integral on a gate (flux of a field).

Consider the case of fields **E** and **H** defined by a curl. Calculating their circulation along a path, they can be linked to the electromotive force in electrostatics [1.51] and in electrokinetics [1.56] or to the magnetomotive force in magnetostatics [1.60].

Likewise, consider the case of fields **B** and **J** defined by a divergence. By calculating their flux, through a surface, it is possible to make the link between the magnetic flux density **B** or the current density **J** and, respectively, the magnetic flux ϕ [1.61] or the current intensity I [1.57].

In order to integrate these source terms into Maxwell's equations, the notion of source fields, support fields and associated potentials will be introduced. The objective is to bring the surface constraints at the boundaries on fields that are by nature volume-based and defined on the entire domain. The introduction of these new source terms then allows for the definition of potentials (new unknowns of the problem) with less constraining boundary conditions. This will significantly simplify the solution of the problem, particularly with numerical methods (see sections 4.3 and 4.4).

For the sake of clarity and in order to alleviate the developments, source fields and associated potentials will be introduced for a general problem. The results obtained will then be used when developing the potential formulations in the case of static and magnetodynamic problems.

Consider a contractible domain Ω of boundary $\Gamma = \Gamma_{n1} \cup \Gamma_{n2} \cup \Gamma_m$ (see Figure 3.1).

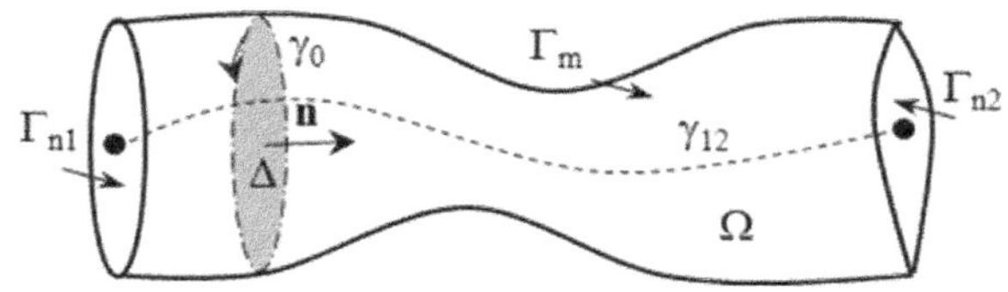

Figure 3.1. *Example of contractible domain Ω with the various notations employed*

Inside the domain, two fields **U** and **V** are defined such that:

$$\mathbf{curlU} = 0 \qquad [3.1]$$

$$\mathrm{div}\mathbf{V} = 0 \qquad [3.2]$$

These two fields can be linked by a behavior law. On the boundary Γ, the boundary conditions are homogeneous and can be written as follows:

$$\Gamma = \Gamma_{n_1} \cup \Gamma_{n_2} \cup \Gamma_m \quad \text{with} \quad \mathbf{V.n}\big|_{\Gamma_m} = 0, \quad \mathbf{U} \wedge \mathbf{n}\big|_{\Gamma_{n_k}} = 0$$
$$k \in \{1, 2\} \qquad [3.3]$$

where Γ_m represents a wall for the field **V**, and Γ_{n1} and Γ_{n2} represent two gates for the field **U**. Considering equations [3.1]–[3.3], **U** and **V** belong to the function spaces:

$$\mathbf{U} \in H_{\Gamma_{n1} \cup \Gamma_{n2}}(\mathbf{curl}0, \Omega) \quad \text{and} \quad \mathbf{V} \in H_{\Gamma_m}(\mathrm{div}0, \Omega) \qquad [3.4]$$

For this problem, two types of source terms are considered:

– The circulation C_s of the field along a path γ_{12} between the two gates Γ_{n1} and Γ_{n2}:

$$C_s = \int_{\gamma_{12}} \mathbf{U.dl} \qquad [3.5]$$

As **U** is curl free, it should be noted that this equation is valid for any path γ_{12}, belonging to the domain Ω linking Γ_{n1} to Γ_{n2}.

– The flux ϕ_v of the field **V** through the gates Γ_{nk}, defined by:

$$\iint_{\Gamma_{nk}} \mathbf{V}.\mathbf{n}dS = \pm\phi_V \quad [3.6]$$

– And in this expression, **n** represents the outward unit normal vector with respect to $\Gamma_{n1} \cup \Gamma_{n2}$.

NOTE.– Let us consider surface Δ in the domain Ω in Figure 3.1. Its contour, denoted by γ_0, belongs to the boundary Γ_m, but cannot be contracted to a point by successive transformations. Therefore, surface Γ_m is not simply connected. Moreover, by moving the contour γ_0 to the limits of the boundary Γ_m, surface Δ may merge with the boundaries Γ_{n1} or Γ_{n2} of the domain. The flux of **V** through the surface Δ is always equal to $\pm\,\phi_v$ (according to the orientation of **n**), as this vector is divergence free; therefore, its flux is conservative. Under these conditions, equation [3.6] can be rewritten as follows:

$$\iint_{\Delta} \mathbf{V}.\mathbf{n}dS = \phi_V \quad [3.7]$$

As an example, in electrokinetics, condition [3.5] means imposing an electromotive force between the two surfaces Γ_{n1} and Γ_{n2}. The field **U** is then the electric field **E**. Imposing condition [3.6] in electrokinetics means imposing the current intensity I through the surface Γ_{n1}. The vector field **V** is then the current density **J**.

It is important to note that these conditions (imposing C_s and ϕ_v) are exclusive, meaning that the two conditions cannot be simultaneously applied. Considering again the case of electrokinetics, it is not possible to impose both an electromotive force across a conductor and the current flowing through it. Likewise, it is not possible to simultaneously impose the current and the voltage across a resistor.

Having defined the problem and the equations to be solved, the next section explores how the notion of source field can be introduced into the entire domain. This field will allow for the representation in an equivalent manner of the conditions imposed on the boundaries.

3.2.1.1. *Source term related to a field defined by a curl*

The main idea is to determine, inside the domain, a source field denoted by $\mathbf{U}_s$, assumed to be known, which verifies the constraints imposed on the boundaries by the vector **U** (see relations [3.3] and [3.5]). It is also assumed that the field $\mathbf{U}_s$

verifies, similar to **U**, equation [3.1]. It should be noted that there are an infinite number of fields $\mathbf{U}_s$ verifying these constraints. Indeed, if a field $\mathbf{U}_s$ is the solution, then $\mathbf{U}_s$ + **grad**p, where p is a scalar field such that p = 0 on Γ_{n1} and Γ_{n2}, is also a solution (see section 2.5).

The field **U** is decomposed into two terms as follows:

$$\mathbf{U} = \mathbf{U}_S + \mathbf{U}' \quad [3.8]$$

In this equation, **U'** becomes the new unknown of the problem, but this time with conditions of null circulation between the two gates Γ_{n1} and Γ_{n2}. Nevertheless, it must still verify the boundary condition [3.3] on Γ_{n1} and Γ_{n2}. Moreover, it can be readily shown that **U'** verifies the same equilibrium equation as the field **U**. The properties of **U'** can finally be stated as follows:

$$\begin{aligned} &\mathbf{curlU}' = 0,\ \int_{\gamma_{12}} \mathbf{U}'.\mathbf{dl} = 0,\ \mathbf{U}' \wedge \mathbf{n}\big|_{\Gamma_{n1}\cup\Gamma_{n2}} = 0 \\ &\text{i.e. } \mathbf{U}' \in \mathrm{H}_{\Gamma_{n1}\cup\Gamma_{n2}}(\mathbf{curl}0,\Omega) \end{aligned} \quad [3.9]$$

As for the properties of the field $\mathbf{U}_s$, they are written as:

$$\begin{aligned} &\mathbf{curlU}_S = 0,\ \int_{\gamma_{12}} \mathbf{U}_S.\mathbf{dl} = \mathrm{C}_S,\ \mathbf{U}_S \wedge \mathbf{n}\big|_{\Gamma_{n1}\cup\Gamma_{n2}} = 0 \\ &\text{i.e. } \mathbf{U}_S \in \mathrm{H}_{\Gamma_{n1}\cup\Gamma_{n2}}(\mathbf{curl}0,\Omega) \end{aligned} \quad [3.10]$$

It can be noted that the constraints imposed on fields $\mathbf{U}_s$ and **U'** are similar to those of field **U** defined by equations [3.1] and [3.3].

The source field $\mathbf{U}_s$ is directly proportional to the circulation C_s. Under these conditions, a support field can be introduced, denoted by $\boldsymbol{\beta}_s$, such that:

$$\mathbf{U}_S = \mathrm{C}_S\boldsymbol{\beta}_S \quad [3.11]$$

where $\boldsymbol{\beta}_s$ must verify the properties of $\mathbf{U}_s$ as follows:

$$\begin{aligned} &\mathbf{curl}\boldsymbol{\beta}_S = 0,\ \int_{\gamma_{12}} \boldsymbol{\beta}_S.\mathbf{dl} = 1, \text{ with } \boldsymbol{\beta}_S \wedge \mathbf{n}\big|_{\Gamma_{n1}\cup\Gamma_{n2}} = 0 \\ &\text{i.e. } \boldsymbol{\beta}_S \in \mathrm{H}_{\Gamma_{n1}\cup\Gamma_{n2}}(\mathbf{curl}0,\Omega) \end{aligned} \quad [3.12]$$

Considering the above-stated properties and the fact that the domain Ω is contractible (see equation [2.32]), $\boldsymbol{\beta}_s$ can be expressed using an "associated" scalar potential denoted by α_s, such that:

$$\boldsymbol{\beta}_s = -\mathbf{grad}\alpha_s, \text{ with } \alpha_s\big|_{\Gamma_{nk}} = \kappa_{nk},\ k \in \{1,2\}, \alpha_s \in H(\mathbf{grad},\Omega) \qquad [3.13]$$

For the choice of constants κ_{nk}, in order to verify the second relation of equation [3.12], a simple solution is to consider:

$$\kappa_{n_1} = 1 \text{ and } \kappa_{n_2} = 0 \qquad [3.14]$$

By grouping equations [3.11] and [3.13], the source field $\mathbf{U}_s$ can also be written as:

$$\mathbf{U}_s = -C_s\mathbf{grad}\alpha_s \qquad [3.15]$$

Similar to $\mathbf{U}_s$, there are an infinite number of fields $\boldsymbol{\beta}_s$ and α_s satisfying equations [3.12] and [3.13], respectively. It is therefore necessary to impose a gauge condition to ensure uniqueness (see section 2.5.2). Section 3.2.3 will provide an example of the calculation of fields $\boldsymbol{\beta}_s$ and α_s.

The equations to be solved can now be written by highlighting the source term C_s:

$$\mathbf{curl}(\mathbf{U}'+C_s\boldsymbol{\beta}_s) = \mathbf{curl}(\mathbf{U}'-C_s\mathbf{grad}\alpha_s) = 0 \qquad [3.16]$$

$$\text{div } \mathbf{V} = 0 \qquad [3.17]$$

It can be noted that the circulation C_s, which was at the beginning a condition imposed on the boundaries, now appears in the equilibrium equation in the form "$C_s\boldsymbol{\beta}_s$" or "$-C_s\mathbf{grad}\alpha_s$", which will be easier to handle in the numerical resolution. As for the boundary conditions of the problem for the new unknown of the problem $\mathbf{U}'$, they remain homogeneous and are stated as follows:

$$\mathbf{U}'\wedge\mathbf{n}\big|_{\Gamma_{nk}} = 0 \text{ with } k \in \{1,2\} \text{ and } \mathbf{V}.\mathbf{n}\big|_{\Gamma_m} = 0 \qquad [3.18]$$

3.2.1.2. *Source term related to a field defined by a divergence*

3.2.1.2.1. Expression of the source field and the support vector field

Consider again the problem illustrated by Figure 3.1 and defined by equations [3.1]–[3.3]. However, the flux ϕ_v is now considered the source term and is related to the field **V** by relation [3.6]. In order to determine a source field, the idea developed in section 3.2.1.1 is applied, which consists of introducing inside the domain a term assumed to be known respecting the constraints imposed on the boundaries. The field **V** is then decomposed into two terms, in the form below:

$$\mathbf{V} = \mathbf{V}_S + \mathbf{V}' \quad [3.19]$$

NOTE.– The function space to which **V** belongs is defined by equation [3.4]. The field **V** therefore has a conservative flux. The same is true for fields $\mathbf{V}_s$ and **V'** which, being built from **V**, have similar properties. Under these conditions, a decision is made to use Δ as a reference surface (see equation [3.7]). It is important to note that this surface can be superimposed to boundaries Γ_{nk} by sliding along surface Γ_m.

Considering equation [3.19], the field **V'** becomes the unknown of the problem and its properties are significantly equivalent to those of field **V**, except on the boundaries. Indeed, the conditions defined in equation [3.6] are transferred to $\mathbf{V}_s$ and the flux of **V'**, through the surfaces Γ_{nk}, is equal to 0. This field is then defined by the properties:

$$\operatorname{div}\mathbf{V}'=0,\ \iint_\Delta \mathbf{V}'.\mathbf{n}ds=0,\ \ \mathbf{V}'.\mathbf{n}\big|_{\Gamma_m}=0 \text{ i.e. } \mathbf{V}'\in H_{\Gamma_m}(\text{div}0,\Omega) \quad [3.20]$$

Then, the field $\mathbf{V}_s$ represents the source term and must verify equation [3.7]. Its properties can be written as follows:

$$\operatorname{div}\mathbf{V}_S=0,\ \iint_\Delta \mathbf{V}_S.\mathbf{n}ds=\pm\phi_V,\ \ \mathbf{V}_S.\mathbf{n}\Big|_{\Gamma_m}=0$$
$$\text{i.e. } \mathbf{V}_S\in H_{\Gamma_m}(\text{div}0,\Omega) \quad [3.21]$$

In this expression, the ± sign before ϕ_v depends on the orientation of the normal vector with respect to the surface Δ.

Since the field source $\mathbf{V}_s$ is proportional to flux ϕ_V, a support field $\boldsymbol{\lambda}_s$ is introduced, such that:

$$\mathbf{V}_S=\phi_V\,\boldsymbol{\lambda}_S \quad [3.22]$$

The properties of the support field $\boldsymbol{\lambda}_s$ are similar to those of $\mathbf{V}_s$ (see equation [3.21]) and are stated as follows:

$$\text{div}\boldsymbol{\lambda}_S = 0, \text{ with } \boldsymbol{\lambda}_S.\mathbf{n}\big|_{\Gamma_m} = 0 \text{ and } \iint_\Delta \boldsymbol{\lambda}_S.\mathbf{n}ds = \pm 1$$
$$\text{i.e. } \boldsymbol{\lambda}_S \in H_{\Gamma_m}(\text{div}0, \Omega) \qquad [3.23]$$

Grouping equations [3.1] and [3.2] with expressions [3.19] and [3.22], the equations to be solved can be written by introducing the source term ϕ_v such that:

$$\mathbf{curlU} = 0 \qquad [3.24]$$

$$\text{div}(\mathbf{V}' + \phi_V \boldsymbol{\lambda}_S) = 0 \qquad [3.25]$$

Similar to section 3.2.1.1, the boundary condition is transferred to a volume source term "$\phi_v\boldsymbol{\lambda}_s$". Finally, the boundary conditions for the unknown of the problem are stated as follows:

$$\mathbf{V}'.\,\mathbf{n}\big|_{\Gamma_m} = 0, \quad \mathbf{U} \wedge \mathbf{n}\big|_{\Gamma_{n1} \cup \Gamma_{n2}} = 0 \qquad [3.26]$$

Since the studied domain Ω is contractible, it is possible to replace $\boldsymbol{\lambda}_s$ by an associated vector potential $\boldsymbol{\chi}_S$. It is nevertheless important to note that the homogeneous boundary conditions of field $\boldsymbol{\lambda}_s$ are imposed on a not simply connected boundary Γ_m (see Figure 3.1). In the context of such a configuration, some precautions must be taken. This will be described in section 3.2.1.2.2.

3.2.1.2.2. Not simply connected boundary: discussion

Figure 3.2a represents the vector field $\boldsymbol{\lambda}_s$ with the not simply connected boundary Γ_m. To introduce the field $\boldsymbol{\chi}_S$, considering the properties of $\boldsymbol{\lambda}_s$ (see equation [3.23]), one solution would be to consider: $\boldsymbol{\chi}_S \in H_{\Gamma m}(\mathbf{curl}, \Omega)$. In this case, $\boldsymbol{\lambda}_s = \mathbf{curl}\boldsymbol{\chi}_s$ having on Γ_m: $\boldsymbol{\chi}_S \wedge \mathbf{n} = 0$. Replacing $\boldsymbol{\lambda}_s$ by $\boldsymbol{\chi}_s$ in the surface integral of equation [3.23], we have:

$$\iint_\Delta \boldsymbol{\lambda}_S.\mathbf{n}ds = \iint_\Delta \mathbf{curl}\ \boldsymbol{\chi}_S.\mathbf{n}ds = \int_{\gamma_0} \boldsymbol{\chi}_S.\mathbf{dl} = 0 \qquad [3.27]$$

This result is in contradiction to properties [3.23] that $\boldsymbol{\lambda}_s$ must meet. It is due to the fact that the contour γ_0 belongs to the boundary Γ_m where the tangential component of $\boldsymbol{\chi}_s$ is zero and that therefore the term $\boldsymbol{\chi}_S.\mathbf{dl}$ is equal to zero. This contradiction is explained by the fact that the boundary Γ_m is not simply connected.

A solution to this constraint, in the presence of a not simply connected boundary, is to enrich the space of definition of $\boldsymbol{\chi}_s$ as follows:

$$\chi_s \in H^{\Delta}_{\Gamma_m}(\mathbf{curl},\Omega) = H_{\Gamma_m}(\mathbf{curl},\Omega) \oplus \mathcal{H}(\Gamma_m) \qquad [3.28]$$

The function space $H_{\Gamma m}(\mathbf{curl}, \Omega)$ relies on relation [2.54]. As for $\mathcal{H}(\Gamma_m)$, it is a function space of finite dimension equal to the number of cuts to make the boundary simply connected. In our case, for Γ_m (see Figure 3.2a), one cut needs to be introduced, and its dimension is therefore equal to one. The properties of the function space $\mathcal{H}(\Gamma_m)$ can then be defined as follows:

$$\mathcal{H}(\Gamma_m) = \Big\{\chi_c \in \mathbf{L}^2(\Omega); \chi_c\big|_{\Omega} = 0; \\ \chi_c \wedge \mathbf{n}\big|_{\Gamma_m} = 0 \;; \oint_{\gamma_0} \chi_c.\mathbf{dl} = \pm 1\Big\} \qquad [3.29]$$

As shown by Figure 3.2a, it should be recalled that γ_0 is a contour belonging to Γ_m that cannot be contracted to a point by continuous transformation.

A possible cut is represented in Figure 3.2b. It is in fact a path linking surfaces Γ_{n1} and Γ_{n2} and cutting the surface Γ_m so that it becomes simply connected. At the passage of the cut, the circulation of $\boldsymbol{\chi}_c$ along a segment crossing it jumps by ± 1 depending on the crossing sense with respect to the orientation of the cut. As an example, Figure 3.2b shows three closed paths γ_0, γ_1 and γ_2. γ_0 is the only one that cannot be contracted to a point by continuous transformation on surface Γ_m. The circulation of contour γ_0 crosses the cut in the direction of the arrows. Under these conditions, the circulation of $\boldsymbol{\chi}_c$ along γ_0 is equal to 1. Consider now two contours γ_1 and γ_2 that can be contracted to a point by continuous transformation on Γ_m. The contour γ_1, belonging to Γ_m, does not cross the cut, the circulation of $\boldsymbol{\chi}_c$ is equal to "0". The same is true for the circulation on contour γ_2 which intersects the cut in one direction, then in the opposite direction. Therefore, the function $\boldsymbol{\chi}_c$ accounts for the topological "singularity" of surface Γ_m.

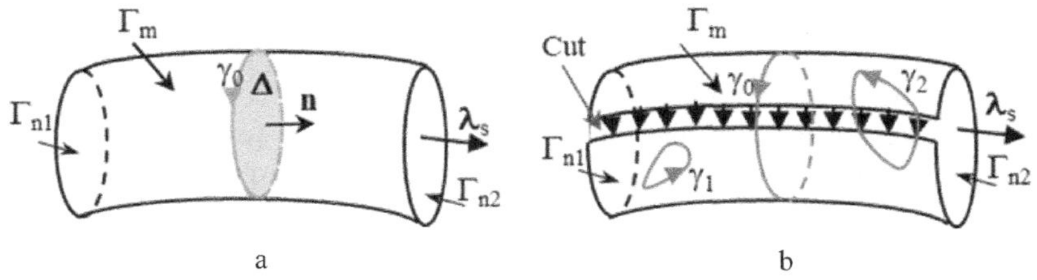

Figure 3.2. *a) Domain with a not simply connected boundary Γ_m; b) introduction of a cut on Γ_m*

Under these conditions, $\boldsymbol{\chi}_s$ can be written as:

$$\chi_s = \chi'_s + \chi_c \quad \text{with:} \chi'_s \in H_{\Gamma_m}(\mathbf{curl}, \Omega) \text{ and } \chi_c \in \mathcal{H}(\Gamma_m) \qquad [3.30]$$

And the expression of $\boldsymbol{\lambda}_s$ is:

$$\lambda_s = \mathbf{curl}\chi_s = \mathbf{curl}(\chi'_s + \chi_c) \qquad [3.31]$$

Consider now the integral form of equation [3.23]. $\boldsymbol{\lambda}_s$ is replaced by $\mathbf{curl}\boldsymbol{\chi}_s$ and the Stokes formula is applied. The following succession of equations can then be written as:

$$\iint_\Delta \lambda_s.\mathbf{n}ds = \iint_\Delta \mathbf{curl}\ \chi_s.\mathbf{n}ds = \int_{\gamma_0} \chi_c.\mathbf{dl} = 1 \qquad [3.32]$$

Finally, it is important to note that $\boldsymbol{\chi}_s$ is not unique and is defined up to a gradient, as shown by equation [2.46]. A gauge condition should therefore be imposed, as proposed in section 2.5.2.

3.2.2. *Source quantities inside the domain*

It is important to note that, for an electrostatic problem, the source term can be a charge density ρ. This charge density leads, through a volume integral, to the quantity Q. Likewise, in magnetostatics, inside the studied domain, there may be an inductor through which flows a current density $\mathbf{J}_0$, see equation [1.65]. It is important to note that, in this case, the global source quantity is the intensity I of the current flowing through the conductors [1.62]. Still in magnetostatics, the existence of permanent magnets in the domain leads to a source term associated with the coercive field or the remanent magnetic flux density (see equations [1.63] and [1.64]). This particular source term will be considered in section 3.5.

In what follows, similar to section 3.2.1, the notion of source fields, support fields and associated potentials will be introduced, to simplify the equations to be solved. In order to maintain a degree of consistency with the previous developments, the same notations defined at the beginning of section 3.2.1 will be used. However, in order to take into account the local quantities, equations [3.1] and [3.2] will be modified by introducing, respectively, the source terms $\mathcal{J}$ and q as follows:

$$\mathbf{curl}\mathbf{U} = \mathcal{J} \qquad [3.33]$$

$$\text{div}\mathbf{V} = q \qquad [3.34]$$

According to the problem dealt with, $\mathcal{J}$ or q will be considered equal to zero. The boundary conditions are those defined by equation [3.3]:

$$\Gamma = \Gamma_{n_1} \cup \Gamma_{n_2} \cup \Gamma_{m}, \text{ with } \mathbf{V}.\mathbf{n}\big|_{\Gamma_m} = 0, \ \mathbf{U} \wedge \mathbf{n}\big|_{\Gamma_{n1} \cup \Gamma_{n2}} = 0 \qquad [3.35]$$

Depending on the constraints on fields **U** and **V**, the boundary conditions can be modified to fit the studied problem.

3.2.2.1. *Local source quantity defined by a curl*

For this study, consider, as an example, the geometry shown in Figure 3.3, whose domain Ω is contractible. In this case, for equation [3.34], we pose $q = 0$. It should be noted that the subdomain Ω_s, of boundary Γ_s, support to the source term, is not simply connected (case of torus). The complementary of Ω_s, in the domain Ω, is denoted by Ω_0. On the one hand, in the studied example and to simplify the developments (see Figure 3.3), there are no gates and therefore $\mathbf{V}.\mathbf{n} = 0$ on the boundary Γ.

Because the vector $\mathcal{J}$ derives from a curl [3.33], it is divergence free. The density $\mathcal{J}$ is defined only in the subdomain Ω_s and verifies on its boundary $\mathcal{J}.\mathbf{n}\,|_{\Gamma s} = 0$. The flux of $\mathcal{J}$ through an arbitrary surface Δ (see Figure 3.3), representing a cross section of the inductor, is denoted by I. Then:

$$\iint_{\Delta} \mathcal{J}.\mathbf{n}dS = I \qquad [3.36]$$

It can be noted that the section of the inductor may vary. Considering equation [3.33] and its domain of definition, the field $\mathcal{J}$ belongs to $H_{\Gamma s}(\text{div}0, \Omega_s)$. By extension, the field $\mathcal{J}$ is set to zero in Ω_0. This extension is possible, as, even though it appears to be a discontinuity of $\mathcal{J}$ at the boundary of the domain Ω_s, there is no discontinuity of the normal component as $\mathcal{J}.\mathbf{n} = 0$ on this surface. Under these conditions, $\mathcal{J}$ is defined throughout the domain Ω, which is contractible. Then, it belongs to H_0 (div0, Ω).

Now we introduce the support field $\boldsymbol{\lambda}_{sl}$ in the form:

$$\mathcal{J} = I\,\boldsymbol{\lambda}_{sl} \qquad [3.37]$$

Relying on the domain of definition of the vector field $\mathcal{J}$, the properties of $\boldsymbol{\lambda}_{sl}$ are deduced from equation [3.23] and can be written as follows:

$$\begin{aligned} &\text{div}\boldsymbol{\lambda}_{sl} = 0, \quad \iint_{\Delta} \boldsymbol{\lambda}_{sl}.\mathbf{n}\,dS = 1, \quad \text{with} \quad \boldsymbol{\lambda}_{sl}.\mathbf{n}\big|_{\Gamma_s} = 0 \\ &\text{i.e.} \quad \boldsymbol{\lambda}_{sl} \in H_0(\text{div}0, \Omega) \end{aligned} \qquad [3.38]$$

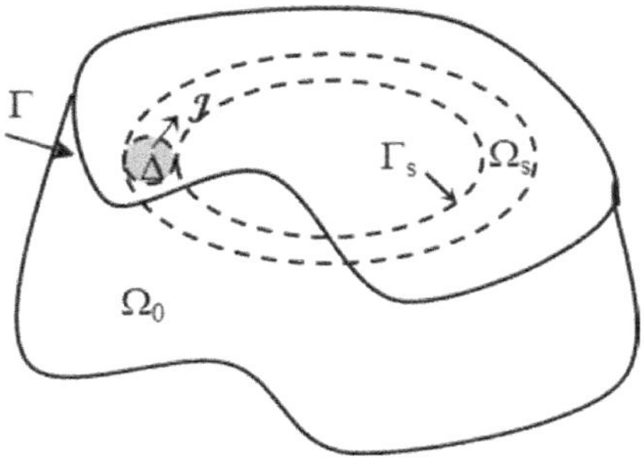

Figure 3.3. *Geometry with a local source quantity $\mathcal{J}$ inside the studied domain*

In equation [3.38], concerning the second condition, similar to equation [3.36], surface Δ represents an arbitrary internal section of the inductor whose contour lies on the boundary Γ_s.

It is important to note that the subdomain Ω_s is not simply connected. But $\mathcal{J}$ and therefore $\boldsymbol{\lambda}_{sl}$ were defined over the whole domain Ω, which is contractible. Considering this definition (see section 2.4.2.1), an associated potential $\boldsymbol{\chi}_{sl}$ is introduced, having the following properties:

$$\mathbf{curl}\chi_{sl} = \boldsymbol{\lambda}_{sl}, \text{ with } \chi_{sl} \wedge \mathbf{n}\big|_{\Gamma} = 0 \text{ i.e. } \chi_{sl} \in H_0(\mathbf{curl}, \Omega) \qquad [3.39]$$

Consider again equation [3.33] and replace the source term $\mathcal{J}$ by its expression as a function of I and $\boldsymbol{\chi}_{sl}$, obtained from equations [3.37] and [3.39]. The following equations can then be written as:

$$\mathbf{curlU} = \mathbf{curl}\, I\,\chi_{sl} \Rightarrow \mathbf{curl}(\mathbf{U} - I\chi_{sl}) = 0 \qquad [3.40]$$

Introducing a new unknown **U'** such that $\mathbf{U}' = \mathbf{U} - \mathrm{I}\boldsymbol{\chi}_{sl}$, equation [3.33] of the initial problem can be written, based on equation [3.40], as follows:

$$\mathbf{curlU}' = 0, \text{ with } \mathbf{U}' \in H(\mathbf{curl}0, \Omega) \qquad [3.41]$$

Dissociating the source term and introducing **U'** will allow us to readily develop potential formulations, as the equation is now homogeneous (the right-hand side term is zero).

3.2.2.2. *Local source quantity defined by a divergence*

Consider again the previous example (see Figure 3.4), but in the presence of a density q defined on Ω_s and of two gates Γ_{n1} and Γ_{n2}. In this case, in equation [3.33], we have $\mathcal{J} = 0$. On the contrary, the local source term q in equation [3.34] is preserved. The studied domain Ω of boundary Γ is the union of Ω_0 and Ω_s. The integration over Ω_s of density q highlights the global integral quantity $\mathcal{Q}$ as follows:

$$\iiint_{\Omega_s} q \, d\tau = \mathcal{Q} \qquad [3.42]$$

Finally, the boundary conditions on the boundary Γ of the domain are defined by equation [3.35].

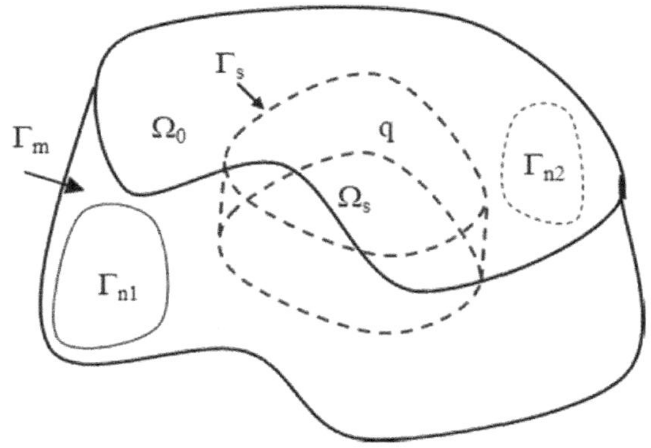

Figure 3.4. *Geometry with a local source quantity, represented by a density q inside the studied domain*

As the distribution of density q is known, its expression can be written as a function of the global quantity $\mathcal{Q}$. In fact, the distribution of density q is transferred on a support scalar term ξ_{sl} such that:

$$q = \mathcal{Q}\,\xi_{sl} \text{ and } \iiint_{\Omega_s} \xi_{sl} d\tau = 1, \text{ with } \xi_{sl} \in L^2(\Omega_s) \qquad [3.43]$$

Outside Ω_s, the density is zero (q = 0), posing $\xi_{sl} = 0$ in Ω_0, which allows for the extension of its definition to the entire domain, hence: $\xi_{sl} \in L^2(\Omega)$.

Let us now introduce a source field $\mathbf{V}_s$, whose boundary conditions are identical to those of $\mathbf{V}$, defined as follows:

$$\text{div}\mathbf{V}_s = Q\,\xi_{sl}, \text{ with } \mathbf{V}_s \in H_{\Gamma_m}(\text{div},\Omega) \quad [3.44]$$

It is then possible to introduce a second support field $\boldsymbol{\eta}_{sl}$ such that:

$$\mathbf{V}_s = Q\,\boldsymbol{\eta}_{sl} \quad [3.45]$$

Grouping equations [3.44] and [3.45], after simplification, the following can be written:

$$\text{div}\,\boldsymbol{\eta}_{sl} = \xi_{sl} \quad [3.46]$$

where $\boldsymbol{\eta}_{sl}$ verifies the following properties:

$$\boldsymbol{\eta}_{sl}.\mathbf{n}\big|_{\Gamma_m} = 0 \text{ i.e. } \boldsymbol{\eta}_{sl} \in H_{\Gamma_m}(\text{div},\Omega) \quad [3.47]$$

It should be noted that there are an infinite number of fields $\boldsymbol{\eta}_{sl}$ verifying conditions [3.46] and [3.47] and uniqueness is imposed by adding a gauge condition (see section 2.5.2).

Based on equations [3.34], [3.43] and [3.46], the following can be written:

$$\text{div}(\mathbf{V} - Q\,\boldsymbol{\eta}_{sl}) = 0 \quad [3.48]$$

Consider now $\mathbf{V}' = \mathbf{V} - Q\boldsymbol{\eta}_{sl}$, with the same boundary conditions applicable for $\mathbf{V}$ and $\boldsymbol{\eta}_{sl}$ being valid for $\mathbf{V}'$. Equation [3.48] can then be rewritten in the form:

$$\text{div}\mathbf{V}' = 0, \text{ with } \mathbf{V}' \in H_{\Gamma_m}(\text{div}0,\Omega) \quad [3.49]$$

The field $\mathbf{V}'$ is therefore divergence free, which, as the following section will show, makes it easy to introduce potential formulations.

3.2.3. *Examples of the calculation of support fields*

Sections 3.2.1 and 3.2.2 referred to source terms represented by support vector fields $\boldsymbol{\beta}_s$, $\boldsymbol{\lambda}_s$, $\boldsymbol{\eta}_{sl}$ and the scalar field ξ_{sl} as well as the associated potentials α_s, $\boldsymbol{\chi}_s$ and

χ_{sl}. There are an infinite number of fields meeting these conditions. As an illustration, this section presents possible analytical solutions for extremely simple cases. The aim is to illustrate our purpose and also the fact that support fields, source fields and associated potentials have an infinite number of solutions. On the contrary, in the case of complex geometries, where Maxwell's equations will be solved using the finite element method, section 4.3.7 proposes general and systematic numerical methods for calculating these fields.

3.2.3.1. *Calculation of a support field β_s and the potential α_s*

The support field $\boldsymbol{\beta}_s$ serves to impose the circulation of a field between two disjoint boundaries of a domain. This field is defined by the relations given in equation [3.12]. As for the associated scalar potential α_s, it must verify equations [3.13].

For the calculation of these two terms, the geometry in Figure 3.5 can be viewed as an example, namely a brick-shaped domain Ω, of boundary $\Gamma = \Gamma_{n1} \cup \Gamma_{n2} \cup \Gamma_m$. The boundaries Γ_{n1} and Γ_{n2} are in the (x,y) plane in z = 0 and z = L, respectively. For the source field, the boundaries Γ_{n1} and Γ_{n2} represent gates and the boundary Γ_m is a wall.

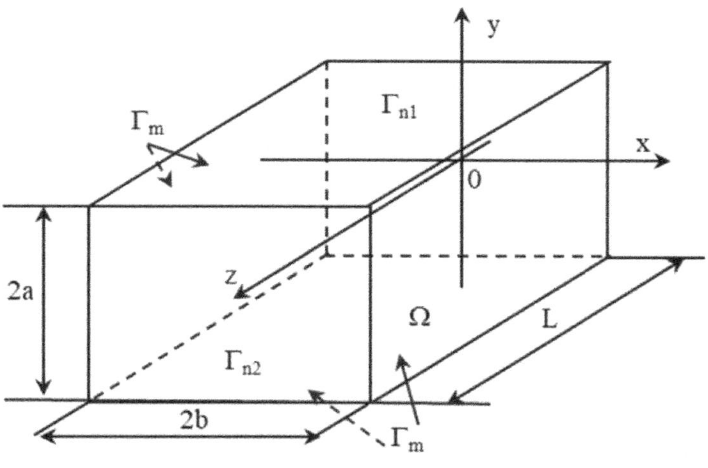

Figure 3.5. *Geometry studied for the calculation of fields β_s and α_s*

As mentioned, there are an infinite number of solutions for the support field $\boldsymbol{\beta}_s$. However, given the simplicity of the studied geometry, it is easy to find an analytical solution. As an example, it is possible to have for vector $\boldsymbol{\beta}_s$:

$$\boldsymbol{\beta}_S = \begin{pmatrix} 0 \\ 0 \\ 1/L \end{pmatrix} \qquad [3.50]$$

This vector field verifies the properties defined in equation [3.12]. As shown by equation [3.13], the scalar potential α_s is defined by its gradient. The integration of equation [3.13] allows us to write the following equation:

$$\alpha_s(z) = -\frac{z}{L} + \mathcal{K} \qquad [3.51]$$

where $\mathcal{K}$ is a constant. If this constant is set to 1, relation [3.14] is verified. Then, $\alpha_s(0) = 1$ and $\alpha_s(L) = 0$.

3.2.3.2. *Calculation of a support field λ_s and of the associated potential χ_s*

For the calculation of fields $\boldsymbol{\lambda}_s$ and $\boldsymbol{\chi}_s$, the same geometry as in Figure 3.5 is used. But now the source term is considered to be an imposed flux ϕ_v flowing through the boundaries Γ_{n1} and Γ_{n2} (see equation [3.21]). The properties of field $\boldsymbol{\lambda}_s$ are given by relations [3.23]. As for the potential $\boldsymbol{\chi}_s$, with the boundary Γ_m being not simply connected, it is defined by relation [3.31]. In the case of the studied geometry, to verify the properties of $\boldsymbol{\lambda}_s$, one solution consists of considering, over the domain Ω, a field defined as follows:

$$\boldsymbol{\lambda}_S = \begin{pmatrix} 0 \\ 0 \\ 1/(4ab) \end{pmatrix} \qquad [3.52]$$

It can be readily verified that $\boldsymbol{\lambda}_s$ is divergence free. Moreover, considering its direction along z, its normal component on the boundary Γ_m is zero. Finally, the surface integral on Γ_{n1} is equal to –1 and to 1 on Γ_{n2}. The field $\boldsymbol{\lambda}_s$ can therefore be used for the studied problem as a support field.

However, based on the above-mentioned expression of $\boldsymbol{\lambda}_s$, it is impossible to deduce an analytical expression of the field $\boldsymbol{\chi}_s$ that verifies the properties defined by relations [3.28]–[3.31]. This difficulty is mainly due to the fact that Γ_m is a not simply connected boundary. Therefore, a cut needs to be introduced.

Relying on the developments proposed in section 3.2.1.2.2, a possibility to cut the boundary Γ_m, to "make it" simply connected, can be found in Figure 3.6. The field $\boldsymbol{\chi}_c$ is zero in the domain and its tangential component is also zero on Γ_m except, as shown in Figure 3.6, at the cut, with a jump of ± 1 depending on the direction in which it is crossed. It is important to note that the field $\boldsymbol{\chi}_c$ is directed along "y". A new field $\boldsymbol{\lambda}_s$ is also defined along the cut (see gray arrow in Figure 3.6), directed along the "z" axis.

Let us now choose an arbitrary surface "Δ_0" (see Figure 3.6), belonging to the domain Ω and the contour γ_0. If the contour γ_0 intersects the cut, then the flux of $\boldsymbol{\lambda}_s$ is equal to ± 1 according to the orientation of γ_0. Considering equations [3.31] and [3.32], $\boldsymbol{\chi}_c$ and $\boldsymbol{\lambda}_s$ have the following properties:

$$\iint_{\Delta_0} \lambda_S.\mathbf{n}dS = \iint_{\Delta_0} \mathbf{curl}\,\chi_S.\mathbf{n}dS = \oint_{\gamma_0} \chi_C.\mathbf{dl} = \pm 1 \qquad [3.53]$$

Otherwise, if there is no intersection between γ_0 and the cut, the flux is equal to zero.

As a conclusion, for the example presented in Figure 3.5, there are two possibilities for the support field. The first one consists of taking as a source term the vector field $\boldsymbol{\lambda}_s$ defined by equation [3.52]. However, in this case, it will not be possible to build a field $\boldsymbol{\chi}_s$ belonging to the spaces defined by equation [3.28]. For the second possibility, the aim is to find a field $\boldsymbol{\chi}_s$ by introducing a cut on the boundary Γ_m, as shown in Figure 3.6, verifying equation [3.31] with a field $\boldsymbol{\chi}_c$ belonging to the space $\mathcal{H}(\Gamma_m)$, defined by relation [3.29].

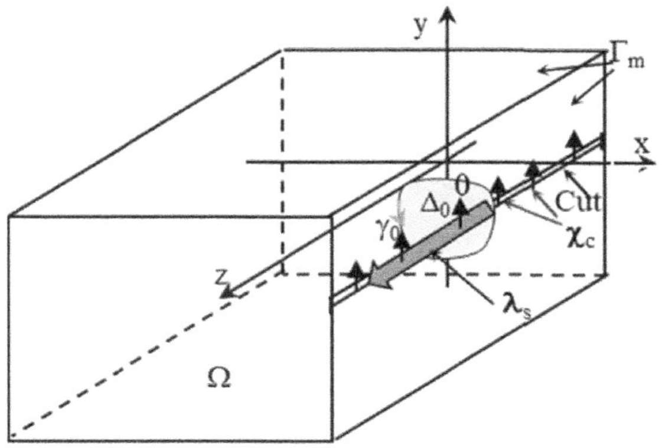

Figure 3.6. *Introduction of a cut in the geometry studied in Figure 3.5*

3.2.3.3. *Local source term: calculation of the support field λ_{sl} and the potential χ_{sl}*

In the case of a source term located inside the domain and defined by a curl, section 3.2.2.1 introduced the support field $\boldsymbol{\lambda}_{sl}$ and the associated potential $\boldsymbol{\chi}_{sl}$. Unlike the sources imposed on the boundaries, we have to account for their geometry.

For our study, consider the relatively simple example presented in Figure 3.7. There is a domain Ω of boundary Γ. At the center of Ω, there is a subdomain Ω_s in the form of square section circular ring (not simply connected domain) support to the source term. The dimensions as well as the orientation of the source $\mathcal{J}$ are defined in Figure 3.7.

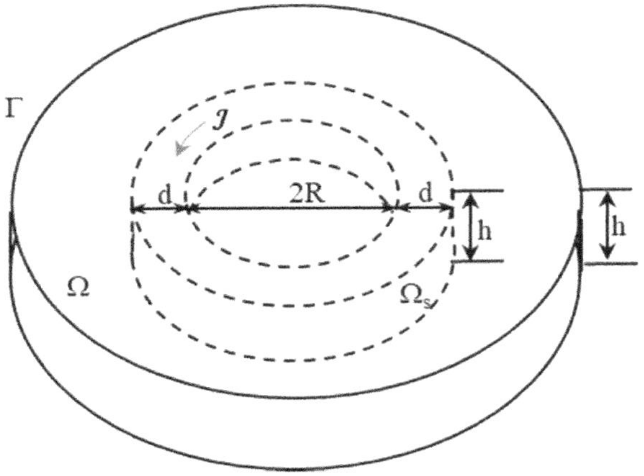

Figure 3.7. *Geometry studied for the calculation of support fields λ_{sl} and χ_{sl}*

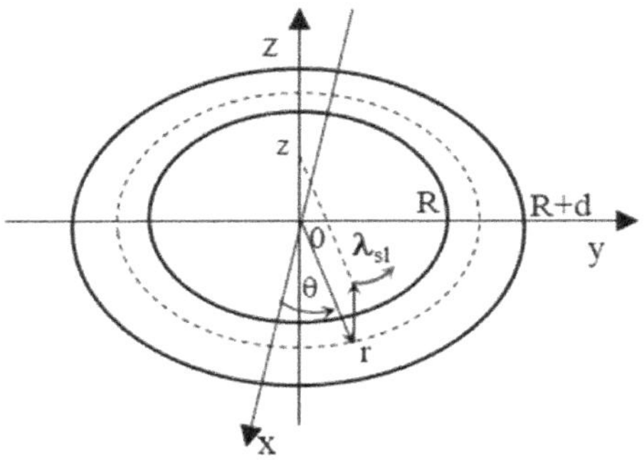

Figure 3.8. *Cylindrical coordinates of support fields λ_{sl}*

As already seen in section 3.2.2.1, in order to address the connectivity problems of the subdomain Ω_s, the domain of definition of $\boldsymbol{\lambda}_{sl}$ and $\boldsymbol{\chi}_{sl}$ is extended to Ω. The support field $\boldsymbol{\lambda}_{sl}$ can be determined by relying on equation [3.38]. In cylindrical coordinates, this vector field (see Figure 3.8) has only one component following θ and can be defined by the following relation for:

$$R \le r \le (R+d),\ 0 \le z \le h,\ \boldsymbol{\lambda}_{sl} = \begin{pmatrix} \lambda_{slr} \\ \lambda_{sl\theta} \\ \lambda_{slz} \end{pmatrix} = \begin{pmatrix} 0 \\ \dfrac{1}{hd} \\ 0 \end{pmatrix} \tag{3.54}$$

$$\text{and } \boldsymbol{\lambda}_{sl} = 0 \text{ elsewhere}$$

Based on the above expression, it can be readily verified that $\boldsymbol{\lambda}_{sl}$ meets the conditions stated in equation [3.38]. It is important to note that, in the complementary of Ω_s in Ω, $\boldsymbol{\lambda}_{sl} = 0$. Figure 3.8 represents the orientation of the support field $\boldsymbol{\lambda}_{sl}$ for $R \le r \le (R + d)$ and $0 \le z \le h$.

Having defined the vector field $\boldsymbol{\lambda}_{sl}$, the objective is to determine $\boldsymbol{\chi}_{sl}$ relying on its definition given by the relations in equation [3.39]. Let us first recall the expression of the curl in cylindrical coordinates, which leads to the expression of $\boldsymbol{\lambda}_{sl}$:

$$\mathbf{curl}\boldsymbol{\chi}_{sl} = \begin{pmatrix} \dfrac{1}{r}\dfrac{\partial \chi_{slz}}{\partial \theta} - \dfrac{\partial \chi_{sl\theta}}{\partial z} \\ \dfrac{\partial \chi_{slr}}{\partial z} - \dfrac{\partial \chi_{slz}}{\partial r} \\ \dfrac{1}{r}\left(\dfrac{\partial (r\chi_{sl\theta})}{\partial r} - \dfrac{\partial \chi_{slr}}{\partial \theta}\right) \end{pmatrix} = \boldsymbol{\lambda}_{sl} = \begin{pmatrix} 0 \\ \dfrac{1}{hd} \\ 0 \end{pmatrix} \tag{3.55}$$

$$\text{for } R \le r \le (R+d) \ ,\ 0 \le z \le h$$

Considering the constraints defined by expression [3.54], a solution for $\boldsymbol{\chi}_{sl}$ can be:

$$\chi_{sl} = \begin{pmatrix} 0 \\ 0 \\ \chi_{slz(r)} \end{pmatrix} \text{ with } 0 \le z \le h:$$

$$\text{for } r < R : \chi_{slz(r)} = \frac{1}{h},$$

$$\text{for } R \le r \le (R+d) : \chi_{slz(r)} = \frac{1}{hd}(-r + (R+d)), \tag{3.56}$$

$$\text{for } (R+d) < r : \chi_{slz(r)} = 0$$

It can be readily verified that the vector $\boldsymbol{\chi}_{ls}$ meets the conditions defined in equation [3.39]. On the contrary, it can be verified that $\boldsymbol{\lambda}_{sl}$ is zero outside Ω_s, which is not the case for the vector $\boldsymbol{\chi}_{ls}$.

3.2.3.4. *Local source term: calculation of the support field ξ_{sl} and the potential η_{sl}*

For the calculation of source terms, in the case of a source located inside the domain and associated with a divergence, the elementary geometry represented in Figure 3.9 is considered. This parallelepipedic domain Ω contains a subdomain Ω_s inside which there is a charge density q. In order to simplify these calculations, this density is assumed to be uniformly distributed. The boundary Γ of the domain is the union of lateral boundaries Γ_{n1} and Γ_{n2} which represent two gates and of the boundary Γ_m associated with a wall. It is important to note that the subdomain Ω_s, also parallelepipedic, has its lower and upper faces in contact with Γ_m, as well as two of its lateral faces. The origin of Cartesian coordinates is located at the center of the domain Ω. For this example, the objective is to calculate the possible solutions for $\boldsymbol{\eta}_{sl}$ and ξ_{sl} linked by equation [3.46].

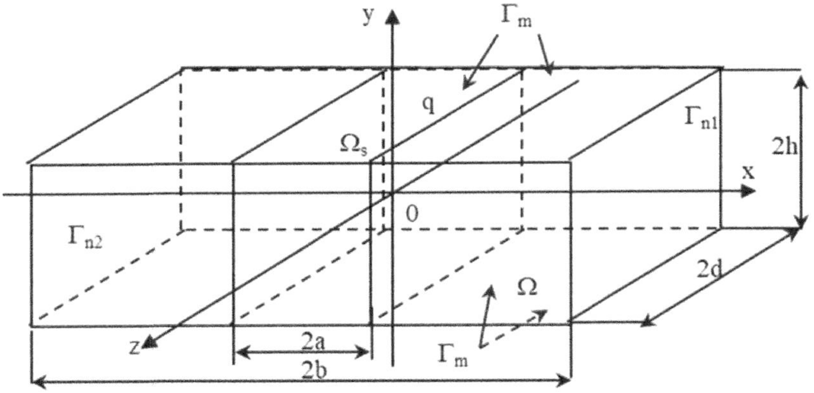

Figure 3.9. *Geometry studied for the calculation of support fields ξ_{sl} and η_{sl}*

For the determination of ξ_{sl}, expression [3.43] is used. As the density q is assumed to be uniformly distributed over the subdomain Ω_s, the expression of ξ_{sl} is given by:

$$\xi_{sl} = \frac{1}{\text{vol}(\Omega_s)} \quad [3.57]$$

where $\mathrm{vol}(\Omega_s)$ represents the volume Ω_s. It is easy to extend the definition of ξ_{sl} to the entire domain Ω by posing $\xi_{sl} = 0$ on the complementary of Ω_s with respect to Ω. Then, $\xi_{sl} \in L^2(\Omega)$.

Concerning the support field $\boldsymbol{\eta}_{sl}$, it is defined by relations [3.46] and [3.47]. Considering the symmetries of the studied geometry and the boundaries Γ_{n1} and Γ_{n2}, which represent gates, it is invariant along y and z. A possible solution involves the definition of this vector field by only one component along x, such that:

$$\eta_{slx} = -\frac{a}{\mathrm{vol}(\Omega_s)} \quad \forall -b \le x < -a,$$
$$\eta_{slx} = \frac{x}{\mathrm{vol}(\Omega_s)} \quad \forall -a \le x \le a \text{ and } \eta_{slx} = \frac{a}{\mathrm{vol}(\Omega_s)} \quad \forall\ a < x < b \qquad [3.58]$$

with the components $\boldsymbol{\eta}_{sly}$ and $\boldsymbol{\eta}_{slz}$ being zero. The divergence of field $\boldsymbol{\eta}_{sl}$, defined by relation [3.58], is equal to ξ_{sl}, as defined by equation [3.46]. Moreover, it can be verified that $\boldsymbol{\eta}_{sl} \in H_{\Gamma m}(\mathrm{div}, \Omega)$.

3.3. Electrostatics

In the case of electrostatics, potential formulations can be developed relying on function spaces defined in Chapter 2. The first notion introduced is the electric scalar potential V, and then the electric vector potential **P**. Then, the Tonti diagram will be obtained.

In order to consider various possibilities, we examine in the first example, close to the one presented in section 1.5.1, the case of source terms (f_s [1.51], then σ_s [1.52]) imposed on the boundary of the domain. Then, a problem with source terms on the boundary of the domain and an internal electrode will be studied.

3.3.1. *Source terms imposed on the boundary of the domain*

The objective of this section is to develop the formulations in terms of scalar and vector potential when the source term is imposed on the boundary of the domain. To this end, the simplified geometry represented in Figure 3.10 will be studied. The domain Ω is contractible and its boundary is denoted by Γ. This example involves two types of boundary conditions. The first one, denoted by Γ_d (see equation [1.39]), represents a wall for the electric displacement field on the boundaries of the domain. The second one, denoted by Γ_{ek} (see equation [1.35]), represents a gate for the electric field. The two boundaries Γ_{e1} and Γ_{e2} behave as perfect electrodes denoted, respectively, by $\mathcal{E}_1$ and $\mathcal{E}_2$. The permittivity of the domain Ω is denoted by ε and may depend on the position.

As already seen in section 1.5.1, it is possible to consider two source terms imposed on the boundaries of the domain. The first one, denoted by f_s, corresponds to the circulation of the electric field between Γ_{e1} and Γ_{e2} (see equation [1.51]). The second one is the total charges $\pm Q_\sigma$ on each of the two electrodes (see equation [1.53]).

It should also be recalled that, in the absence of volume charges, the equations of electrostatics are stated as follows:

$$\mathbf{curlE} = 0 \quad [3.59]$$

$$\mathrm{div}\mathbf{D} = 0 \quad [3.60]$$

with the electric behavior law [1.19] and the boundary conditions defined by equations [1.35] and [1.39] written as:

$$\Gamma = \Gamma_{e1} \cup \Gamma_{e2} \cup \Gamma_d \text{ with } \mathbf{D.n}\big|_{\Gamma_d} = 0,\ \mathbf{E} \wedge \mathbf{n}\big|_{\Gamma_{ek}} = 0, \text{ with } k \in \{1,2\} \quad [3.61]$$

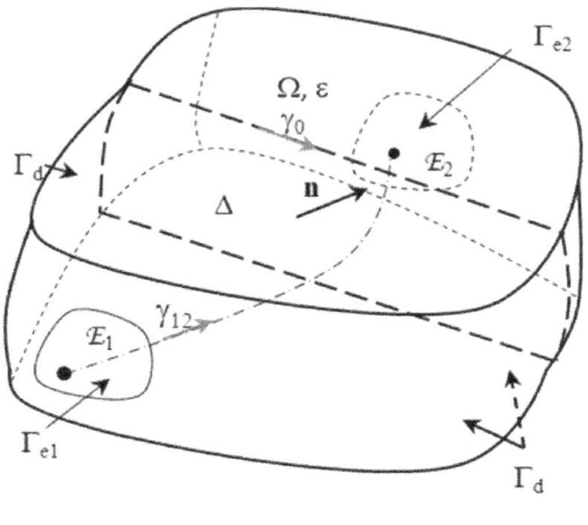

Figure 3.10. *Geometry studied in electrostatics when the source terms are imposed on the boundary*

Considering equations [3.59]–[3.61], the fields **E** and **D** belong to the following function spaces:

$$\mathbf{E} \in H_{\Gamma_{e1} \cup \Gamma_{e2}}(\mathbf{curl}0, \Omega) \quad \text{and} \quad \mathbf{D} \in H_{\Gamma_d}(\text{div}0, \Omega) \qquad [3.62]$$

3.3.1.1. *Scalar potential V formulation*

The electric scalar potential formulation is well suited to the case when the circulation of the electric field (source term f_s) is imposed. On the contrary, when the source term is directly related to the electric displacement field, which is the case when a surface charge density is imposed on the electrodes, it is then necessary to introduce an additional equation, resulting from an energy balance.

3.3.1.1.1. Imposed circulation of the electric field f_s

This configuration corresponds to a global quantity imposed on the boundary via a source term defined by a curl. We will therefore introduce into the local equations the term f_s following the steps proposed in section 3.2.1.1. As shown in equation [3.8], the approach involves the decomposition of the electric field into two terms as follows:

$$\mathbf{E} = \mathbf{E}_s + \mathbf{E}' \qquad [3.63]$$

where $\mathbf{E}_s$ represents the source field linked to the circulation f_s and **E**' is the unknown of the problem. As the source term is supported by $\mathbf{E}_s$, the circulation of **E**' between the gates Γ_{e1} and Γ_{e2} is equal to zero. Relying on equation [3.9], the properties of field **E**' are written as:

$$\mathbf{curl}\mathbf{E}' = 0, \ \int_{\gamma_{12}} \mathbf{E}'.\mathbf{dl} = 0, \ \mathbf{E}' \wedge \mathbf{n}\big|_{\Gamma_{e1} \cup \Gamma_{e2}} = 0 \ \text{and} \ \mathbf{E}' \in H_{\Gamma_{e1} \cup \Gamma_{e2}}(\mathbf{curl}0, \Omega) \qquad [3.64]$$

Similarly, based on equation [3.10], the properties of $\mathbf{E}_s$ are given by the relations:

$$\mathbf{curl}\mathbf{E}_s = 0, \ \int_{\gamma_{12}} \mathbf{E}_s.\mathbf{dl} = f_s, \ \mathbf{E}_s \wedge \mathbf{n}\big|_{\Gamma_{e1} \cup \Gamma_{e2}} = 0 \ \text{and} \ \mathbf{E}_s \in H_{\Gamma_{e1} \cup \Gamma_{e2}}(\mathbf{curl}0, \Omega) \qquad [3.65]$$

The field $\mathbf{E}_s$ can be expressed (see equation [3.11]) as a function of f_s using a support field, denoted by $\boldsymbol{\beta}_e$:

$$\mathbf{E}_s = f_s \boldsymbol{\beta}_e \qquad [3.66]$$

Since the support vector field is proportional to $\mathbf{E}_s$, then for $\boldsymbol{\beta}_e$, the properties defined by equation [3.12] can be written as:

$$\mathbf{curl}\boldsymbol{\beta}_e = 0,\ \int_{\gamma_{12}} \boldsymbol{\beta}_e.\mathbf{dl} = 1, \text{ with } \boldsymbol{\beta}_e \wedge \mathbf{n}\big|_{\Gamma_{e1}\cup\Gamma_{e2}} = 0,\ \boldsymbol{\beta}_e \in H_{\Gamma_{e1}\cup\Gamma_{e2}}(\mathbf{curl}0,\Omega) \qquad [3.67]$$

Given its properties and the fact that the studied domain is contractible, $\boldsymbol{\beta}_e$ can be expressed as a function of an associated scalar potential α_e using relations [3.13] and [3.14]. Under these conditions, for the given example, we have:

$$\boldsymbol{\beta}_e = -\mathbf{grad}\alpha_e \text{ with } \alpha_e\big|_{\Gamma_{e1}} = 1,\ \alpha_e\big|_{\Gamma_{e2}} = 0,\ \alpha_e \in H(\mathbf{grad},\Omega) \qquad [3.68]$$

Then, the source field $\mathbf{E}_s$ has the form:

$$\mathbf{E}_s = -f_s\mathbf{grad}\,\alpha_e \qquad [3.69]$$

Relying on equations [3.63] and [3.66], the source term is introduced in the expression of the electric field $\mathbf{E}$, and therefore equation [3.59] can be written as:

$$\mathbf{curl}(f_s\boldsymbol{\beta}_e + \mathbf{E}') = 0 \qquad [3.70]$$

where the source field "$f_s\boldsymbol{\beta}_e$" is assumed to be known. The unknown of the problem is then the field $\mathbf{E}'$.

Concerning the field $\mathbf{E}'$, taking into account the properties stated in equation [3.64], it can be expressed by means of an electric scalar potential, which is denoted by V. As the tangential component of $\mathbf{E}'$ is zero on surfaces Γ_{e1} and Γ_{e2}, this potential is constant and may take, respectively, the values V_1 and V_2. Moreover, the circulation of $\mathbf{E}'$ between two electrodes being equal to zero (see relation [3.64]), we have to impose $V_1 = V_2$. Finally, in order to obtain a unique solution (see section 2.5.2.1), the value of the electric scalar potential at one point of the domain must be fixed. Under these conditions, consider: $V_1 = V_2 = 0$, which corresponds to Dirichlet boundary conditions. The expression of $\mathbf{E}'$ can then be written as follows:

$$\mathbf{E}' = -\mathbf{grad}V \text{ with } V \in H_{\Gamma_{e1}\cup\Gamma_{e2}}(\mathbf{grad},\Omega) \qquad [3.71]$$

Based on equations [3.63], [3.66] and [3.71], the expression of the electric field is:

$$\mathbf{E} = f_s\boldsymbol{\beta}_e - \mathbf{grad}V \qquad [3.72]$$

The electric displacement field results from equation [3.72] and the behavior law [1.19]:

$$\mathbf{D} = \varepsilon(f_s\boldsymbol{\beta}_e - \mathbf{grad}V) \tag{3.73}$$

NOTE.– It can be noted that, on the boundary Γ_d, the wall-type condition (see equation [3.61]) for the electric displacement field is reflected by a Neumann boundary condition for the scalar potential V.

Replacing in equation [3.60] the electric displacement field by its expression given in equation [3.73], the scalar potential electrostatic formulation can be written as follows:

$$\text{div}(\varepsilon(f_s\boldsymbol{\beta}_e - \mathbf{grad}V)) = 0 \tag{3.74}$$

Replacing the support field $\boldsymbol{\beta}_e$ by the associated scalar potential α_e, equation [3.74] can be rewritten as follows:

$$\text{div}(\varepsilon(f_s\mathbf{grad}\alpha_e + \mathbf{grad}V)) = 0 \tag{3.75}$$

We have thus developed the electric scalar potential formulation when the circulation of the electric field **E**, inside the domain, is imposed between two electrodes.

3.3.1.1.2. Charges Q_σ imposed on the electrodes

With the scalar potential formulation, imposing the total charges Q_σ requires a complementary development. The objective is to express Q_σ on the electrodes as a function of the electric scalar potential and the circulation f_s, which then becomes an unknown of the problem. This requires an energy balance.

In the case of electrostatics, the expression of the stored energy W_e is:

$$W_e = \frac{1}{2}\iiint_\Omega \mathbf{E}.\mathbf{D}d\tau \tag{3.76}$$

If the electric field is replaced by its expression given in equation [3.72], the following can be written:

$$W_e = \frac{1}{2}\iiint_\Omega (-\mathbf{grad}\,V + f_s\boldsymbol{\beta}_e).\mathbf{D}d\tau \tag{3.77}$$

Let us analyze the first term of the volume integral. Applying relation [2.23] for vector operators, we have:

$$-\iiint_{\Omega}\mathbf{grad}\mathrm{V}.\mathbf{D}d\tau=\iiint_{\Omega}\mathrm{V}\mathrm{div}\mathbf{D}d\tau-\oiint_{\Gamma_{e1}\cup\Gamma_{e2}\cup\Gamma_{d}}\mathrm{V}\mathbf{D}.\mathbf{n}\,d\tau=0 \qquad [3.78]$$

This term is equal to zero, as the divergence of **D** is zero (see equation [3.60]), as well as to the surface integral, taking into account the boundary conditions of V on Γ_{ek} (see equation [3.71]) and that of **D** on Γ_d (see equation [3.61]). Under these conditions, equation [3.77] has the form:

$$\mathrm{W_e}=\frac{1}{2}\iiint_{\Omega}f_s\boldsymbol{\beta}_e.\mathbf{D}d\tau \qquad [3.79]$$

The stored energy can also be expressed via global quantities f_s and Q_σ. Then, the following relation can be written:

$$\mathrm{W_e}=\frac{1}{2}f_s Q_\sigma=\frac{1}{2}\iiint_{\Omega}f_s\boldsymbol{\beta}_e.\mathbf{D}d\tau \qquad [3.80]$$

By identification, the expression of the total charges Q_σ can be deduced:

$$Q_\sigma=\iiint_{\Omega}\boldsymbol{\beta}_e.\mathbf{D}d\tau \qquad [3.81]$$

This expression allows us to write the total charges Q_σ as a function of the support field $\boldsymbol{\beta}_e$ and the electric displacement field **D**. Hence, by replacing the electric displacement field by its expression given in equation [3.73], we obtain:

$$Q_\sigma=\iiint_{\Omega}\varepsilon\boldsymbol{\beta}_e.(f_s\boldsymbol{\beta}_e-\mathbf{grad}\mathrm{V})d\tau \qquad [3.82]$$

When writing the equation of the electrostatic problem, presented in Figure 3.10, with the electric scalar potential formulation, and given that the source term is the total charges Q_σ, the circulation f_s becomes an unknown. The system of equations to be solved is composed of relations [3.75] and [3.82] having as unknown the scalar potential V and circulation f_s.

It should be noted that equation [3.81] can also be used to calculate the total charges Q_σ from the scalar potential V when the circulation is imposed.

3.3.1.2. *Vector potential P formulation*

The example in Figure 3.10 is also used in the case of the vector potential formulation, with the two source terms f_s and Q_σ imposed on the boundaries of the domain. The vector potential formulation is well suited to the source term corresponding to the total charges Q_σ on the electrodes. On the contrary, when the circulation of the electric field f_s is applied as the source term, the developments are not straightforward. An additional equation should then be added that can be obtained from an energy balance. The objective is to express the circulation f_s as a function of the total charges Q_σ and of the electric vector potential.

3.3.1.2.1. Total charges Q_σ imposed on the electrodes

For our example, when the total charges Q_σ are imposed on the boundaries Γ_{e1} and Γ_{e2}, the expression of the source term is given by equation [1.53]. To introduce this term in the local forms of the equations, the approach proposed in section 3.2.1.2.1 will be followed. Therefore, as proposed in relation [3.19], in order to introduce the source field, the electric displacement field is decomposed as follows:

$$\mathbf{D} = \mathbf{D}_s + \mathbf{D}' \qquad [3.83]$$

In this expression, $\mathbf{D}_s$ represents the source field produced by the total charges Q_σ and $\mathbf{D}'$ is the new unknown of the problem. The absolute value of the flux of $\mathbf{D}_s$, on the boundaries Γ_{e1} and Γ_{e2}, is equal to the charges Q_σ. Therefore, the flux of $\mathbf{D}'$, on these boundaries, is equal to zero. Relying on equation [3.20], in the context of our problem, the properties of field $\mathbf{D}'$ can be written as follows:

$$\mathrm{div}\mathbf{D}' = 0,\ \mathbf{D}'.\mathbf{n}\big|_{\Gamma_d} = 0,\ \iint_{\Gamma_{ek}} \mathbf{D}'.\mathbf{n}ds = 0 \ \text{ and } \ \mathbf{D}' \in H_{\Gamma_d}(\mathrm{div}0,\Omega) \qquad [3.84]$$

As for the properties of the source field $\mathbf{D}_s$, they are deduced from equation [3.21] and, for our example, they are written as follows:

$$\mathrm{div}\mathbf{D}_s = 0,\ \iint_{\Gamma_{ek}} \mathbf{D}_s.\mathbf{n}ds = \pm Q_\sigma,\ \mathbf{D}_s.\mathbf{n}\Big|_{\Gamma_d} = 0$$
$$\text{i.e. } \mathbf{D}_s \in H_{\Gamma_d}(\mathrm{div}0,\Omega) \qquad [3.85]$$

NOTE.– This configuration is similar to the notes concerning equations [3.6] and [3.19]. The fluxes of fields $\mathbf{D}'$ and $\mathbf{D}_s$ are conservative and any surface Δ whose contour γ_0 (see Figure 3.10) lies on the boundary Γ_d, can replace the surface integral on Γ_{ek}.

Since the source field $\mathbf{D}_s$ is proportional to the total charges Q_σ, it can be expressed using a support field, denoted by $\boldsymbol{\lambda}_e$, such that:

$$\mathbf{D}_s = Q_\sigma \boldsymbol{\lambda}_e \tag{3.86}$$

The properties of $\boldsymbol{\lambda}_e$ are equivalent to those of $\mathbf{D}_s$ (see equation [3.85]) and can be written as follows:

$$\text{div}\boldsymbol{\lambda}_e = 0 \text{ and } \boldsymbol{\lambda}_e.\mathbf{n}\big|_{\Gamma_d} = 0, \iint_{\Gamma_{ek}} \boldsymbol{\lambda}_e.\mathbf{n}ds = \pm 1, \ \boldsymbol{\lambda}_e \in H_{\Gamma_d}(\text{div}0, \Omega) \tag{3.87}$$

As for $\boldsymbol{\lambda}_e$, considering its properties and the fact that the domain Ω is contractible, it can be expressed by means of the vector potential $\boldsymbol{\chi}_e$. However, precautions are required when building the field $\boldsymbol{\chi}_e$, since the boundary Γ_d is not simply connected. To address this difficulty, a cut along this boundary is introduced (see section 3.2.1.2.2). The field $\boldsymbol{\chi}_e$ then belongs to the function space $H^{\Delta}_{\Gamma d}(\mathbf{curl}, \Omega)$ defined by relation [3.28] and it verifies:

$$\boldsymbol{\chi}_e \wedge \mathbf{n}\big|_{\Gamma_d} = 0, \ \int_{\gamma_0} \boldsymbol{\chi}_e.\mathbf{dl} = \pm 1 \tag{3.88}$$

where γ_0 is the contour of the surface Δ defined in Figure 3.10.

The support field $\boldsymbol{\lambda}_e$ can then be expressed as a function of $\boldsymbol{\chi}_e$ as follows:

$$\boldsymbol{\lambda}_e = \mathbf{curl}\boldsymbol{\chi}_e \tag{3.89}$$

Grouping equations [3.60], [3.83] and [3.86], the following can be written as:

$$\text{div}(Q_\sigma \boldsymbol{\lambda}_e + \mathbf{D}') = 0 \tag{3.90}$$

where the source field "Q_σ $\boldsymbol{\lambda}_e$" is assumed to be known. On the contrary, the unknown of the problem is now the field $\mathbf{D}'$. Relying on the properties of $\mathbf{D}'$ defined in equation [3.84], it can be expressed as a function of an electric vector potential, denoted by $\mathbf{P}$, in the following form:

$$\mathbf{D}' = \mathbf{curlP}, \ \mathbf{P} \wedge \mathbf{n}\big|_{\Gamma_d} = 0 \text{ i.e. } \mathbf{P} \in H_{\Gamma_d}(\mathbf{curl}, \Omega) \tag{3.91}$$

As indicated in section 2.5.2, to have a unique solution to the electric vector potential, a gauge condition must be imposed. Based on equations [3.83], [3.86] and [3.91], the electric displacement field can be written as follows:

$$\mathbf{D} = \mathbf{curlP} + Q_\sigma \boldsymbol{\lambda}_e \tag{3.92}$$

Similarly, using the behavior law [1.19], the electric field takes the form:

$$\mathbf{E} = \varepsilon^{-1}(\mathbf{curlP} + Q_\sigma \boldsymbol{\lambda}_e) \quad [3.93]$$

When writing the electric field by means of the electric vector potential **P** and the source term "$Q_\sigma\boldsymbol{\lambda}_e$", equation [3.60] is automatically verified. The solution to the problem must now simply verify the behavior law [3.93] and the equilibrium equation [3.59].

Grouping equations [3.59] and [3.93], the following equation can be written as:

$$\mathbf{curl}(\varepsilon^{-1}(\mathbf{curlP} + Q_\sigma \boldsymbol{\lambda}_e)) = 0 \quad [3.94]$$

Or still by introducing the source field $\boldsymbol{\chi}_e$ (see equation [3.89]), then we have:

$$\mathbf{curl}(\varepsilon^{-1}(\mathbf{curlP} + Q_\sigma \mathbf{curl}\boldsymbol{\chi}_e)) = 0 \quad [3.95]$$

These equations represent the vector potential formulation of the electrostatic problem when the source term is the total charges Q_σ on the electrodes Γ_{e1} and Γ_{e2}. The steps to be taken to solve these equations are to express the source fields $\boldsymbol{\lambda}_e$ or $\boldsymbol{\chi}_e$, and then to calculate the vector potential **P**. Then, the electric displacement field **D** and the electric field **E** can be deduced.

3.3.1.2.2. Imposed circulation f_s of the electric field

Still with the case of the formulation in terms of the vector potential **P** and the problem in Figure 3.10, let us now consider that the source term is the circulation f_s of the electric field between the boundaries Γ_{e1} and Γ_{e2}. As indicated in the introduction to section 3.3.1.2, in this case, the charge Q_σ becomes an unknown of the problem. An additional equation should therefore be provided allowing for the expression of the source term f_s as a function of the charge Q_σ and of the vector potential **P**. This requires an energy balance.

In the expression of the electrostatic energy W_e (see equation [3.76]), the electric displacement field is replaced by its expression given in equation [3.92]. Then:

$$W_e = \frac{1}{2}\iiint_\Omega \mathbf{E}.(\mathbf{curlP} + Q_\sigma \boldsymbol{\lambda}_e)d\tau \quad [3.96]$$

Let us now consider the first term of the volume integral to which the formula [2.27] related to vector operators is applied. We then obtain the following:

$$\iiint_{\Omega} \mathbf{E}.\mathbf{curlP}d\tau = \iiint_{\Omega} \mathbf{P}.\mathbf{curlE}d\tau - \oiint_{\Gamma} (\mathbf{P} \wedge \mathbf{n}).\mathbf{E}dS \qquad [3.97]$$

Considering the properties of the electric field **E** (see equation [3.59]), the volume integral, on the right-hand side of the equality, is zero. Concerning the surface integral, on the boundary Γ of the domain, it can be decomposed into two terms as follows:

$$\oiint_{\Gamma} (\mathbf{P} \wedge \mathbf{n}).\mathbf{E}dS = \iint_{\Gamma_{e1} \cup \Gamma_{e2}} (\mathbf{P} \wedge \mathbf{n}).\mathbf{E}dS + \iint_{\Gamma_d} (\mathbf{P} \wedge \mathbf{n}).\mathbf{E}dS \qquad [3.98]$$

The integral on Γ_{e1} or Γ_{e2} is equal to zero as, according to the properties of the mixed product, the tangential component of the electric field is zero on these boundaries. As for the second term, it is also zero, considering the properties of the vector potential **P** on the boundary Γ_d (see equation [3.91]). Under these conditions, the expression of the electrostatic energy is:

$$W_e = \frac{1}{2} \iiint_{\Omega} Q_{\sigma} \mathbf{E}.\boldsymbol{\lambda}_e d\tau \qquad [3.99]$$

If the electrostatic energy is expressed as a function of global quantities f_s and Q_{σ} (see equation [3.80]), then:

$$\frac{1}{2} f_s Q_{\sigma} = \frac{1}{2} \iiint_{\Omega} Q_{\sigma} \mathbf{E}.\boldsymbol{\lambda}_e d\tau \qquad [3.100]$$

After simplification, the source term f_s can be expressed as a function of **E** and $\boldsymbol{\lambda}_e$, hence:

$$f_s = \iiint_{\Omega} \boldsymbol{\lambda}_e.\mathbf{E}d\tau \qquad [3.101]$$

or by replacing the electric field **E** with its expression in equation [3.93], the following can be written:

$$f_s = \iiint_{\Omega} \varepsilon^{-1} \boldsymbol{\lambda}_e.(\mathbf{curlP} + Q_{\sigma}\boldsymbol{\lambda}_e)d\tau \qquad [3.102]$$

Let us now introduce the source potential $\boldsymbol{\chi}_e$ (see equation [3.89]). Then, equation [3.102] can be rewritten in the following form:

$$f_s = \iiint_\Omega \varepsilon^{-1}\boldsymbol{\lambda}_e.(\mathbf{curlP} + Q_\sigma \mathbf{curl}\boldsymbol{\chi}_e)d\tau \qquad [3.103]$$

In conclusion, for the electric vector potential formulation, when the source term is the circulation f_s of the electric field, the unknowns are the electric vector potential **P** and the total charges Q_σ. Then, the system of equations defined by relations [3.94] and [3.102] or [3.95] and [3.103] must be solved.

It should be noted that equation [3.101] can also be used to calculate the circulation f_s from the vector potential **P** when the total charge is imposed.

3.3.1.3. *Summary tables*

This section presents a summary of the equations to be solved for the example in Figure 3.10.

Table 3.1 summarizes the equations for the scalar potential formulation when the source term is the circulation f_s of the electric field or the total charges Q_σ. The table also contains support fields and function spaces to which they belong.

Electrostatics (studied domain)		
Scalar potential V formulation	Source term: circulation of the electric field f_s	
		Decomposition of the electric field: $\mathbf{E} = \mathbf{E}_s + \mathbf{E}'$
	Source field support two possibilities: $\boldsymbol{\beta}_e$ or α_e	$\mathbf{E}_s = f_s\boldsymbol{\beta}_e,\ \boldsymbol{\beta}_e \in H_{\Gamma_{e1}\cup\Gamma_{e2}}(\mathbf{curl}0,\Omega)$
		$\mathbf{E}_s = -f_s\mathbf{grad}\alpha_e\ ,\ \alpha_e \in H(\mathbf{grad}\Omega)$
	Properties of the unknown E' and introduction of potential V	$\mathbf{E}' \in H_{\Gamma_{e1}\cup\Gamma_{e2}}(\mathbf{curl}0,\Omega)$
		$\mathbf{E}' = -\mathbf{grad}V\ ,\ V \in H_{\Gamma_{e1}\cup\Gamma_{e2}}(\mathbf{grad},\Omega)$
	Equation to be solved: two possible forms depending on $\boldsymbol{\beta}_e$ or α_e	$\mathrm{div}(\varepsilon(f_s\boldsymbol{\beta}_e - \mathbf{grad}V)) = 0$
		$\mathrm{div}(\varepsilon(f_s\mathbf{grad}\alpha_e + \mathbf{grad}V)) = 0$
	Source term: total charges Q_σ	
	f_s becomes an unknown; an additional equation is needed	
	$Q_\sigma = \iiint_\Omega \boldsymbol{\beta}_e.\mathbf{D}d\tau$	

Table 3.1. *Summary of the equations to be solved in electrostatics for the scalar potential formulation (see Figure 3.10)*

For the same source terms (Q_σ and f_s), Table 3.2 presents the equations to be solved for a vector potential formulation.

Electrostatics (studied domain)		
Vector potential P formulation	Source term: total charges Q_σ	
	Source field support two possibilities: $\boldsymbol{\lambda}_e$ or $\boldsymbol{\chi}_e$ with a cut on not simply connected Γ_d	Decomposition of the electric field: $\mathbf{D} = \mathbf{D}_s + \mathbf{D}'$
		$\mathbf{D}_S = Q_\sigma \lambda_e$, $\lambda_e \in H_{\Gamma_d}(\text{div}0, \Omega)$
		$\mathbf{D}_s = Q_\sigma \mathbf{curl}\chi_e$, $\chi_e \in H^{\Delta}_{\Gamma_d}(\mathbf{curl}, \Omega)$
	Properties of the unknown **D'** and introduction of the vector potential **P**	$\mathbf{D}' \in H_{\Gamma_d}(\text{div}0, \Omega)$
		$\mathbf{D}' = \mathbf{curlP}$, $\mathbf{P} \in H_{\Gamma_d}(\mathbf{curl}, \Omega)$
	Equation to be solved: two possible forms depending on $\boldsymbol{\lambda}_e$ or $\boldsymbol{\chi}_e$	$\mathbf{curl}\ (\varepsilon^{-1}(\mathbf{curlP} + Q_\sigma \lambda_e)) = 0$
		$\mathbf{curl}\ (\varepsilon^{-1}(\mathbf{curlP} + Q_\sigma \mathbf{curl}\ \chi_e)) = 0$
	Source term: circulation of the electric field f_s	
	Q becomes an unknown; an additional equation is needed	
	$f_S = \iiint_\Omega \lambda_e.\mathbf{E}d\tau$	

Table 3.2. *Summary of equations to be solved in electrostatics for the vector potential formulation (see Figure 3.10)*

3.3.2. *Internal electrode*

This section again uses the example in Figure 3.10, adding inside the domain, as shown in Figure 3.11, a subdomain $\Omega_{\mathcal{E}}$, namely an internal electrode, denoted by $\mathcal{E}_3$, of boundary Γ_{e3}. It should be recalled that at electrostatic equilibrium, the electric field is zero inside an electrode and the surface charge density is σ_s. Under these conditions, the electric field is normal to the surface. The studied domain then relates to $\Omega' = \Omega - \Omega_{\mathcal{E}3}$, it is therefore simply connected with disconnected boundary. For this problem, the source term can be the circulation of the electric field f_s between the boundaries Γ_{ek} (with $k \in \{1,2,3\}$), equation [1.51] or the electric flux on the gates [1.53].

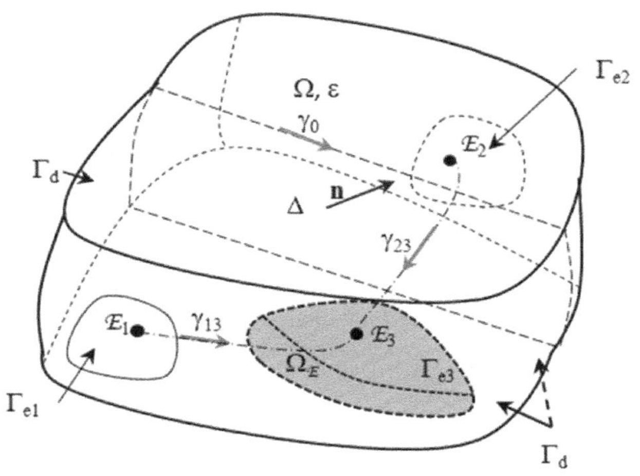

Figure 3.11. *Geometry studied in electrostatics in the case of an internal electrode*

The equations to be solved are written as follows:

$$\mathbf{curlE} = 0 \quad [3.104]$$

$$\text{div}\mathbf{D} = 0 \quad [3.105]$$

With the dielectric behavior law [1.19] and the boundary conditions defined by equations [1.35] and [1.39] which, in this case, are written as:

$$\begin{aligned} &\Gamma = \Gamma_d \cup \Gamma_{e1} \cup \Gamma_{e2} \cup \Gamma_{e3} \\ &\text{with : } \mathbf{D.n}\big|_{\Gamma_d} = 0,\ \mathbf{E} \wedge \mathbf{n}\big|_{\Gamma_{ek}} = 0,\ \ k \in \{1,2,3\} \end{aligned} \quad [3.106]$$

Considering the above equations, the fields **E** and **D** are defined in the following function spaces:

$$\mathbf{E} \in H_{\Gamma_{e1} \cup \Gamma_{e2} \cup \Gamma_{e3}}(\mathbf{curl}0, \Omega') \ \text{ and } \ \mathbf{D} \in H_{\Gamma_d}(\text{div}0, \Omega') \quad [3.107]$$

Let us now define the various possible source terms. This can be the circulation of the electric field **E** between two electrodes. Then, for electrodes i, j, the circulation f_{ij} is:

$$\int_{\gamma_{ij}} \mathbf{E.dl} = f_{ij} \ \text{ with } i \neq j \text{ and } i, j \in \{1,2,3\} \quad [3.108]$$

It is important to note that with the electric field being curl free, the path γ_{ij}, linking two electrodes can be arbitrary since in this case the circulation does not depend on the path followed. Moreover, in the presence of three electrodes, it is possible to define three circulations of the electric field. Nevertheless, two are sufficient, given that the electric field **E** is curl free, its circulation along a closed contour is zero. This leads to the following property:

$$f_{12} + f_{23} + f_{31} = 0 \qquad [3.109]$$

The source term can also be the total charges Q_k on the electrodes of boundaries Γ_{ek}. Then, we have:

$$\iint_{\Gamma_{ek}} \mathbf{D.n}\mathrm{dS} = Q_k \quad \text{with} \quad k \in \{1,2,3\} \qquad [3.110]$$

In this case, Gauss' law (see equation [1.50]) leads to:

$$Q_1 + Q_2 + Q_3 = 0 \qquad [3.111]$$

Similar to the case of circulations f_{ij}, this relation shows that the three values of charge are not independent.

The following section develops the potential (V and **P**) formulations by considering, for the source terms, the following possibilities:

– The two circulations of the electric field, f_{13} and f_{23}, are imposed. The third circulation f_{12} can be deduced using relation [3.109]. The charges Q_k are unknown.

– The charges Q_1 and Q_2 are imposed, and in this case Q_3 is fixed by equation [3.111]. The circulations f_{ij} are unknown.

– Hybrid source terms are imposed, namely a circulation of the electric field and a total charge on an electrode. For example, f_{13} and the total charge Q_2 are imposed. In this case, the unknowns are f_{23} and Q_1 (f_{12} and Q_3 are then obtained by [3.109] and [3.111]).

3.3.2.1. *Scalar potential V formulation*

As indicated in section 3.3.1.1, when the source terms are the circulations of the electric field, the scalar potential formulation is perfectly adapted. Our approach will therefore focus first on this case, and then on how to consider, as the source term, the total charges on the electrodes.

3.3.2.1.1. Imposed circulations of the electric field f_{13} and f_{23}

When the circulation of the electric field is imposed, as mentioned above (see equation [3.72]), it is expressed as a function of the scalar potential V and a support field $\boldsymbol{\beta}_e$. As our example holds two source terms, f_{13} and f_{23}, two support fields are introduced (see equations [3.63] and [3.67]), such that:

$$\mathbf{E} = f_{13}\boldsymbol{\beta}_{13} + f_{23}\boldsymbol{\beta}_{23} + \mathbf{E}', \text{ with } \mathbf{E}' \in H_{\Gamma_{e1}\cup\Gamma_{e2}\cup\Gamma_{e3}}(\mathbf{curl}0, \Omega') \quad [3.112]$$

The support fields $\boldsymbol{\beta}_{13}$ and $\boldsymbol{\beta}_{23}$ are defined by the following relations:

$$\begin{aligned}&\mathbf{curl}\,\boldsymbol{\beta}_{ij} = 0, \ \int_{\gamma_{kj}} \boldsymbol{\beta}_{ij}.\mathbf{dl} = \delta_{ki} \quad \boldsymbol{\beta}_{ij} \wedge \mathbf{n}\Big|_{\Gamma_{e1}\cup\Gamma_{e2}\cup\Gamma_{e3}} = 0 \\ &\boldsymbol{\beta}_{ij} \in H_{\Gamma_{e1}\cup\Gamma_{e2}\cup\Gamma_{e3}}(\mathbf{curl}0, \Omega') \ \ (i,k) \in \{1.2\} \text{ and } j = 3\end{aligned} \quad [3.113]$$

Considering the above properties, and since the studied domain is simply connected with a disconnected boundary, the support fields $\boldsymbol{\beta}_{ij}$ can be expressed as a function of scalar potentials α_{ij} (see section 2.4.2.3). Based on relations [3.13] and [3.14], we can write:

$$\begin{aligned}&\boldsymbol{\beta}_{ij} = -\mathbf{grad}\alpha_{ij} \ \text{ and } \alpha_{ij}\Big|_{\Gamma_{ek}} = \delta_{ki} \quad \alpha_{ij} \in H(\mathbf{grad}, \Omega') \\ &\text{with } i \in \{1,2\}, \ k \in \{1,2,3\}, j = 3\end{aligned} \quad [3.114]$$

The properties of the field **E**', introduced in equation [3.112], are identical to those given by relation [3.64]. An electric scalar potential V can therefore be defined (see equation [3.71]). On the gates Γ_{ek}, the tangential component of field **E**' is zero. Therefore, for the scalar potential V, these gates represent equipotential surfaces whose values may be V_k = Constant with k = {1, 2, 3}. Nevertheless, the circulation of the electric field, between various electrodes, is imposed by the support fields $\boldsymbol{\beta}_{ij}$ associated with circulations f_{ij}. The circulation of the field **E**' is then equal to "0" between the surfaces Γ_{ek}. Under these conditions, we pose V = 0 on the three gates, and therefore the gauge condition can be imposed (see section 2.5.2.1). The field **E**' is then written as:

$$\mathbf{E}' = -\mathbf{grad}V \text{ with } V \in H_{\Gamma_{e1}\cup\Gamma_{e2}\cup\Gamma_{e3}}(\mathbf{grad}, \Omega') \quad [3.115]$$

Gathering equations [3.112] and [3.115], the electric field is written as:

$$\mathbf{E} = f_{13}\boldsymbol{\beta}_{13} + f_{23}\boldsymbol{\beta}_{23} - \mathbf{grad}\ V \quad [3.116]$$

Based on this equation and the behavior law [1.19], the electric displacement field can be written as:

$$\mathbf{D} = \varepsilon(f_{13}\boldsymbol{\beta}_{13} + f_{23}\boldsymbol{\beta}_{23} - \mathbf{grad}\mathrm{V}) \qquad [3.117]$$

To obtain the scalar potential formulation, in the presence of the two source terms f_{13} and f_{23}, the electric displacement field is replaced in equation [3.105] by its expression given by equation [3.117]. The equation to be solved is then obtained in the following form:

$$\mathrm{div}(\varepsilon(f_{13}\boldsymbol{\beta}_{13} + f_{23}\boldsymbol{\beta}_{23} - \mathbf{grad}\mathrm{V})) = 0 \qquad [3.118]$$

Replacing the support fields $\boldsymbol{\beta}_{ij}$ by the associated potentials α_{ij} (see equation [3.114]), equation [3.118] is written as:

$$\mathrm{div}(\varepsilon(f_{13}\mathbf{grad}\alpha_{13} + f_{23}\mathbf{grad}\alpha_{23} + \mathbf{grad}\mathrm{V})) = 0 \qquad [3.119]$$

3.3.2.1.2. Total charges Q_1 and Q_2 imposed on the electrodes

In the formulation developed in section 3.3.2.1.1, the source terms are the circulations of the electric field between the electrodes. The charges, carried by the electrodes, are not explicitly present. If instead of circulations, charges Q_1 and Q_2 should be imposed, respectively, on electrodes $\mathcal{E}_1$ and $\mathcal{E}_2$, the formulation [3.118] can be used. In this case, circulations f_{13} and f_{23} become unknowns and two new equations should be added. These equations can be obtained from an energy balance, as in section 3.3.1.1.2.

Based on global quantities, namely the circulations f_{13} and f_{23} and the total charges Q_1 and Q_2, the electrostatic energy can be expressed by the following equation:

$$\mathrm{W_e} = \frac{1}{2} f_{13}\mathrm{Q}_1 + \frac{1}{2} f_{23}\mathrm{Q}_2 \qquad [3.120]$$

On the contrary, if the energy W_e is expressed based on local quantities (see equation [3.76]), replacing the electric field by its expression given in equation [3.116] yields:

$$\mathrm{W_e} = \frac{1}{2} \iiint_{\Omega'} (f_{13}\boldsymbol{\beta}_{13} + f_{23}\boldsymbol{\beta}_{23} - \mathbf{grad}\mathrm{V}).\mathbf{D}d\tau \qquad [3.121]$$

As shown in section 3.3.1.1.2 (see equation [3.78]), the contribution of the term related to "**grad**V.**D**" is zero. Indeed, a similar development can be readily obtained taking into consideration equations [3.105] and [3.107]. After simplification, by gathering equations [3.120] and [3.121], the following can be written as:

$$\frac{1}{2} f_{13} Q_1 + \frac{1}{2} f_{23} Q_2 = \frac{1}{2} \iiint_{\Omega'} (f_{13} \boldsymbol{\beta}_{13} + f_{23} \boldsymbol{\beta}_{23}).\mathbf{D} d\tau \qquad [3.122]$$

This equation is true, irrespective of the values of f_{13} and f_{23}. Under these conditions, posing $f_{13} = 1$ and $f_{23} = 0$, the expression of Q_1 results quite naturally as follows:

$$Q_1 = \iiint_{\Omega'} \boldsymbol{\beta}_{13}.\mathbf{D} d\tau \qquad [3.123]$$

Similarly, considering $f_{13} = 0$ and $f_{23} = 1$, the expression of Q_2 is obtained:

$$Q_2 = \iiint_{\Omega'} \boldsymbol{\beta}_{23}.\mathbf{D} d\tau \qquad [3.124]$$

If the electric displacement field **D** is replaced by its expression given by equation [3.117], the following expressions are obtained for Q_1 and Q_2:

$$Q_1 = \iiint_{\Omega'} \varepsilon \boldsymbol{\beta}_{13}.(f_{13} \boldsymbol{\beta}_{13} + f_{23} \boldsymbol{\beta}_{23} - \mathbf{grad} V) d\tau \qquad [3.125]$$

$$Q_2 = \iiint_{\Omega'} \varepsilon \boldsymbol{\beta}_{23}.(f_{13} \boldsymbol{\beta}_{13} + f_{23} \boldsymbol{\beta}_{23} - \mathbf{grad} V) d\tau \qquad [3.126]$$

In conclusion, if the charges Q_1 and Q_2 are known, the circulations f_{13} and f_{23} become the unknowns of the problem. In this case, the system of equations composed of equation [3.118] and relations [3.125] and [3.126] should be solved.

3.3.2.1.3. Hybrid source terms: circulation and total charges

When imposing a circulation and also a total charge on an electrode, taking into account the developments of sections 3.3.2.1.1 and 3.3.2.1.2, it is relatively simple to write the equation. In fact, a system of equations is built, consisting of relation [3.118] to which, depending on the imposed source terms, equation [3.125] or [3.126] is added. If the sources are f_{13} and Q_2, the expression of Q_2 (see equation [3.126]) is added to equation [3.118]. On the contrary, if the source terms are the circulation f_{23} and the total charges Q_1, then equation [3.118] is completed by the expression of Q_1 given by equation [3.125].

3.3.2.2. *Vector potential P formulation*

When the total charges Q_1 and Q_2 are imposed, the vector potential formulation is naturally obtained. This case will be discussed first, and then the focus will be on how the circulations f_{13} and f_{23} of the electric field can be introduced as source terms. Finally, section 3.3.2.2.3 will address the case of complementary hybrid source terms.

It is important to note that, for the vector potential formulation, with this example holding three electrodes, the developments are similar to those presented in section 3.3.1.2.1.

3.3.2.2.1. Total charges Q_1 and Q_2 imposed on the electrodes

When the total charges Q_1 and Q_2 are, respectively, on the electrodes $\mathcal{E}_1$ and $\mathcal{E}_2$, the electric displacement field can be expressed using [3.83], but with two source fields, as follows:

$$\mathbf{D} = \mathbf{D}_{s1} + \mathbf{D}_{s2} + \mathbf{D}' \qquad [3.127]$$

In this expression, $\mathbf{D}_{s1}$ and $\mathbf{D}_{s2}$ represent the two source fields due to the charges Q_1 and Q_2 on the two electrodes $\mathcal{E}_1$ and $\mathcal{E}_2$. As for the field $\mathbf{D}'$, it represents the new unknown of the problem. Under these conditions, the properties of $\mathbf{D}'$ are stated (see equation [3.84]) as follows:

$$\begin{aligned}&\mathrm{div}\mathbf{D}' = 0, \iint_{\Gamma_{ek}} \mathbf{D}'.\mathbf{n}\mathrm{dS} = 0 \text{ with } k \in \{1,2,3\}, \ \mathbf{D}'.\mathbf{n}\big|_{\Gamma_d} = 0\\ &\text{i.e. } \mathbf{D}' \in H_{\Gamma_d}(\mathrm{div}0,\Omega')\end{aligned} \qquad [3.128]$$

The source fields have properties similar to those of the electric displacement field, but they take into account the constraints on the total charges imposed on the electrodes $\mathcal{E}_k$.

The properties of the source field $\mathbf{D}_{s1}$ are written as follows:

$$\begin{aligned}&\mathrm{div}\mathbf{D}_{s1} = 0, \iint_{\Gamma_{e1}} \mathbf{D}_{s1}.\mathbf{n}\mathrm{dS} = Q_1, \iint_{\Gamma_{e3}} \mathbf{D}_{s1}.\mathbf{n}\mathrm{dS} = -Q_1,\\ &\text{and, } \mathbf{D}_{s1}.\mathbf{n}\big|_{\Gamma_{e2}\cup\Gamma_d} = 0\\ &\text{i.e. } \mathbf{D}_{s1} \in H_{\Gamma_{e2}\cup\Gamma_d}(\mathrm{div}0,\Omega')\end{aligned} \qquad [3.129]$$

and for $\mathbf{D}_{s2}$, we have:

$$\mathrm{div}\mathbf{D}_{s2} = 0,\ \iint_{\Gamma_{e2}} \mathbf{D}_{s2}.\mathbf{n}dS = Q_2,\ \iint_{\Gamma_{e3}} \mathbf{D}_{s2}.\mathbf{n}dS = -Q_2,$$
$$\text{and, } \mathbf{D}_{s2}.\mathbf{n}\big|_{\Gamma_{e1}\cup\Gamma_d} = 0 \qquad [3.130]$$
$$\text{i.e. } \mathbf{D}_{s2} \in H_{\Gamma_{e1}\cup\Gamma_d}(\mathrm{div}0,\Omega')$$

Support fields $\boldsymbol{\lambda}_{13}$ and $\boldsymbol{\lambda}_{23}$ are now introduced, which are defined as follows:

$$\mathrm{div}\boldsymbol{\lambda}_{13} = 0,\ \iint_{\Gamma_{e1}} \boldsymbol{\lambda}_{13}.\mathbf{n}dS = 1,\ \iint_{\Gamma_{e3}} \boldsymbol{\lambda}_{13}.\mathbf{n}dS = -1$$
$$\boldsymbol{\lambda}_{13}.\mathbf{n}\big|_{\Gamma_{e2}\cup\Gamma_d} = 0 \text{ i.e. } \boldsymbol{\lambda}_{13} \in H_{\Gamma_{e2}\cup\Gamma_d}(\mathrm{div}0,\Omega') \qquad [3.131]$$

$$\mathrm{div}\boldsymbol{\lambda}_{23} = 0,\ \iint_{\Gamma_{e2}} \boldsymbol{\lambda}_{23}.\mathbf{n}dS = 1,\ \iint_{\Gamma_{e3}} \boldsymbol{\lambda}_{23}.\mathbf{n}dS = -1$$
$$\boldsymbol{\lambda}_{23}.\mathbf{n}\big|_{\Gamma_{e1}\cup\Gamma_d} = 0 \text{ i.e. } \boldsymbol{\lambda}_{23} \in H_{\Gamma_{e1}\cup\Gamma_d}(\mathrm{div}0,\Omega') \qquad [3.132]$$

By identifying the properties of the source fields $\mathbf{D}_{s1}$ and $\mathbf{D}_{s2}$ with those of the support fields $\boldsymbol{\lambda}_{13}$ and $\boldsymbol{\lambda}_{23}$, we have:

$$\mathbf{D}_{s1} = Q_1\boldsymbol{\lambda}_{13}\ ,\quad \mathbf{D}_{s2} = Q_2\boldsymbol{\lambda}_{23} \qquad [3.133]$$

If in equation [3.127] the source fields are replaced by their expression provided in equation [3.133], the electric displacement field is written as:

$$\mathbf{D} = Q_1\boldsymbol{\lambda}_{13} + Q_2\boldsymbol{\lambda}_{23} + \mathbf{D}' \qquad [3.134]$$

The divergence operator is now applied to this equation. Then, the following can be written as:

$$\mathrm{div}\mathbf{D} = Q_1\mathrm{div}\boldsymbol{\lambda}_{13} + Q_2\mathrm{div}\boldsymbol{\lambda}_{23} + \mathrm{div}\mathbf{D}' \qquad [3.135]$$

Based on this equation and considering the properties of the support fields given in equations [3.131] and [3.132], as well as those of **D'** field defined in equation [3.128], it can be noted that equation [3.105] and the boundary conditions on **D** are verified.

Based on the properties of **D'**, see [3.128], the notion of potential can be introduced. Nevertheless, as the domain is simply connected with a disconnected boundary, some precautions must be taken (see section 2.4.2.3). It should be noted that, for our study, there is only one cavity. As shown by equation [2.41], a vector potential **P** and a field **h** are introduced, which makes it possible to consider the fact that the studied domain is not contractible. Under these conditions, the field **D'** can be written in the following form:

$$\mathbf{D}' = \mathbf{curlP} + \mathrm{K}\mathbf{h} \;\; \text{with} \;\; \mathbf{P} \in \mathrm{H}_{\Gamma_d}(\mathbf{curl}, \Omega') \;\; \text{and} \;\; \mathbf{h} \in \mathcal{H}^2(\Omega') \qquad [3.136]$$

In this expression, the function space $\mathcal{H}^2(\Omega')$ is defined by equation [2.39] and K is a constant to be determined. To this end, let us calculate the flux of the electric displacement field **D** through the external surface of the domain Ω. Based on equations [3.134] and [3.136] and expression [3.110], we can write the following:

$$\oiint_{\Gamma_d \cup \Gamma_{e1} \cup \Gamma_{e2}} (Q_1 \boldsymbol{\lambda}_{13} + Q_2 \boldsymbol{\lambda}_{23} + \mathbf{curlP} + \mathrm{K}\mathbf{h}).\mathbf{n}\mathrm{dS} = Q_1 + Q_2 \qquad [3.137]$$

Based on the properties of the support fields $\boldsymbol{\lambda}_{13}$ and $\boldsymbol{\lambda}_{23}$ (see equations [3.131] and [3.132]), it can be deduced that their surface integrals on $\Gamma_d \cup \Gamma_{e1} \cup \Gamma_{e2}$ are equal to 1. The function space to which the term **curlP** belongs implies that the surface integral is zero on Γ_d. The same is true on the boundaries Γ_{e1} and Γ_{e2} due to the constraints that **D'** must meet (see equation [3.128]). Finally, the integral on the external surface of the basis function **h** is equal to 1 (Bossavit 1988). Under these conditions, in order to verify the conservation equation [3.137], we obtain K = 0.

NOTE.– In fact, constant K makes it possible to consider the total charges inside the domain $\Omega_{\mathcal{E}}$ (see Figure 3.11). As it is an electrode, there are by definition zero charges inside.

Equation [3.136] is then written as:

$$\mathbf{D}' = \mathbf{curlP} \;\; \text{with} \;\; \mathbf{P} \in \mathrm{H}_{\Gamma_d}(\mathbf{curl}, \Omega') \qquad [3.138]$$

Under these conditions, based on equation [3.134], we have:

$$\mathbf{D} = Q_1 \boldsymbol{\lambda}_{13} + Q_2 \boldsymbol{\lambda}_{23} + \mathbf{curlP} \qquad [3.139]$$

Using the induction **D**, defined by the above expression, the electric field can be expressed via the behavior law [1.19] as follows:

$$\mathbf{E} = \varepsilon^{-1}(Q_1\boldsymbol{\lambda}_{13} + Q_2\boldsymbol{\lambda}_{23} + \mathbf{curlP}) \qquad [3.140]$$

If the above expression of **E** is introduced in equation [3.104], the equation to be solved is:

$$\mathbf{curl}(\varepsilon^{-1}(Q_1\boldsymbol{\lambda}_{13} + Q_2\boldsymbol{\lambda}_{23} + \mathbf{curlP})) = 0 \qquad [3.141]$$

which represents the vector potential formulation, when the source terms are the total charges Q_1 and Q_2 on the electrodes $\mathcal{E}_1$ and $\mathcal{E}_2$ and in the presence of an electrode ($\mathcal{E}_3$) inside the domain.

3.3.2.2.2. Imposed circulations of the electric field f_{13} and f_{23}

Consider now, as source terms, the circulations f_{13} and f_{23}. The equation to be solved is still [3.141], but the total charges Q_1 and Q_2 on the electrodes become unknowns. Two new equations should then be introduced in order to build a complete system of equations. To this end, as in section 3.3.1.2.2, an energy balance is written.

In equation [3.76], the electric displacement field is replaced by its expression, provided in equation [3.139]. The following can then be written as:

$$W_e = \frac{1}{2}\iiint_{\Omega'}(\mathbf{curl\ P} + Q_1\boldsymbol{\lambda}_{13} + Q_2\boldsymbol{\lambda}_{23}).\mathbf{E}d\tau \qquad [3.142]$$

Let us now consider the first term of the volume integral. Using formula [2.27], related to vector operators, the following can be written as:

$$\iiint_{\Omega'}\mathbf{E.curlP}d\tau = \iiint_{\Omega'}\mathbf{curlE.P}d\tau + \oiint_{\Gamma_{e1}\cup_{e2}\cup_{e3}\cup\Gamma_d}(\mathbf{E}\wedge\mathbf{n}).\mathbf{P}d\tau \qquad [3.143]$$

The first right-hand side term is equal to zero (see equation [3.104]). The same is true for the surface integral, as the tangential component of **E** is equal to zero on the boundaries Γ_{e1}, Γ_{e2} and Γ_{e3} as well as the tangential component of **P** on Γ_d (see equation [3.136]). Equation [3.142] is then written as:

$$W_e = \frac{1}{2}\iiint_{\Omega'}(Q_1\boldsymbol{\lambda}_{13} + Q_2\boldsymbol{\lambda}_{23}).\mathbf{E}d\tau \qquad [3.144]$$

This relation must be equal to [3.120] for all the values of Q_1 and Q_2. This leads to:

$$f_{13} = \iiint_{\Omega'} \boldsymbol{\lambda}_{13}.\mathbf{E}d\tau \ , \quad f_{23} = \iiint_{\Omega'} \boldsymbol{\lambda}_{23}.\mathbf{E}d\tau \qquad [3.145]$$

If the electric field is replaced by its expression provided in equation [3.140], then f_{13} is written as follows:

$$f_{13} = \iiint_{\Omega'} \boldsymbol{\lambda}_{13}.\varepsilon^{-1}(Q_1\boldsymbol{\lambda}_{13} + Q_2\boldsymbol{\lambda}_{23} + \mathbf{curlP})d\tau \qquad [3.146]$$

and f_{23}:

$$f_{23} = \iiint_{\Omega'} \boldsymbol{\lambda}_{23}.\varepsilon^{-1}(Q_1\boldsymbol{\lambda}_{13} + Q_2\boldsymbol{\lambda}_{23} + \mathbf{curlP})d\tau \qquad [3.147]$$

In conclusion, for the problem studied with the vector potential formulation and when the source terms are the circulations f_{13} and f_{23}, the unknowns of the problem are the vector potential **P** and the charges Q_1 and Q_2. The system to be solved is then composed of equations [3.141], [3.146] and [3.147].

3.3.2.2.3. Hybrid source terms: circulation and total charges

Similar to the approach for the scalar potential formulation, two complementary source terms are now imposed, one circulation and one total charge.

In this case, writing the equation is relatively simple. If the source terms are the total charges Q_1 and the circulation f_{23}, relation [3.146] should be added to equation [3.141]. On the contrary, if the source terms are f_{13} and Q_2, then relation [3.147] should be added to equation [3.141].

3.3.3. *Tonti diagram*

This section focuses on obtaining the Tonti diagram in electrostatics based on Figure 2.13. Then, we have a succession of function spaces with imposed boundary conditions and we place the various terms defined in sections 3.3.1 and 3.3.2.

Besides the physical quantities **E**, **D** and ρ (see Figure 3.12), there are also the source fields $\mathbf{E}_s$ and $\mathbf{D}_s$ as well as potentials V and **P**.

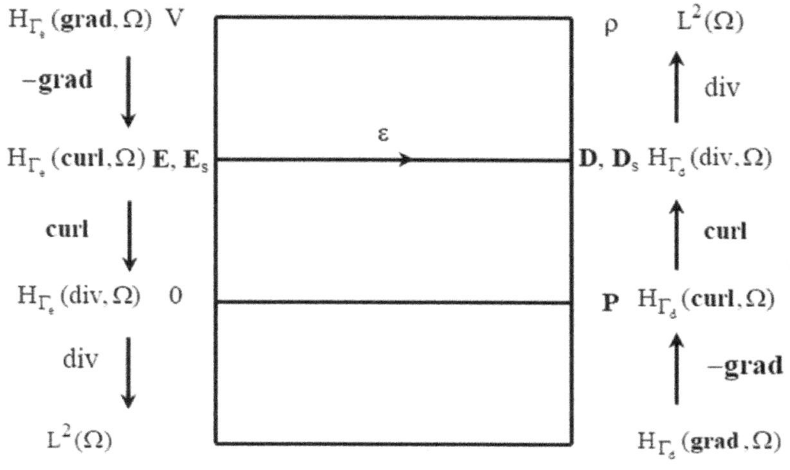

Figure 3.12. *Tonti diagram in electrostatics*

3.4. Electrokinetics

In electrokinetics, if a general problem is considered, the studied geometry may have the form presented in Figure 1.10 with two possible types of source terms: the electromotive force "e" and the current density flux "I". However, to make the developments less cumbersome while maintaining a certain generality, a simplified geometry is considered. Nevertheless, in section 3.4.2, we will show how to address the case of a multisource problem.

3.4.1. *Elementary geometry*

The studied geometry, namely a section of a conductor, is represented in Figure 3.13. Its conductivity, which may depend on the position, will be denoted by σ and its boundary Γ comprises three surfaces, denoted by Γ_{e1}, Γ_{e2} and Γ_j such that:

$$\Gamma = \Gamma_{e1} \cup \Gamma_{e2} \cup \Gamma_j \qquad [3.148]$$

On the two surfaces Γ_{e1} and Γ_{e2}, the tangential component of the electric field is zero, namely a gate-type boundary condition [1.35]. These two surfaces, considered gates, are therefore in contact with perfect conductors (the tangential component of the electric field is zero). On the contrary, the wall-type surface Γ_j can be considered in contact with a perfect insulator. Then, the condition given by equation [1.38] is verified for the current density.

Let us note that the studied domain Ω, which is limited to the conductor, is a contractible domain. On the contrary, the boundary Γ_j is not simply connected.

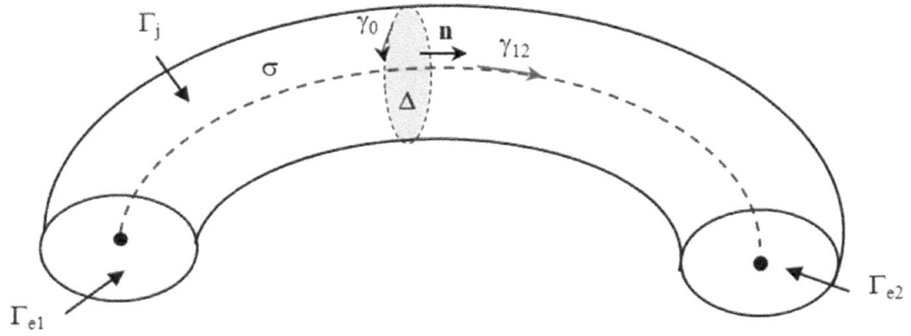

Figure 3.13. *Simplified geometry studied for electrokinetics*

For this example, there are two possibilities for the source term (see section 1.5.2):

– the first one is an electromotive force, denoted by "e", imposed between the two boundaries Γ_{e1} and Γ_{e2} (see equation [1.56]);

– the second one consists of imposing the flux "I" of the current density on the boundaries Γ_{e1} and Γ_{e2} (see equation [1.57]).

In the case of electrokinetics, the initial problem is governed by equations [1.58] and [1.59] written as follows:

$$\mathbf{curl}\mathbf{E} = 0 \quad [3.149]$$

$$\mathrm{div}\mathbf{J} = 0 \quad [3.150]$$

which are completed by the electric behavior law [1.20] and the boundary conditions [1.35] and [1.38] as follows:

$$\Gamma = \Gamma_j \cup \Gamma_{e1} \cup \Gamma_{e2} \text{ with } \mathbf{J.n}\big|_{\Gamma_j} = 0,\ \mathbf{E} \wedge \mathbf{n}\big|_{\Gamma_{ek}} = 0 \quad \text{with } k \in \{1,2\} \quad [3.151]$$

Based on the above constraints, the electric field and the current density are defined in the function spaces:

$$\mathbf{E} \in H_{\Gamma_{e1} \cup \Gamma_{e2}}(\mathbf{curl0}, \Omega) \text{ and } \mathbf{J} \in H_{\Gamma_j}(\mathrm{div}0, \Omega) \quad [3.152]$$

Sections 3.4.1.1 and 3.4.1.2 will develop the formulations in terms of the scalar potential and the vector potential when for each of them an electromotive force and the current density flux are imposed. It is important to note that, in the case of electrokinetics, the equations to be solved are equivalent to those encountered in electrostatics when a source term is imposed on the boundaries of the domain (see section 3.3.1). Therefore, the developments given in the following are very similar.

3.4.1.1. *Scalar potential V formulation*

3.4.1.1.1. Imposed electromotive force

The electric scalar potential formulation is very well suited when an electromotive force "e" is imposed between the surfaces Γ_{e1} and Γ_{e2} (see equation [1.56]). To account for the source term "e" in the local equations, a source field is introduced by decomposing the electric field into two terms (see section 3.2.1.1):

$$\mathbf{E} = \mathbf{E}_s + \mathbf{E}' \qquad [3.153]$$

where $\mathbf{E}_s$ represents a known source field depending on the electromotive force and $\mathbf{E}'$ is the unknown of the problem.

The properties of $\mathbf{E}_s$ are close to those of the electric field. Based on the general case, presented in section 3.2.1.1, they are stated (see equation [3.10]) as follows:

$$\mathbf{curlE}_s = 0, \int_{\gamma_{12}} \mathbf{E}_s.\mathbf{dl} = e \text{ , } \mathbf{E}_s \wedge \mathbf{n}\big|_{\Gamma_{e1}\cup\Gamma_{e2}} = 0$$
$$\text{i.e. } \mathbf{E}_s \in H_{\Gamma_{e1}\cup\Gamma_{e2}}(\mathbf{curl}0,\Omega) \qquad [3.154]$$

As the field $\mathbf{E}_s$ is curl free, the integral is true for any path γ_{12}, belonging to the domain Ω, linking the gates Γ_{e1} and Γ_{e2}.

The properties of the field $\mathbf{E}'$ are similar to those of the electric field $\mathbf{E}$, except for its circulation between the two gates Γ_{e1} and Γ_{e2} which is equal to zero. Indeed (see equation [3.154]), this constraint is supported by $\mathbf{E}_s$.

The properties of $\mathbf{E}'$ are then written as follows (see equation [3.9]):

$$\mathbf{curlE}' = 0, \int_{\gamma_{12}} \mathbf{E}'.\mathbf{dl} = 0, \ \mathbf{E}' \wedge \mathbf{n}\big|_{\Gamma_{e1}\cup\Gamma_{e2}} = 0$$
$$\text{i.e. } \mathbf{E}' \in H_{\Gamma_{e1}\cup\Gamma_{e2}}(\mathbf{curl}0,\Omega) \qquad [3.155]$$

Based on equation [3.11], the source field $\mathbf{E}_s$ can be defined using the electromotive force "e" and a support vector field $\boldsymbol{\beta}_e$, such that:

$$\mathbf{E}_S = e\boldsymbol{\beta}_e \qquad [3.156]$$

the vector field $\boldsymbol{\beta}_e$ having the following properties:

$$\begin{aligned} &\mathbf{curl}\ \boldsymbol{\beta}_e = 0,\ \boldsymbol{\beta}_e \wedge \mathbf{n}\big|_{\Gamma_{e1}\cup\Gamma_{e2}} = 0 \text{ and } \int_{\gamma_{12}} \boldsymbol{\beta}_e.\mathbf{dl} = 1 \\ &\text{i.e. : } \boldsymbol{\beta}_e \in H_{\Gamma_{e1}\cup\Gamma_{e2}}(\mathbf{curl}0, \Omega) \end{aligned} \qquad [3.157]$$

Considering the function space to which it belongs, the field $\boldsymbol{\beta}_e$ can be defined by means of an associated scalar potential, which will be denoted by "α_e", as follows:

$$\begin{aligned} &\boldsymbol{\beta}_e = -\mathbf{grad}\alpha_e, \qquad \alpha_e\big|_{\Gamma_{e1}} = \alpha_{e1}, \alpha_e\big|_{\Gamma_{e2}} = \alpha_{e2}, \alpha_{e1} - \alpha_{e2} = 1, \\ &\alpha_e \in H(\mathbf{grad}, \Omega') \end{aligned} \qquad [3.158]$$

NOTE.– For the choice of the constants α_{e1} and α_{e2}, similar to the approach in the case of electrostatics, a simple solution involves taking $\alpha_{e1} = 1$ on Γ_{e1} and $\alpha_{e2} = 0$ on Γ_{e2}.

The field **E'** also belongs to the function space $H_{\Gamma e1\cup\Gamma e2}(\mathbf{curl}0, \Omega)$. Therefore, it can be defined using an electric scalar potential V (see equation [2.21]), such that:

$$\mathbf{E}' = -\mathbf{grad}V \text{ with } V \in H_{\Gamma e1\cup\Gamma e2}(\mathbf{grad}, \Omega)\big| \qquad [3.159]$$

In this expression, homogeneous conditions are chosen for V on Γ_{e1} and Γ_{e2} to make sure that the circulation of **E'** is equal to zero. It is important to note that fixing a value of potential V, in this case zero, makes it possible to impose the gauge condition and therefore the uniqueness. If in expression [3.153] the field $\mathbf{E}_s$ is replaced by equation [3.156] and **E'** by equation [3.159], the electric field can be written as follows:

$$\mathbf{E} = e\boldsymbol{\beta}_e - \mathbf{grad}V \qquad [3.160]$$

As for the current density, using the behavior law [1.20], it is written as:

$$\mathbf{J} = \sigma(e\boldsymbol{\beta}_e - \mathbf{grad}V) \qquad [3.161]$$

Finally, if in equation [3.150] the current density is replaced by the above expression, we can write:

$$\text{div}(\sigma(\mathrm{e}\boldsymbol{\beta}_\mathrm{e} - \mathbf{grad}\mathrm{V})) = 0 \qquad [3.162]$$

which corresponds to the scalar potential formulation of an electrokinetics problem when the electromotive force "e" is imposed. Furthermore, $\boldsymbol{\beta}_e$ can be replaced by its expression defined in equation [3.158], which leads to:

$$\text{div}(\sigma(\mathrm{e}\,\mathbf{grad}\alpha_\mathrm{e} + \mathbf{grad}\mathrm{V})) = 0 \qquad [3.163]$$

Solving equation [3.162] or [3.163] leads to obtaining the scalar potential V. The electric field **E** is expressed using equation [3.160] and the current density can be obtained via equation [3.161]. The current density can also be expressed using the associated scalar potential α_e as follows:

$$\mathbf{J} = -\sigma(\mathrm{e}\,\mathbf{grad}\alpha_\mathrm{e} + \mathbf{grad}\mathrm{V}) \qquad [3.164]$$

3.4.1.1.2. Flux of imposed current density

Let us now consider as a source term the flux of current density "I" defined by equation [1.57]. The electric scalar potential formulation will therefore be centered on this source term, which does not naturally appear in the equations. To this end, expression [3.162] is kept, but the electromotive force "e" is now an unknown of the problem. A new equation accounting for the current intensity "I" should therefore be introduced. To deduce it, a power balance is written.

In its classical form, depending on the distribution of the electric field and the current density, the expression of the power dissipated in a conducting domain is:

$$\mathrm{P} = \iiint_\Omega \mathbf{E}.\mathbf{J}d\tau \qquad [3.165]$$

Replacing the electric field by its expression given in equation [3.160] yields:

$$\mathrm{P} = \iiint_\Omega (\mathrm{e}\boldsymbol{\beta}_\mathrm{e} - \mathbf{grad}\,\mathrm{V}).\mathbf{J}d\tau \qquad [3.166]$$

The second term of this integral can be written using the formula related to vector operators [2.23] in the following form:

$$-\iiint_\Omega \mathbf{grad}\mathrm{V}.\mathbf{J}d\tau = \iiint_\Omega \mathrm{V}\text{div}\mathbf{J}d\tau - \oiint_{\Gamma_{e1}\cup\Gamma_{e2}\cup\Gamma_j} \mathrm{V}\mathbf{J}.\mathbf{n}d\tau = 0 \qquad [3.167]$$

This equation is equal to zero, as the divergence of **J** is equal to zero (see equation [3.150]) and, considering the boundary conditions, the surface integral is also equal to zero ($V = 0$ on Γ_{e1} and Γ_{e2} [3.159] and $\mathbf{J}.\mathbf{n} = 0$ on Γ_j [3.150]). Under these conditions, equation [3.166] takes the form:

$$P = \iiint_\Omega e\boldsymbol{\beta}_e.\mathbf{J}d\tau \tag{3.168}$$

The power dissipated in the conductor can also be expressed using global quantities, as the electromotive force "e" across it and the current intensity I. Then, we have:

$$P = eI = \iiint_\Omega e\boldsymbol{\beta}_e.\mathbf{J}d\tau \tag{3.169}$$

By identification, the current I can be very easily deduced:

$$I = \iiint_\Omega \boldsymbol{\beta}_e.\mathbf{J}d\tau \tag{3.170}$$

Replacing the current density **J** by its expression given in equation [3.164], we obtain:

$$I = -(\iiint_\Omega \sigma\mathbf{grad}V.\boldsymbol{\beta}_e d\tau + \iiint_\Omega e\sigma\mathbf{grad}\alpha_e.\boldsymbol{\beta}_e d\tau) \tag{3.171}$$

Solving a problem of electrokinetics with the electric scalar potential formulation, when the source term is the current density flux, amounts to solving the system of equations formed of expressions [3.163] and [3.171]. In this system of equations, the unknowns are then the electric scalar potential V and the electromotive force "e" imposed across the conductor.

3.4.1.2. *Vector potential T formulation*

Similar to the scalar potential formulation, for the vector potential formulation, the studied case imposes as the source term either an electromotive force or the current density flux. The first to be studied is the case where the source term is the current density flux that is naturally imposed in the vector potential formulation.

3.4.1.2.1. Imposed current density flux

For this problem, equations [3.149], [3.150] and [1.20] must be solved. The boundary conditions on the boundary of the domain are defined in equation [3.151] and the source term "I", corresponding to the current density flux, is given by relationship [1.57]. As the current density is defined by means of the divergence operator, in order to introduce the source term in the local form of the equations, the procedure presented in section 3.2.1.2 will be used. The current density **J** is decomposed (see equation [3.19]) in the form:

$$\mathbf{J} = \mathbf{J}_S + \mathbf{J}' \quad [3.172]$$

In this expression, the field **J'** becomes the unknown of the problem and the field $\mathbf{J}_s$ makes it possible to take into account the source term imposed on the boundaries Γ_{e1} and Γ_{e2}. The properties of $\mathbf{J}_s$ are given by the relations:

$$\text{div}\mathbf{J}_S = 0,\ \iint_{\Gamma_{ek}} \mathbf{J}_S.\mathbf{n}ds = \pm I\ k \in \{1,2\},\ \mathbf{J}_S.\mathbf{n}\Big|_{\Gamma_j} = 0$$
$$\text{i.e. } \mathbf{J}_S \in H_{\Gamma_j}(\text{div}0,\Omega) \quad [3.173]$$

It is important to note that the current density $\mathbf{J}_s$ having a conservative flux, the above-mentioned surface integral is valid for any surface Δ whose contour lies on the boundary Γ_j (see Figure 3.13).

Since the current density $\mathbf{J}_s$ is proportional to the current intensity I, a support field $\boldsymbol{\lambda}_I$ is introduced such that:

$$\mathbf{J}_S = I\boldsymbol{\lambda}_I \quad [3.174]$$

Under these conditions, the properties of $\boldsymbol{\lambda}_I$ are identical to those of $\mathbf{J}_s$ and can be stated as follows:

$$\text{div}\boldsymbol{\lambda}_I = 0,\ \boldsymbol{\lambda}_I.\mathbf{n}\big|_{\Gamma_j} = 0 \text{ and } \iint_{\Delta} \boldsymbol{\lambda}_I.\mathbf{n}ds = \pm 1$$
$$\text{i.e., } \boldsymbol{\lambda}_I \in H_{\Gamma_j}(\text{div}0,\Omega) \quad [3.175]$$

In this expression, the surface Δ, whose contour is denoted by γ_0, lies on the boundary Γ_j (see Figure 3.13). It should be recalled that by sliding the contour γ_0 on this boundary, the surface Δ can be superposed with the boundaries Γ_{e1} and Γ_{e2}.

Since the flux I of the current density on the boundaries Γ_{ek} of the domain is now supported by $\mathbf{J}_s$, $\mathbf{J}'$ is defined by the following relations:

$$\text{div}\mathbf{J}'=0,\ \iint_{\Gamma_{ek}} \mathbf{J}'.\mathbf{n}ds = 0,\ \text{with } k\in\{1,2\},\ \mathbf{J}'.\mathbf{n}\big|_{\Gamma_j}=0$$
$$\text{i.e. } \mathbf{J}'\in H_{\Gamma_j}(\text{div}0,\Omega) \quad [3.176]$$

As the field $\mathbf{J}'$ is divergence free, Γ_{ek} ($k\in\{1,2\}$) can be replaced by any surface Δ whose contour belongs to Γ_j.

Since the domain Ω is contractible and considering the function space to which the support field $\boldsymbol{\lambda}_I$ belongs, the latter can be expressed based on an associated vector potential $\boldsymbol{\chi}_I$. However, precautions must be taken when building $\boldsymbol{\chi}_I$, as the boundary Γ_j is not simply connected. To this end, a cut is introduced along Γ_j (see section 3.2.1.2.2). Based on equation [3.175], the properties of the potential $\boldsymbol{\chi}_I$ are written as:

$$\boldsymbol{\lambda}_I = \mathbf{curl}\boldsymbol{\chi}_I,\ \boldsymbol{\chi}_I\wedge\mathbf{n}\big|_{\Gamma_j}=0 \text{ and } \int_{\gamma_0}\boldsymbol{\chi}_I.\mathbf{dl}=\pm 1$$
$$\text{i.e. } \boldsymbol{\chi}_I\in H^{\Delta}_{\Gamma_j}(\mathbf{curl},\Omega) \quad [3.177]$$

where γ_0 represents any contour supported by Γ_j that cannot be contracted to a point by successive transformations.

After having defined the source term $\mathbf{J}_s$, $\mathbf{J}'$ must be expressed. Its properties are given by relations [3.176] or $\mathbf{J}'\in H_{\Gamma j}(\text{div}0,\Omega)$. As the domain Ω is contractible, $\mathbf{J}'$ can be expressed using an electric vector potential (see equation [2.31]), which is denoted by $\mathbf{T}$, such that:

$$\mathbf{J}'=\mathbf{curl}\mathbf{T}\quad \mathbf{T}\in H_{\Gamma_j}(\mathbf{curl},\Omega) \quad [3.178]$$

NOTE.– The vector potential $\mathbf{T}$ is not unique. It is defined up to a gradient (see equation [2.46]). In order to have a unique solution, a gauge condition [2.50] or [2.51] must be imposed.

Based on equations [3.172], [3.174] and [3.178], the following can be deduced:

$$\mathbf{J}=I\boldsymbol{\lambda}_I+\mathbf{curl}\mathbf{T} \quad [3.179]$$

This equation and the behavior law [1.20] lead to the expression of the electric field:

$$\mathbf{E} = \sigma^{-1}(I\boldsymbol{\lambda}_I + \mathbf{curl}\mathbf{T}) \qquad [3.180]$$

Using the associated vector potential $\boldsymbol{\chi}_I$ (see equation [3.177]), the electric field can also be written as follows:

$$\mathbf{E} = \sigma^{-1}(\mathbf{curl}\mathbf{T} + I\mathbf{curl}\boldsymbol{\chi}_I) \qquad [3.181]$$

If the electric field is replaced in equation [3.149] by its expression given in equation [3.180], we obtain:

$$\mathbf{curl}(\sigma^{-1}(\mathbf{curl}\mathbf{T} + I\boldsymbol{\lambda}_I)) = 0 \qquad [3.182]$$

The electric field defined by equation [3.181] can also be replaced in equation [3.149]. Then we obtain:

$$\mathbf{curl}(\sigma^{-1}(\mathbf{curl}\mathbf{T} + I\mathbf{curl}\boldsymbol{\chi}_I)) = 0 \qquad [3.183]$$

which is the electrokinetics formulation in terms of the electric vector potential having as a source term the flux of density of current I.

This problem can be solved in two steps. The first one is to determine the support field $\boldsymbol{\lambda}_I$ verifying equation [3.175] or the associated vector potential $\boldsymbol{\chi}_I$ (see equations [3.177]). Knowing that the determination of $\boldsymbol{\chi}_I$ can be complex due to the topology of the surface Γ_j, which is not simply connected, the support field $\boldsymbol{\lambda}_I$ is generally preferred. Once the source field is calculated, equation [3.182] or [3.183] is solved with $\mathbf{T} \in H_{\Gamma j}(\mathbf{curl}, \Omega)$.

3.4.1.2.2. Imposed electromotive force

Let us now focus on the vector potential formulation having as a source term the electromotive force imposed between the surfaces Γ_{e1} and Γ_{e2}. Introducing this source term, with the vector potential formulation, is not natural. Indeed, in this case, current I becomes an unknown of the problem. To address this issue, equation [3.182] or [3.183] is kept and we look for an additional expression of the electromotive force as a function of quantities I and **T**. A system of equations is thus obtained. To this end, a power balance is written.

Let us go back to the expression of power (see equation [3.165]) in which the current density is replaced by its expression given in equation [3.179]. This yields:

$$P = \iiint_{\Omega} \mathbf{E}.\mathbf{curlT} d\tau + \iiint_{\Omega} I\boldsymbol{\lambda}_I.\mathbf{E} d\tau \qquad [3.184]$$

Let us consider the first integral term of equation [3.184] to which the formula [2.27] related to vector operators is applied. Then, the following can be written as:

$$\iiint_{\Omega} \mathbf{E}.\mathbf{curlT} d\tau = \iiint_{\Omega} \mathbf{curlE}.\mathbf{T} d\tau + \oiint_{\Gamma_{e1} \cup \Gamma_{e2} \cup \Gamma_j} (\mathbf{E} \wedge \mathbf{n}).\mathbf{T} d\tau \qquad [3.185]$$

Considering equation [3.149], the first integral term on the right is equal to zero. It can be readily shown that the second term is also equal to zero. Indeed, the surface integral is decomposed into three terms, namely Γ_{e1}, Γ_{e2} and Γ_j. On Γ_{e1} and Γ_{e2}, the surface integral is zero, considering the properties of **E** (see equation [3.151]). The same is true for the integral on Γ_j due to the properties of **T** (see equation [3.178]). The power dissipated in the domain Ω only depends on the second term of equation [3.184] as follows:

$$P = \iiint_{\Omega} I\boldsymbol{\lambda}_I.\mathbf{E} d\tau \qquad [3.186]$$

Expressing power as a function of global quantities, namely the electromotive force "e" and the flux of the current density "I", equation [3.186] is written as:

$$eI = \iiint_{\Omega} I\boldsymbol{\lambda}_I.\mathbf{E} d\tau \qquad [3.187]$$

Simplifying by "I", the following expression of the electromotive force is obtained:

$$e = \iiint_{\Omega} \boldsymbol{\lambda}_I.\mathbf{E} d\tau \qquad [3.188]$$

This equation can be rewritten by replacing the electric field by its expression given in equation [3.180] as follows:

$$e = \iiint_{\Omega} \sigma^{-1}\boldsymbol{\lambda}_I.\mathbf{curlT} d\tau + \iiint_{\Omega} I\sigma^{-1}\boldsymbol{\lambda}_I.\boldsymbol{\lambda}_I d\tau \qquad [3.189]$$

The expression of the electromotive force "e" is obtained as a function of the electric vector potential **T** and the current intensity I.

In order to solve this problem, with the electromotive force as a source term, assume that the vector field $\boldsymbol{\lambda}_I$ is known (see equation [3.175]). In this case, the system of equations to be solved, whose unknowns are the electric vector potential **T** and the current intensity I, has the following form:

$$\mathbf{curl}(\sigma^{-1}\mathbf{curlT}) + \mathbf{curl}(\sigma^{-1}\boldsymbol{\lambda}_I I) = 0$$

$$\iiint_\Omega \sigma^{-1}\boldsymbol{\lambda}_I.\mathbf{curlT}d\tau + \iiint_\Omega I\sigma^{-1}\boldsymbol{\lambda}_I.\boldsymbol{\lambda}_I d\tau = 0 \qquad [3.190]$$

For the problem to be complete, the gauge condition and the boundary conditions for **T** should be added.

3.4.1.3. *Summarizing tables*

<table>
<tr><th colspan="3">Electrokinetics (studied domain)</th></tr>
<tr><td rowspan="12">Scalar potential V formulation</td><td colspan="2">Source term: electromotive force e</td></tr>
<tr><td rowspan="3">Source field support two possibilities: $\boldsymbol{\beta}_e$ or α_e</td><td>Decomposition of the electric field: $\mathbf{E} = \mathbf{E}_s + \mathbf{E}'$</td></tr>
<tr><td>$\mathbf{E}_S = e\boldsymbol{\beta}_e, \ \boldsymbol{\beta}_e \in H_{\Gamma_{e1} \cup \Gamma_{e2}}(\mathbf{curl}0, \Omega)$</td></tr>
<tr><td>$\mathbf{E}_S = -e\mathbf{grad}\alpha_e, \ \alpha_e \in H(\mathbf{grad}, \Omega)$</td></tr>
<tr><td rowspan="2">Properties of the unknown E' and introduction of potential V</td><td>$\mathbf{E}' \in H_{\Gamma_{e1} \cup_{e2}}(\mathbf{curl}0, \Omega)$</td></tr>
<tr><td>$\mathbf{E}' = -\mathbf{grad}V, \ V \in H_{\Gamma_{e1} \cup \Gamma_{e2}}(\mathbf{grad}, \Omega)$</td></tr>
<tr><td rowspan="2">Equation to be solved: two possible forms as a function of $\boldsymbol{\beta}_e$ or α_e</td><td>$\mathrm{div}(\sigma(e\boldsymbol{\beta}_e - \mathbf{grad}V)) = 0$</td></tr>
<tr><td>$\mathrm{div}(\sigma(e\mathbf{grad}\alpha_e + \mathbf{grad}V)) = 0$</td></tr>
<tr><td colspan="2">Source term: current intensity I</td></tr>
<tr><td colspan="2">e becomes an unknown; an additional equation is needed</td></tr>
<tr><td colspan="2">$I = \iiint_\Omega \boldsymbol{\beta}_e.\mathbf{J}d\tau$</td></tr>
</table>

Table 3.3. *Summary of equations to be solved in electrokinetics for the scalar potential formulation (see Figure 3.13)*

<table>
<tr><th colspan="3">Electrokinetics (studied domain)</th></tr>
<tr><td rowspan="11">Vector potential T formulation</td><td colspan="2">Source term: current intensity I</td></tr>
<tr><td rowspan="3">Source field support two possibilities: $\boldsymbol{\lambda}_I$ or $\boldsymbol{\chi}_I$ with a cut on Γ_j not simply connected</td><td>Decomposition of the electric field: $\mathbf{J} = \mathbf{J}_s + \mathbf{J}'$</td></tr>
<tr><td>$\mathbf{J}_S = I\boldsymbol{\lambda}_I,\ \boldsymbol{\lambda}_I \in H_{\Gamma_j}(\mathrm{div}0, \Omega)$</td></tr>
<tr><td>$\mathbf{J}_s = I\,\mathbf{curl}\boldsymbol{\chi}_I,\ \boldsymbol{\chi}_I \in H^{\Delta}_{\Gamma_j}(\mathbf{curl}, \Omega)$</td></tr>
<tr><td rowspan="2">Properties of the unknown J' and introduction of the vector potential T</td><td>$\mathbf{J}' \in H_{\Gamma_j}(\mathrm{div}0, \Omega)$</td></tr>
<tr><td>$\mathbf{J}' = \mathbf{curl}\mathbf{T},\ \mathbf{T} \in H_{\Gamma_j}(\mathbf{curl}, \Omega)$</td></tr>
<tr><td rowspan="2">Equation to be solved: two possible forms depending on $\boldsymbol{\lambda}_I$ or $\boldsymbol{\chi}_I$</td><td>$\mathbf{curl}(\sigma^{-1}(\mathbf{curl}\mathbf{T} + I\boldsymbol{\lambda}_I)) = 0$</td></tr>
<tr><td>$\mathbf{curl}(\sigma^{-1}(\mathbf{curl}\mathbf{T} + I\,\mathbf{curl}\boldsymbol{\chi}_I)) = 0$</td></tr>
<tr><td colspan="2">Source term: electromotive force e</td></tr>
<tr><td colspan="2">I becomes an unknown; an additional equation is needed</td></tr>
<tr><td colspan="2">$e = \iiint_\Omega \boldsymbol{\lambda}_I . \mathbf{E} d\tau$</td></tr>
</table>

Table 3.4. *Summary of equations to be solved in electrokinetics for the vector potential formulation (see Figure 3.13)*

3.4.2. *Multisource case*

Section 3.4.1 only considered two gates on which either an electromotive force or the current intensity was imposed. The approach can be generalized to a set-up with N boundaries of Γ_{ek} type (N gates). The source terms can be electromotive forces (circulation of the electric field between two gates), current intensity (flux of current density) or still a combination of the two.

For the electric field, which is curl free (see equation [3.149]), the path γ_{ij}, linking two gates i and j, can be arbitrary, since in this case the circulation is independent of the path followed. Nevertheless, the paths on which the circulation (the electromotive force) is imposed should not form a closed loop. As there are N gates, the maximum number of conditions to be imposed on the circulations (electromotive forces) is N – 1.

Similarly, as **J** is divergence free (see equation [3.150]), the sum of the fluxes of current density, imposed across the gates, must be equal to zero. Therefore, the maximum number of independent values of current I that can be imposed is N – 1.

Conditions on the circulations and the fluxes of current density can also be imposed simultaneously. Likewise, the number of these conditions must be equal to N – 1 and respect the above constraints.

In order to illustrate these various possibilities, the example to be studied is that of Figure 3.14, composed of three gates Γ_{e1}, Γ_{e2} and Γ_{e3}.

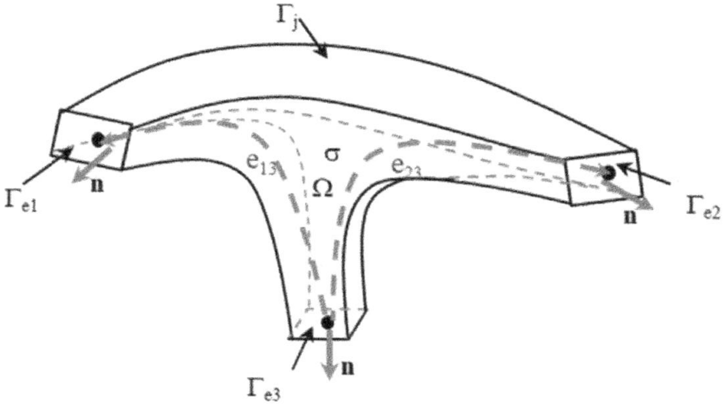

Figure 3.14. *Electrokinetics: example of multisource geometry with 3 gates (N = 3)*

The studied domain Ω, of the conductivity σ, is contractible, of the boundary Γ, such that:

$$\Gamma = \Gamma_{e1} \cup \Gamma_{e2} \cup \Gamma_{e3} \cup \Gamma_j \qquad [3.191]$$

It can be noted that the boundary Γ_j is not simply connected. The boundary conditions on various boundaries are equivalent to those given in equation [3.151]. The fields **E** and **J** are governed by equations [3.149] and [3.150] and the electric behavior law [1.20]. In this example, in the presence of three gates, two source terms must be imposed. As shown in Figure 3.14, the source terms can be, for example, the electromotive forces e_{13} and e_{23}, the currents I_1 and I_2 or a condition on an electromotive force and a current.

In what follows, in reference to sections 3.4.1.1 and 3.4.1.2, the two formulations in terms of the scalar potential V and the vector potential **T** will be built. Considering the equations to be solved, a certain similarity with the electrostatic problem studied in section 3.3.2 can be noted. This will serve as a reference when building the support fields and associated potentials.

3.4.2.1. *Scalar potential V formulation*

As a first step, the source terms are considered to be the electromotive forces e_{13} and e_{23}. Based on equation [3.153], two source fields $\mathbf{E}_{s1}$ and $\mathbf{E}_{s2}$ are evidenced, and they can be expressed, respectively, using two support fields $\boldsymbol{\beta}_{13}$ and $\boldsymbol{\beta}_{23}$ (see equation [3.157]) such that:

$$\begin{aligned}&\mathbf{curl}\,\boldsymbol{\beta}_{ij} = 0, \int_{\gamma_{kj}} \boldsymbol{\beta}_{ij}.\mathbf{dl} = \delta_{ki} \quad \boldsymbol{\beta}_{ij} \wedge \mathbf{n}\Big|_{\Gamma_{e1}\cup\Gamma_{e2}\cup\Gamma_{e3}} = 0\\&\boldsymbol{\beta}_{ij} \in H_{\Gamma_{e1}\cup\Gamma_{e2}\cup\Gamma_{e3}}(\mathbf{curl}0,\Omega') \ (i,k)\in\{1.2\} \text{ and } j=3\end{aligned} \tag{3.192}$$

Since the studied domain is contractible, the support fields $\boldsymbol{\beta}_{ij}$ can be expressed as a function of associated scalar potentials defined by relations [3.13] and [3.14]. Then, we have:

$$\begin{aligned}&\boldsymbol{\beta}_{ij} = -\mathbf{grad}\alpha_{ij} \text{ and } \alpha_{ij}\Big|_{\Gamma_{ek}} = \delta_{ki} \quad \alpha_{ij} \in H(\mathbf{grad},\Omega')\\&\text{with } i \in \{1,2\}, \ k\in\{1,2,3\}, j=3\end{aligned} \tag{3.193}$$

Following the same approach as in section 3.4.1.1.1, let us introduce the electric scalar potential V (see equation [3.159]). The electric field can then be expressed using support fields, electromotive forces and scalar potential V in the following form:

$$\mathbf{E} = e_{13}\boldsymbol{\beta}_{13} + e_{23}\boldsymbol{\beta}_{23} - \mathbf{grad}V, \text{ with } V\in H_{\Gamma_{e1}\cup\Gamma_{e2}\cup\Gamma_{e3}}(\mathbf{grad},\Omega)\Big| \tag{3.194}$$

As for the current density, considering equation [3.150] and the boundary conditions defined by equations [3.191] and [3.151], they belong to $H_{\Gamma j}(\text{div}0, \Omega)$. Its expression is obtained using the behavior law [1.20] as follows:

$$\mathbf{J} = \sigma(e_{13}\boldsymbol{\beta}_{13} + e_{23}\boldsymbol{\beta}_{23} - \mathbf{grad}V) \text{ with } \mathbf{J}\in H_{\Gamma_j}(\text{div}0,\Omega) \tag{3.195}$$

Applying equation [3.147] to the expression of the current density, the scalar potential formulation is written as:

$$\text{div}\sigma(e_{13}\boldsymbol{\beta}_{13} + e_{23}\boldsymbol{\beta}_{23} - \mathbf{grad}V) = 0 \tag{3.196}$$

The two support fields $\boldsymbol{\beta}_{13}$ and $\boldsymbol{\beta}_{23}$ can be expressed (see equation [3.193]), by means of the associated scalar potentials α_{13} and α_{23}. In this case, equation [3.196] has the following form:

$$\text{div}\sigma(e_{13}\mathbf{grad}\alpha_{13} + e_{23}\mathbf{grad}\alpha_{23} + \mathbf{grad}V) = 0 \tag{3.197}$$

If instead of the electromotive forces, the currents I_1 and I_2 are imposed on the gates Γ_{e1} and Γ_{e2}, as shown in section 3.4.1.1.2, the electromotive forces become the unknowns of the problem. To obtain a full equation system, two additional equations must be added by means of a power balance. Based on equation [3.170], it can be noted that current I is obtained by integrating, over the entire domain Ω, the scalar product of the current density $\mathbf{J}$ and of the support field $\boldsymbol{\beta}_e$. There are two support fields $\boldsymbol{\beta}_{13}$ and $\boldsymbol{\beta}_{23}$ in our application. Applying an approach similar to the one developed in the case of electrostatics (see section 3.3.2.1.2), it can be deduced that the current I_1 is obtained by integrating the scalar product of $\mathbf{J}$ (see equation [3.195]) and $\boldsymbol{\beta}_{13}$ as follows:

$$I_1 = \iiint_\Omega \sigma\boldsymbol{\beta}_{13}.(e_{13}\boldsymbol{\beta}_{13} + e_{23}\boldsymbol{\beta}_{23} - \mathbf{grad}\,V)d\tau \qquad [3.198]$$

Similarly, for current I_2, we have:

$$I_2 = \iiint_\Omega \sigma\boldsymbol{\beta}_{23}.(e_{13}\boldsymbol{\beta}_{13} + e_{23}\boldsymbol{\beta}_{23} - \mathbf{grad}\,V)d\tau \qquad [3.199]$$

The system of equations to be solved is then composed of equations [3.196], [3.198] and [3.199].

Let us now consider that the source terms are a combination of an electromotive force and a current, namely e_{13} and I_2. In order to solve equation [3.196], we have two unknowns: the scalar potential V and the electromotive force e_{23}. To obtain a full equation system, equation [3.199] is added.

3.4.2.2. *Vector potential T formulation*

For the vector potential formulation, we rely on section 3.4.1.2. As a first step, let us consider as source terms the currents I_1 and I_2.

Based on equation [3.172], two source current densities $\mathbf{J}_{s1}$ and $\mathbf{J}_{s2}$ are introduced and their expressions use the two support fields $\boldsymbol{\lambda}_{13}$ and $\boldsymbol{\lambda}_{23}$. The approach used for building these two fields is similar to that for the field $\boldsymbol{\lambda}_I$ of equation [3.175]. This yields:

$$\begin{gathered}\mathrm{div}\boldsymbol{\lambda}_{13} = 0,\ \iint_{\Gamma_{e1}} \boldsymbol{\lambda}_{13}.\mathbf{n}dS = 1,\ \iint_{\Gamma_{e3}} \boldsymbol{\lambda}_{13}.\mathbf{n}dS = -1 \\ \boldsymbol{\lambda}_{13}.\mathbf{n}\big|_{\Gamma_{e2}\cup\Gamma_j} = 0 \ \text{ i.e. } \boldsymbol{\lambda}_{13} \in H_{\Gamma_{e2}\cup\Gamma_j}(\mathrm{div},\Omega)\end{gathered} \qquad [3.200]$$

$$\operatorname{div}\boldsymbol{\lambda}_{23} = 0, \iint_{\Gamma_{e2}} \boldsymbol{\lambda}_{23}.\mathbf{n}dS = 1, \iint_{\Gamma_{e3}} \boldsymbol{\lambda}_{23}.\mathbf{n}dS = -1,$$
$$\boldsymbol{\lambda}_{23}.\mathbf{n}\big|_{\Gamma_{e1}\cup\Gamma_j} = 0 \text{ i.e. } \boldsymbol{\lambda}_{23} \in H_{\Gamma_{e1}\cup\Gamma_j}(\operatorname{div},\Omega) \quad [3.201]$$

Using these support fields and introducing the electric vector potential **T** (see equation [3.178]), the current density is written as:

$$\mathbf{J} = I_1\boldsymbol{\lambda}_{13} + I_2\boldsymbol{\lambda}_{23} + \mathbf{curl}\mathbf{T} \text{ with } \mathbf{T} \in H_{\Gamma_j}(\mathbf{curl},\Omega) \quad [3.202]$$

For the electric field, equation [3.149] and the boundary conditions defined by equations [3.191] and [3.151] show that it belongs to $H_{\Gamma e1\cup\Gamma e2\cup\Gamma e3}\mathbf{curl}0, \Omega)$. Its expression is obtained by means of the behavior law [1.20], i.e.:

$$\mathbf{E} = \sigma^{-1}(I_1\boldsymbol{\lambda}_{13} + I_2\boldsymbol{\lambda}_{23} + \mathbf{curl}\mathbf{T}) \text{ with } \mathbf{E} \in H_{\Gamma_{e1}\cup\Gamma_{e2}\cup\Gamma_{e3}}(\mathbf{curl}0,\Omega) \quad [3.203]$$

The vector potential formulation of this problem is then obtained by applying relation [3.149] to the electric field as follows:

$$\mathbf{curl}(\sigma^{-1}(I_1\boldsymbol{\lambda}_{13} + I_2\boldsymbol{\lambda}_{23} + \mathbf{curl}\mathbf{T})) = 0 \quad [3.204]$$

Considering the properties of the support fields $\boldsymbol{\lambda}_{13}$ and $\boldsymbol{\lambda}_{23}$ (see equations [3.200] and [3.201]), it is possible to introduce the potentials $\boldsymbol{\chi}_{ij}$, as shown by equation [3.177]. However, given that the boundary Γ_j is not simply connected, some precautions must be taken (see section 3.2.1.2.2).

Let us now consider that the source terms are the electromotive forces e_{13} and e_{23}. The current densities I_1 and I_2 become the unknowns of the problem. It is therefore necessary to impose two additional equations. Similar to section 3.4.1.2.2, these two equations are obtained by means of a power balance. In this section, equation [3.188] shows that the electromotive force is expressed by integrating, over the domain Ω, the scalar product of the electric field with a support field. Applied to our example, the expression of the electromotive force e_{13}, with $\boldsymbol{\lambda}_{13}$ as the support field, is:

$$e_{13} = \iiint_{\Omega} \boldsymbol{\lambda}_{13}.\sigma^{-1}(I_1\boldsymbol{\lambda}_{13} + I_2\boldsymbol{\lambda}_{23} + \mathbf{curl}\mathbf{T})d\tau \quad [3.205]$$

As for the electromotive force e_{23}, it is written similarly, with $\boldsymbol{\lambda}_{23}$ as the support field:

$$e_{23} = \iiint_{\Omega} \boldsymbol{\lambda}_{23}.\sigma^{-1}(I_1\boldsymbol{\lambda}_{13} + I_2\boldsymbol{\lambda}_{23} + \mathbf{curl}\mathbf{T})d\tau \quad [3.206]$$

The system of equations to be solved is therefore composed of equations [3.204], [3.205] and [3.206].

Finally, consider the case of a combination of source terms of different natures, for example, the electromotive force e_{13} and the current I_2. We have to solve equation [3.204] but, besides the vector potential **T**, the current I_1 is unknown. Under these conditions, to obtain a full equation system, equation [3.205] is added.

3.4.3. *Tonti diagram*

First of all, the physical quantities, namely **J**, **E** and also the conductivity σ linking them (see equation [1.20]), are positioned in the diagram. Then, similar to **E** and **J**, the source fields $\mathbf{E}_s$, $\mathbf{J}_s$ and the fields **E'** and **J'** can be placed on the diagram as well as the two potentials V and **T** so that they verify, respectively, equations [3.159] and [3.178].

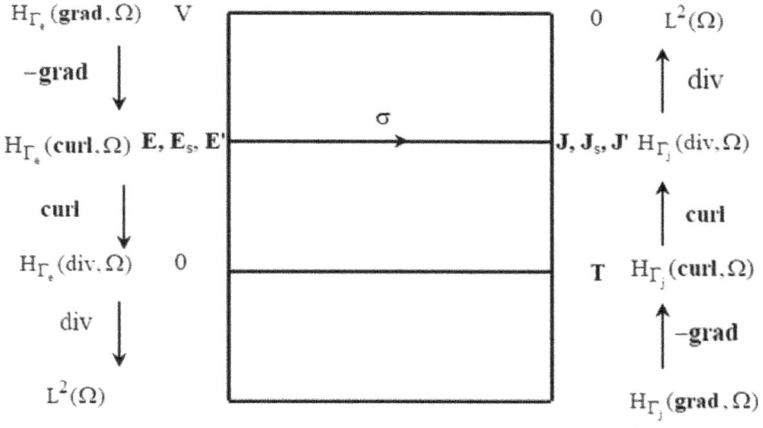

Figure 3.15. *Tonti diagram in the case of electrokinetics*

3.5. Magnetostatics

3.5.1. *Studied problems*

Similar to electrostatics and electrokinetics, two (scalar and vector) potential formulations will be developed for magnetostatics. These developments will be achieved for the four types of source terms presented in section 1.5.3, namely the magnetomotive force, the magnetic flux, a permanent magnet and an inductor through which a known current density flows. Similar to electrostatics and electrokinetics, depending on the source term used, the scalar or vector potential formulation will be introduced naturally. For the other configurations, an additional equation resulting from a magnetic energy balance needs to be introduced.

It can also be noted that if the source terms are imposed on the boundaries of the domain, in this case the magnetomotive force and the magnetic flux, the developments required to obtain the potential formulations are equivalent to those introduced in electrostatics and electrokinetics.

3.5.2. *Scalar potential φ formulation*

3.5.2.1. *Imposed magnetomotive force*

For the study of a set-up in which the source term is a magnetomotive force, let us consider the relatively simple example, represented in Figure 3.16. The domain Ω is contractible, its boundary Γ being $\Gamma = \Gamma_{h1} \cup \Gamma_{h2} \cup \Gamma_b$. The boundary Γ_b, defined by relation [1.37], represents a wall for the magnetic flux density and the boundaries Γ_{h1} and Γ_{h2}, defined by relation [1.36], represent the gates. As shown in Chapter 1 (see equation [1.60]), the magnetomotive force "f_m" is defined by the circulation of the magnetic field between the boundaries Γ_{h1} and Γ_{h2} of the domain Ω along the path γ_{12}.

The equations to be solved, see for example Figure 3.16, are written based on equations [1.65] and [1.66], but in the absence of the current density, in the following form:

$$\mathbf{curlH} = 0 \qquad [3.207]$$

$$\mathrm{div}\mathbf{B} = 0 \qquad [3.208]$$

We should add the magnetic behavior law [1.26] and the boundary conditions [1.36] and [1.37] which, for the studied problem, are defined by the following expressions:

$$\Gamma = \Gamma_b \cup \Gamma_{h1} \cup \Gamma_{h2} \text{ with } \mathbf{B.n}\big|_{\Gamma_b} = 0,\ \mathbf{H} \wedge \mathbf{n}\big|_{\Gamma_{hk}} = 0$$
$$\text{with } k \in \{1,2\} \qquad [3.209]$$

It should be noted that the boundary Γ_b is not simply connected.

Based on the above equations, the function spaces associated with the magnetic field $\mathbf{H}$ and with the magnetic flux density $\mathbf{B}$ can be readily deduced. Then, we have:

$$\mathbf{H} \in H_{\Gamma_{h1} \cup \Gamma_{h2}}(\mathbf{curl}0, \Omega) \text{ and } \mathbf{B} \in H_{\Gamma_b}(\mathrm{div}0, \Omega) \qquad [3.210]$$

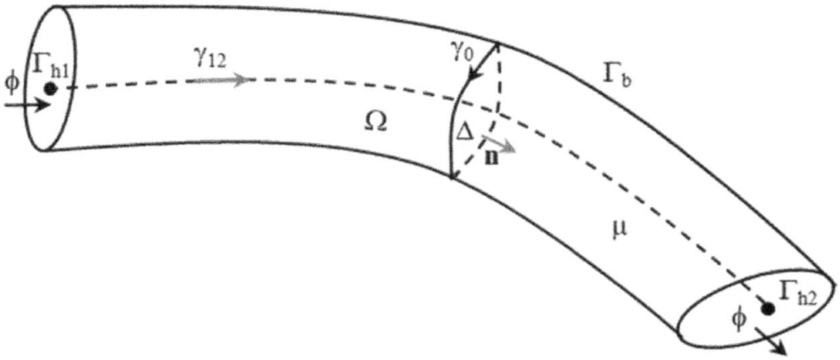

Figure 3.16. *Geometry studied for magnetostatics having as source terms the flux ϕ or the magnetomotive force f_m*

It should be recalled that when the source term is linked to a field defined by a circulation (see section 3.2.1.1), the scalar potential formulation is naturally introduced. Under these conditions, in order to introduce the source term, in the local forms of the equations, the magnetic field **H** is decomposed into two terms (see equation [3.8]) as follows:

$$\mathbf{H} = \mathbf{H}_s + \mathbf{H}' \tag{3.211}$$

where $\mathbf{H}_s$ is the source field due to the magnetomotive force, a priori known, and $\mathbf{H}'$ is the new unknown of the problem. The source field $\mathbf{H}_s$ is defined based on the properties of field **H** as follows:

$$\mathbf{curl}\mathbf{H}_s = 0, \ \int_{\gamma_{12}} \mathbf{H}_s.\mathbf{dl} = f_m, \ \mathbf{H}_s \wedge \mathbf{n}\big|_{\Gamma_{h1} \cup \Gamma_{h2}} = 0$$
$$\text{i.e. } \mathbf{H}_s \in H_{\Gamma_{h1} \cup \Gamma_{h2}}(\mathbf{curl}0, \Omega) \tag{3.212}$$

The field $\mathbf{H}_s$ is proportional to the magnetomotive force "f_m". It can therefore be expressed by means of a support field, denoted by $\boldsymbol{\beta}_s$ (see equation [3.11]), such that:

$$\mathbf{H}_s = f_m \, \boldsymbol{\beta}_s \tag{3.213}$$

where $\boldsymbol{\beta}_s$ has the properties equivalent to those defined in the general case (see equation [3.12]), therefore:

$$\mathbf{curl}\boldsymbol{\beta}_s = 0, \int_{\gamma_{12}} \boldsymbol{\beta}_s.\mathbf{dl} = 1, \text{ with } \boldsymbol{\beta}_s \wedge \mathbf{n}\big|_{\Gamma_{h1}\cup\Gamma_{h2}} = 0$$
$$\text{i.e. } \boldsymbol{\beta}_s \in H_{\Gamma_{h1}\cup\Gamma_{h2}}(\mathbf{curl}0,\Omega) \qquad [3.214]$$

For a contractible studied domain, taking into account its properties, the field $\boldsymbol{\beta}_s$ can be expressed by means of a scalar potential "α_s" (see equation [3.13]). Then, the source field $\mathbf{H}_s$ has the following form:

$$\mathbf{H}_s = -f_m\mathbf{grad}\alpha_s \text{ with } \alpha_s\big|_{\Gamma_{e1}} = 1 \text{ and } \alpha_s\big|_{\Gamma_{e2}} = 0 \qquad [3.215]$$

Having defined the source field, the aim is to determine the field **H'**, the new unknown of the problem. The properties of **H'** are deduced from the magnetic field **H** and from the source field $\mathbf{H}_s$ and have the following form:

$$\mathbf{curlH}' = 0, \int_{\gamma_{12}} \mathbf{H}'.\mathbf{dl} = 0, \mathbf{H}' \wedge \mathbf{n}\big|_{\Gamma_{h1}\cup\Gamma_{h2}} = 0$$
$$\text{i.e. } \mathbf{H}' \in H_{\Gamma_{h1}\cup\Gamma_{h2}}(\mathbf{curl}0,\Omega) \qquad [3.216]$$

For a contractible domain, **H'** can be expressed using the magnetic scalar potential, which will be denoted by φ. In order to have a circulation of **H'** between Γ_{h1} and Γ_{h2} equal to zero and a unique solution, the scalar potential φ is set to zero on these two surfaces. Under these conditions, the scalar potential is defined by the properties:

$$\mathbf{H}' = -\mathbf{grad}\varphi \text{ with } \varphi \in H_{\Gamma_{h1}\cup\Gamma_{h2}}(\mathbf{grad},\Omega) \qquad [3.217]$$

Gathering equations [3.211], [3.213] and [3.217], the magnetic field has the following form:

$$\mathbf{H} = f_m\boldsymbol{\beta}_s - \mathbf{grad}\varphi \qquad [3.218]$$

Based on this equation and using the behavior law [1.26], the magnetic flux density is written as:

$$\mathbf{B} = \mu(f_m\boldsymbol{\beta}_s - \mathbf{grad}\varphi) \qquad [3.219]$$

If the magnetic flux density is replaced in equation [3.208] by expression [3.219], we obtain:

$$\mathrm{div}(\mu(f_m \boldsymbol{\beta}_s - \mathbf{grad}\varphi)) = 0 \qquad [3.220]$$

Furthermore, $\boldsymbol{\beta}_s$ can be replaced by its expression as a function of α_s, defined in equation [3.215]. Under these conditions, equation [3.220] can be rewritten as follows:

$$\mathrm{div}(\mu(f_m \mathbf{grad}\alpha_s + \mathbf{grad}\varphi)) = 0 \qquad [3.221]$$

The solution to equation [3.220] or [3.221] makes it possible to determine the magnetic scalar potential φ. Relation [3.218] is used in order to calculate the magnetic field **H**, knowing φ. Then, in order to determine the magnetic flux density **B**, equation [3.219] is used. The scalar source term α_s introduced into equation [3.215] can also be used. Then, we obtain:

$$\mathbf{B} = -\mu(\mathbf{grad}\varphi + f_m \mathbf{grad}\alpha_s) \qquad [3.222]$$

It can be noted that, in this case, this equation is similar to relation [3.164], obtained in the case of electrokinetics.

3.5.2.2. *Imposed magnetic flux density*

To illustrate the case when the source term is the magnetic flux ϕ (see equation [1.61]), the example in Figure 3.16 is considered. Imposing the flux ϕ is more difficult in scalar potential formulation. Indeed, as developed in the previous section, in the case of formulations in electrostatics and electrokinetics, an additional equation is required. In the present case, the objective is to express the flux ϕ as a function of the magnetomotive force f_m and of the scalar potential φ. To obtain this expression, an energy balance must be written.

The following section presents the developments for a linear behavior law of materials. This result is nevertheless true when the magnetic behavior law is not linear.

In the linear case, the magnetic energy is written as:

$$W_{mag} = \frac{1}{2}\iiint_{\Omega} \mathbf{B}.\mathbf{H}\,d\tau \qquad [3.223]$$

Replacing the magnetic field **H** by its expression given in equation [3.218], we have:

$$W_{mag} = \frac{1}{2} \iiint_{\Omega} \mathbf{B}.(f_m \boldsymbol{\beta}_s - \mathbf{grad}\varphi) d\tau \qquad [3.224]$$

Let us consider the second term of the integral. Using formula [2.23], related to vector operators, we can write:

$$-\frac{1}{2} \iiint_{\Omega} \mathbf{B}.\mathbf{grad}\varphi d\tau = \frac{1}{2} \iiint_{\Omega} \varphi \mathrm{div}\mathbf{B} d\tau - \frac{1}{2} \oiint_{\Gamma_b \cup \Gamma_{h1} \cup \Gamma_{h2}} \varphi \mathbf{B}.\mathbf{n} d\tau \qquad [3.225]$$

It can be easily shown that this integral is equal to zero. Indeed, for the first term, the divergence of **B** is zero (see equation [3.208]). For the second term, it can be readily shown that, taking into account the boundary conditions, the surface integral is also zero (on Γ_b, $\mathbf{B}.\mathbf{n} = 0$ and on $\Gamma_{h1} \cup \Gamma_{h2}$, $\varphi = 0$). Under these conditions, equation [3.224] has the form:

$$W_{mag} = \frac{1}{2} \iiint_{\Omega} f_m \boldsymbol{\beta}_s.\mathbf{B} d\tau \qquad [3.226]$$

The magnetic energy can also be expressed using the global quantities f_m and ϕ. The following can be written:

$$W_{mag} = \frac{1}{2} f_m \phi = \frac{1}{2} \iiint_{\Omega} f_m \boldsymbol{\beta}_s.\mathbf{B} d\tau \qquad [3.227]$$

By identification, the expression of the magnetic flux is obtained:

$$\phi = \iiint_{\Omega} \mathbf{B}.\boldsymbol{\beta}_s d\tau \qquad [3.228]$$

By replacing **B** by its expression, given in equation [3.222]:

$$\phi = -\iiint_{\Omega} \mu(\mathbf{grad}\,\varphi + f_m \mathbf{grad}\,\alpha_s).\boldsymbol{\beta}_s d\tau \qquad [3.229]$$

Under these conditions, when the source term is the magnetic flux, with the scalar potential formulation, the magnetostatic problem to be solved has the form of the following system of equations:

$$\mathrm{div}(\mu(f_m \boldsymbol{\beta}_s - \mathbf{grad}\varphi)) = 0$$

$$\phi = -\iiint_{\Omega} (\mu \boldsymbol{\beta}_s \cdot \mathbf{grad}\,\varphi + f_m \mu \boldsymbol{\beta}_s \cdot \mathbf{grad}\,\alpha_s) d\tau \qquad [3.230]$$

where the unknowns are the scalar potential φ and the magnetomotive force "f_m".

3.5.2.3. *Imposed current density*

For the analysis of a magnetostatics problem, having as a source term the current density $\mathbf{J}_0$, let us consider the example in Figure 3.17. We have an iron core coil, consisting of a multiwire inductor winding, through which flows a current I. The magnetic permeability of the iron core is denoted by μ_1. Knowing the intensity of the current I in the inductor, the current density $\mathbf{J}_0$ is given by the expression [1.62]. The studied domain Ω, of boundary Γ, holds the coil Ω_s of boundary Γ_s, the iron core, all being immersed in an air box of permeability μ_0. On the external boundary Γ, the boundary conditions are of wall type for the magnetic flux density (see equation [1.37]). Finally, it should be noted that the domain defined by the coil (subdomain Ω_s) is not simply connected and its permeability is equal to μ_0.

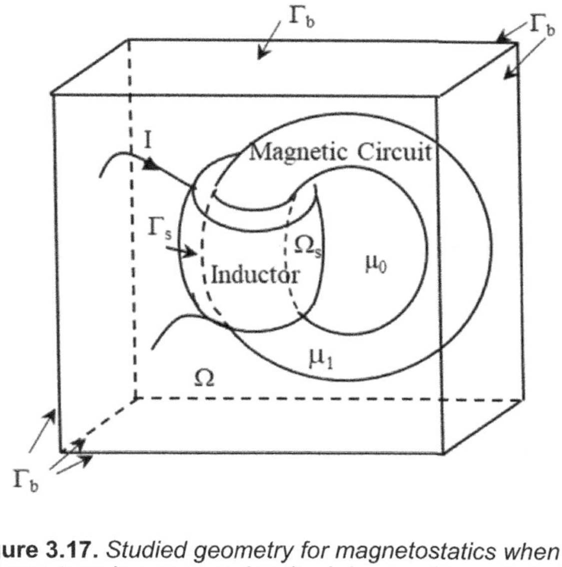

Figure 3.17. *Studied geometry for magnetostatics when the source term is a current density J_0 imposed in an inductor*

In the case of magnetostatics, having as a source term a current density, the equations to be solved are written as follows:

$$\mathbf{curlH} = \mathbf{J}_0 \qquad [3.231]$$

$$\text{div}\mathbf{B} = 0 \qquad [3.232]$$

with the magnetic behavior law [1.26] and the Γ_b type of boundary condition over the entire boundary of the domain, which is:

$$\mathbf{B.n}\big|_{\Gamma} = 0 \qquad [3.233]$$

Based on these equations, the function spaces, to which belong the magnetic field **H** and the magnetic flux density **B**, can be deduced as follows:

$$\mathbf{H} \in H(\mathbf{curl}, \Omega) \text{ and } \mathbf{B} \in H_0(\text{div}0, \Omega) \qquad [3.234]$$

For this problem, the configuration is that of a source term inside a domain associated with a curl. The formulation relies on the results of section 3.2.2.1. Let us first focus on the current density $\mathbf{J}_0$, which is a divergence free vector field. It is proportional to the current intensity I and can be expressed using a support field $\boldsymbol{\lambda}_I$ (see equation [3.37]) in the form:

$$\mathbf{J}_0 = I\boldsymbol{\lambda}_I \qquad [3.235]$$

In reference to the field $\boldsymbol{\lambda}_s$ introduced in section 3.2.2.1, $\boldsymbol{\lambda}_I$ is defined in the inductor and, by extension, it is set to zero in the remaining domain. Let us recall that this extension is possible, as we have $\mathbf{J}_0.\mathbf{n} = 0$ on the boundary of the inductor thus providing the continuity of the normal component of $\mathbf{J}_0$ throughout the domain. This process allows for the definition of $\boldsymbol{\lambda}_I$ throughout the domain Ω that is contractible. The issue related to the non-connectedness of the subdomain Ω_s for the construction of the associated vector potential $\boldsymbol{\chi}_I$ (see equation [3.39]) vanishes. Under these conditions, the properties of $\boldsymbol{\lambda}_I$ are stated (see equation [3.38]) in the following form:

$$\text{div}\boldsymbol{\lambda}_I = 0, \ \iint_{\Delta} \boldsymbol{\lambda}_I.\mathbf{n}dS = 1, \ \boldsymbol{\lambda}_I.\mathbf{n}\big|_{\Gamma_s} = 0$$
$$\text{i.e. } \boldsymbol{\lambda}_I \in H_0(\text{div}0, \Omega) \qquad [3.236]$$

In this expression, Δ represents the cross-section of the conductors, perpendicular to the direction of the current density.

The support field $\boldsymbol{\lambda}_I$ belongs to the space $H_0(\text{div}0, \Omega)$, which can therefore be expressed by means of a potential $\boldsymbol{\chi}_I$ (see equation [3.39]) whose properties are:

$$\mathbf{curl}\boldsymbol{\chi}_I = \boldsymbol{\lambda}_I, \ \boldsymbol{\chi}_I \wedge \mathbf{n}\big|_{\Gamma} = 0 \text{ i.e. } \boldsymbol{\chi}_I \in H_0(\mathbf{curl}, \Omega) \qquad [3.237]$$

For the developments, a source field $\mathbf{H}_s$ is again introduced. In the studied example, it is defined using the current density $\mathbf{J}_0$ via the curl operator. Then, based on equations [3.235] and [3.237], the following succession of equations can be written as:

$$\mathbf{curl H}_s = \mathbf{J}_0 = \mathrm{I}\boldsymbol{\lambda}_\mathrm{I} = \mathrm{I}\mathbf{curl}\boldsymbol{\chi}_\mathrm{I} \qquad [3.238]$$

Based on equation [3.231], introducing the field $\mathbf{H}_s$ and using equation [3.238], a field $\mathbf{H}'$ can be defined as follows:

$$\mathbf{curl}(\mathbf{H} - \mathbf{H}_s) = \mathbf{curl}(\mathbf{H} - \mathrm{I}\boldsymbol{\chi}_\mathrm{I}) = \mathbf{curl H}' = 0 \qquad [3.239]$$

In this expression, the following can be deduced by identification:

$$\mathbf{H} = \mathbf{H}' + \mathbf{H}_s = \mathbf{H}' + \mathrm{I}\boldsymbol{\chi}_\mathrm{I} \qquad [3.240]$$

It should be noted that this is the expression of the decomposition of field $\mathbf{H}$ given by equation [3.211]. The properties of $\mathbf{H}'$ can be deduced from these relations as follows:

$$\mathbf{curl H}' = 0, \Rightarrow \mathbf{H}' \in \mathrm{H}(\mathbf{curl}0, \Omega) \qquad [3.241]$$

Considering the function space to which $\mathbf{H}'$ belongs, where the studied domain is contractible, equation [2.32] yields:

$$\mathbf{H}' = -\mathbf{grad}\varphi, \quad \text{with } \varphi \in \mathrm{H}(\mathbf{grad}, \Omega) \qquad [3.242]$$

where φ represents the magnetic scalar potential to which a gauge condition must be added, namely setting the potential in a point of the domain. Based on relations [3.240] and [3.242], the magnetic field is written as:

$$\mathbf{H} = \mathrm{I}\boldsymbol{\chi}_\mathrm{I} - \mathbf{grad}\varphi \qquad [3.243]$$

Using this expression, with the behavior law [1.26], the magnetic flux density is written as:

$$\mathbf{B} = \mu(\mathrm{I}\boldsymbol{\chi}_\mathrm{I} - \mathbf{grad}\varphi) \qquad [3.244]$$

Using equation [3.244], equation [3.232] is written as:

$$\mathrm{div}(\mu(\mathrm{I}\boldsymbol{\chi}_\mathrm{I} - \mathbf{grad}\varphi)) = 0 \qquad [3.245]$$

To which we should add the boundary conditions on the boundary Γ_b of the domain as follows:

$$\mu(\mathrm{I}\chi_\mathrm{I} - \mathbf{grad}\varphi).\mathrm{n}\Big|_{\Gamma_b} = 0 \qquad [3.246]$$

Equation [3.245] represents the magnetic scalar potential formulation of a magnetostatics problem when the source term is an imposed current density flowing through an inductor located inside the domain.

3.5.2.3.1. Expression of the flux Φ in a coil

To determine the total magnetic flux Φ in a coil, the magnetic energy will first be expressed. This is given by equation [3.223]. Replacing the magnetic field by expression [3.243], we obtain:

$$W_{mag} = \frac{1}{2}\iiint_\Omega \mathbf{B}.(-\mathbf{grad}\varphi + \mathrm{I}\chi_\mathrm{I})d\tau \qquad [3.247]$$

As already seen in section 3.5.2.2, the first term of the integral is equal to zero. Indeed, in this configuration, using equation [2.23], it can be decomposed as follows:

$$-\frac{1}{2}\iiint_\Omega \mathbf{B}.\mathbf{grad}\varphi d\tau = \frac{1}{2}\iiint_\Omega \varphi \mathrm{div}\mathbf{B} d\tau - \frac{1}{2}\oiint_{\Gamma_b} \varphi\mathbf{B}.\mathbf{n}dS = 0 \qquad [3.248]$$

with div$\mathbf{B} = 0$ and the boundary conditions on the boundary, namely $\mathbf{B}.\mathbf{n} = 0$ on Γ_b. Considering this result, if now the magnetic energy is also expressed using the global quantities I and Φ, we then have:

$$W_{mag} = \frac{1}{2}\mathrm{I}\Phi = \frac{1}{2}\mathrm{I}\iiint_\Omega \mathbf{B}.\chi_\mathrm{I} d\tau \qquad [3.249]$$

By identification, the expression of the flux in the coil is written as:

$$\Phi = \iiint_\Omega \mathbf{B}.\chi_\mathrm{I} d\tau \qquad [3.250]$$

This expression allows not only for the calculation of the total flux Φ in the coil but also its imposition. It is sufficient then to couple equations [3.245] and [3.250]. In this case, the current I becomes an unknown of the problem.

3.5.2.4. *Permanent magnet as the source term*

Consider the domain Ω represented in Figure 3.18. It consists of three subdomains (Ω_0, Ω_1, Ω_{PM}), which are, respectively, air, a ferromagnetic material and a permanent magnet. The boundary conditions on the boundary are of type Γ_b. In the air, a magnetic permeability equal to μ_0 is considered. For the ferromagnetic subdomain, a permeability μ_1 is considered. The permanent magnet will be represented by a magnetic permeability μ_A and a coercive field $\mathbf{H}_c$ (see equation [1.64]). In order to simplify the developments, the domain of definition of field $\mathbf{H}_c$ is extended to the entire domain Ω considering it equal to zero in Ω-Ω_{PM}.

In this case, the equations to be solved have the form:

$$\mathbf{curlH} = 0 \tag{3.251}$$

$$\mathrm{div}\mathbf{B} = 0 \tag{3.252}$$

having as the behavior law of the domains Ω_0 and Ω_1, equation [1.26] and relation [1.64] for the permanent magnet. Considering the Γ_b type boundary conditions on the entire boundary, the function spaces associated with field $\mathbf{H}$ and with magnetic flux density $\mathbf{B}$ are given by:

$$\mathbf{H} \in H(\mathbf{curl}0, \Omega) \quad \text{and} \quad \mathbf{B} \in H_0(\mathrm{div}0, \Omega) \tag{3.253}$$

Based on the function space to which the field $\mathbf{H}$ belongs and the fact that the domain Ω is contractible, we can introduce (see equation [2.32]) the magnetic scalar potential, such that:

$$\mathbf{H} = -\mathbf{grad}\varphi \quad \text{with} \quad \varphi \in H(\mathbf{grad}, \Omega) \tag{3.254}$$

Using the scalar potential, equation [3.251] can be automatically verified. The behavior law [1.64] can be used to express the magnetic flux density in the following form:

$$\mathbf{B} = \mu(\mathbf{H} - \mathbf{H}_c) \tag{3.255}$$

where μ takes the value μ_0, μ_1 and μ_A depending on the considered subdomain and $\mathbf{H}_c$ which is zero in the domain Ω-Ω_{PM}.

Under these conditions, grouping equations [3.252], [3.254] and [3.255], we obtain the following for the domain Ω:

$$\mathrm{div}(\mu(\mathbf{grad}\varphi + \mathbf{H}_c)) = 0 \tag{3.256}$$

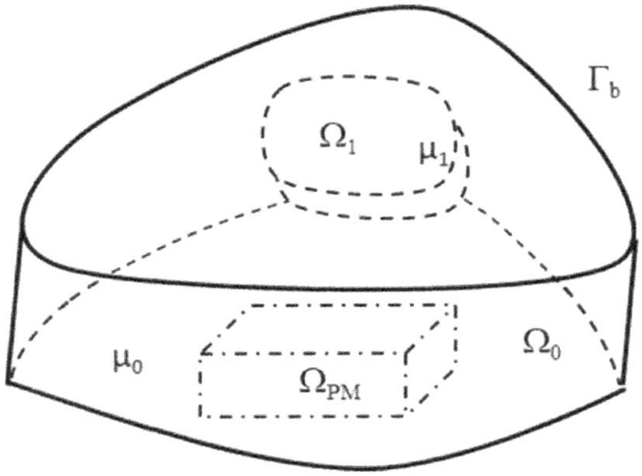

Figure 3.18. *Studied geometry for magnetostatics when the source term is a permanent magnet*

NOTE.– For many applications, the coercive field $\mathbf{H}_c$ is considered constant in the permanent magnet. This leads to discontinuities of its normal component on some parts of its boundary. Consider the example in Figure 3.19, where the permanent parallelepipedic magnet (domain Ω_{PM}) is immersed in the domain Ω, such that $\Omega = \Omega_{PM} \cup \Omega_0$. It can be noted that the normal component of $\mu\mathbf{H}_c$ is discontinuous on the boundary Γ_{An}, since this component is zero in Ω_0, and therefore, the divergence of this term is not defined on the entire domain. In fact, only the divergence of the term $\mu\mathbf{grad}\varphi + \mu\mathbf{H}_c$ is defined. Therefore, the source term $\mu\mathbf{H}_c$ cannot be extracted from the divergence operator. This will no longer be a difficulty when the weighted residual method is introduced in section 4.4.5.1, as the work will be conducted on integral formulations.

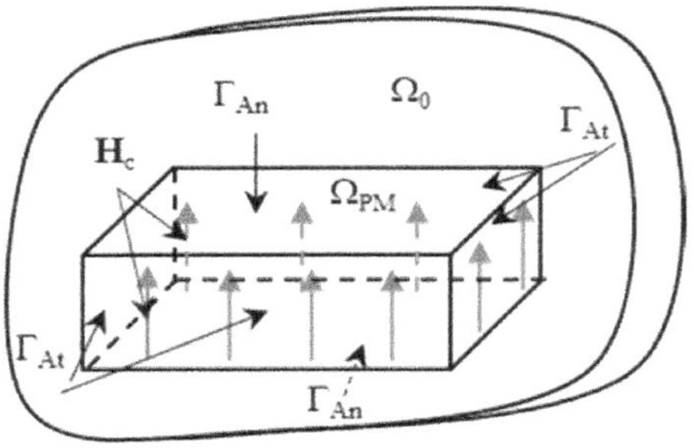

Figure 3.19. *Study of a permanent magnet: discontinuity of the normal component of the coercive field on the boundary Γ_{An}*

Finally, on the boundary of the domain Ω, of Γ_b type, the boundary condition is written as:

$$-\mu \mathbf{grad}\,\varphi.\mathbf{n}\big|_{\Gamma} = 0 \qquad [3.257]$$

Moreover, the system of equations is well posed, provided that a gauge condition is imposed on the scalar potential.

3.5.2.5. *General case*

This section develops the scalar potential formulation in the presence of several source terms, as shown in Figure 3.20. It is not difficult to simultaneously impose source terms of various natures, considering the linearity of vector operators and boundary conditions. The following developments are therefore true, even though the behavior laws are not linear. However, concerning the magnetomotive force f_m and the magnetic flux ϕ, imposed on the boundaries $\Gamma_{\Gamma h1 \cup \Gamma h2}$, only one of these two terms should be considered at a time. In what follows, let us first consider as source terms the current density $\mathbf{J}_0$, a permanent magnet represented by means of the coercive field $\mathbf{H}_c$ and a magnetomotive force f_m. The next step will be to replace f_m by the magnetic flux ϕ.

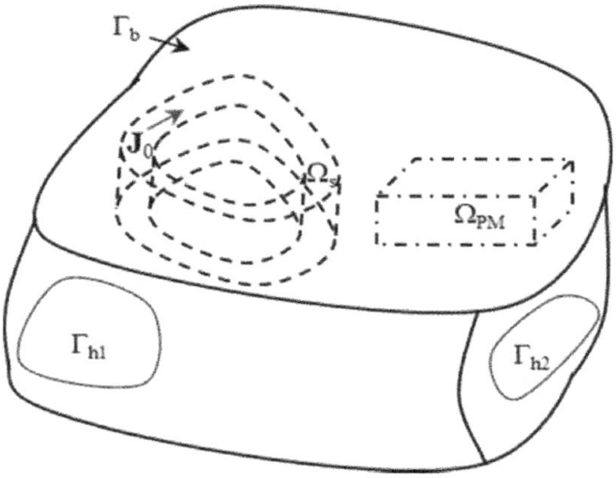

Figure 3.20. *Magnetostatics problem: studied geometry in the general case*

Maxwell's equations to be solved in this context are written as:

$$\mathbf{curlH} = \mathbf{J}_0 \qquad [3.258]$$

$$\text{div}\mathbf{B} = 0 \qquad [3.259]$$

For this example, the boundary conditions on the boundary Γ have the following form:

$$\Gamma = \Gamma_b \cup \Gamma_{h1} \cup \Gamma_{h2},\ \mathbf{B.n}\big|_{\Gamma_b} = 0,\ \mathbf{H} \wedge \mathbf{n}\big|_{\Gamma_{hk}} = 0, \text{ and } k \in \{1,2\} \qquad [3.260]$$

Irrespective of whether the magnetomotive force or the magnetic flux is imposed, the function spaces to which the magnetic field **H** and the magnetic flux density **B** belong are defined by:

$$\mathbf{H} \in H_{\Gamma_{h1} \cup \Gamma_{h2}}(\mathbf{curl}, \Omega) \text{ and } \mathbf{B} \in H_{\Gamma_b}(\text{div}0, \Omega) \qquad [3.261]$$

3.5.2.5.1. Source terms: f_s, J_0, H_c

As developed in sections 3.2.5.1–3.2.5.4, the magnetic field can be decomposed into several terms (see equation [3.211]). In the presence of a magnetomotive force and a source current density, we have:

$$\mathbf{H} = \mathbf{H}' + \mathbf{H}_{sf} + \mathbf{H}_{sI} \qquad [3.262]$$

where **H'** represents the unknown of the problem, $\mathbf{H}_{sf}$ is the source term due to the magnetomotive force and $\mathbf{H}_{sI}$ is the source term due to the current density $\mathbf{J}_0$ in the inductor. It should be recalled that the flux of the current density $\mathbf{J}_0$ is equal to the intensity I (see equation [1.62]). If $\mathbf{H}_{sf}$ is replaced in equation [3.262] by its expression given in equation [3.215], $\mathbf{H}_{sI}$ by its expression given in equation [3.240] and $\mathbf{H}' \in H_{\Gamma_{h1} \cup \Gamma_{h2}}(\mathbf{curl}, \Omega)$ by equation [3.217], the magnetic field is written as:

$$\mathbf{H} = -(\mathbf{grad}\varphi + f_m \mathbf{grad}\alpha_s - I\boldsymbol{\chi}_I) \text{ with } \varphi \in H_{\Gamma_{h1} \cup \Gamma_{h2}}(\mathbf{grad}, \Omega) \qquad [3.263]$$

It is important to note that the function space to which φ belongs, as well as the properties of the scalar potential α_s [3.215] and of the support field $\boldsymbol{\chi}_I$ [3.237], require the tangential component of the magnetic field to be zero on the boundaries $\Gamma_{h1} \cup \Gamma_{h2}$. The contribution due to the permanent magnet appears through the behavior law [3.255], when the magnetic flux density is expressed:

$$\mathbf{B} = -\mu(\mathbf{grad}\varphi + f_m \mathbf{grad}\alpha_s - I\boldsymbol{\chi}_I) - \mu \mathbf{H}_c \qquad [3.264]$$

where μ depends on the position according to the subdomains (Ω_0, Ω_1 and Ω_{PM}) and may take the values μ_0, μ_1 or μ_{PM}. When the source terms are f_m, $\mathbf{J}_0$ and in the presence of a permanent magnet, the scalar potential formulation results by applying equation [3.259] to the expression of the magnetic flux density as follows:

$$\text{div}(\mu(\mathbf{grad}\varphi + f_m \mathbf{grad}\alpha_s - I\boldsymbol{\chi}_I + \mathbf{H}_c)) = 0 \qquad [3.265]$$

This equation can be rewritten by replacing the associated potential α_s by the support field $\boldsymbol{\beta}_s$, see [3.13], which yields:

$$\mathrm{div}(\mu(\mathbf{grad}\varphi - f_m \boldsymbol{\beta}_s - I\boldsymbol{\chi}_I + \mathbf{H}_c)) = 0 \qquad [3.266]$$

3.5.2.5.2. Source terms: ϕ, J_0, H_c

Imposing the magnetic flux ϕ on the boundaries Γ_{h1} and Γ_{h2} (see equation [1.61]) does not modify equation [3.265]. On the contrary, the magnetomotive force f_m that was a source term becomes, as seen in section 3.5.2.2, an additional unknown of the problem. To obtain a full equation system, an additional equation is required in which the magnetic flux on the boundaries Γ_{h1} and Γ_{h2} appears. To this end, relation [3.228] can be used. If in this equation **B** is replaced by its expression (see equation [3.264]), we obtain:

$$\phi = \iiint_{\Omega} \mu(-\boldsymbol{\beta}_s.\mathbf{grad}\,\varphi - f_m \boldsymbol{\beta}_s.\mathbf{grad}\,\alpha_s + I\boldsymbol{\beta}_s.\boldsymbol{\chi}_I - \boldsymbol{\beta}_s.\mathbf{H}_c)d\tau \qquad [3.267]$$

This equation can also be rewritten, posing $\boldsymbol{\beta}_s = -\,\mathbf{grad}\alpha_s$, as follows:

$$\phi = \iiint_{\Omega} \mu(-\boldsymbol{\beta}_s.\mathbf{grad}\,\varphi + f_m \boldsymbol{\beta}_s.\boldsymbol{\beta}_s + I\boldsymbol{\beta}_s.\boldsymbol{\chi}_I - \boldsymbol{\beta}_s.\mathbf{H}_c)d\tau \qquad [3.268]$$

The system of equations composed of relations [3.265] and [3.268] then has to be solved, where the unknowns are the scalar potential φ and the magnetomotive force f_m. The gauge condition on φ and the boundary conditions on the boundaries of the domain must be added to this system.

3.5.3. *Vector potential A formulation*

Similar to the approach for the scalar potential formulation, for the vector potential formulation the equations to be solved will be introduced by considering the four source terms separately, and then the general case will be dealt with.

To facilitate the reasoning, the first case to be studied is that in which the source term is the magnetic flux ϕ imposed on the boundaries of a domain (see section 3.5.3.1). This source term is very well adapted to the vector potential formulation. The next section focuses on how to impose a magnetomotive force "f_m" (see section 3.5.3.2), the current "I" in a stranded inductor (see section 3.5.3.3) and the case of a permanent magnet (see section 3.5.3.4). Finally, the general case with several source terms in the studied domain will be considered.

3.5.3.1. *Imposed magnetic flux*

The development of this formulation uses the example in Figure 3.16. The equations to be solved are those given in equations [3.207] and [3.208] with the magnetic behavior law [1.26] and the Γ_h and Γ_b type boundary conditions (see equation [3.209]). To solve these equations, the flux ϕ across the boundaries Γ_{h1} and Γ_{h2} will be introduced into the local equations. It should be recalled (see equation [1.61]) that the flux is expressed using the magnetic flux density **B**. As the latter is defined by the divergence operator, the procedure proposed in section 3.2.1.2 will be used.

First of all, the magnetic flux density is decomposed into two terms as follows:

$$\mathbf{B} = \mathbf{B}_s + \mathbf{B}' \qquad [3.269]$$

In this expression, $\mathbf{B}_s$ represents the source field created by the flux ϕ imposed on the boundaries Γ_{h1} and Γ_{h2}. The properties of this field (see equation [3.21]) are given by the relations:

$$\mathrm{div}\mathbf{B}_s = 0, \iint_{\Gamma_{hk}} \mathbf{B}_s.\mathbf{n}ds = \pm\phi, \ \mathbf{B}_s.\mathbf{n}\Big|_{\Gamma_b} = 0$$
$$\text{with } k \in \{1,2\} \text{ i.e. } \mathbf{B}_s \in H_{\Gamma_b}(\mathrm{div}0,\Omega) \qquad [3.270]$$

Concerning **B'** on Γ_b, condition [1.37] is obviously applicable. On the contrary, considering equation [3.270], through Γ_{h1} and Γ_{h2}, the flux is equal to zero. Finally, using equation [3.208], the properties of **B'** are stated as follows:

$$\mathrm{div}\,\mathbf{B}' = 0, \iint_{\Gamma_{hk}} \mathbf{B}'.\mathbf{n}ds = 0, \ \mathbf{B}'.\mathbf{n}\big|_{\Gamma_b} = 0$$
$$\text{with } k \in \{1,2\} \text{ i.e. } \mathbf{B}' \in H_{\Gamma_b}(\mathrm{div}0,\Omega) \qquad [3.271]$$

In order to express the source field $\mathbf{B}_s$, the support vector field denoted by $\boldsymbol{\lambda}_\phi$ is introduced (see equation [3.22]), such that:

$$\mathbf{B}_s = \boldsymbol{\lambda}_\phi \phi \qquad [3.272]$$

Similar to $\boldsymbol{\lambda}_s$ (see section 3.2.1.2.1), the properties of $\boldsymbol{\lambda}_\phi$ are similar to $\mathbf{B}_s$, which are written as:

$$\mathrm{div}\boldsymbol{\lambda}_\phi = 0 \text{ with } \iint_{\Delta} \boldsymbol{\lambda}_\phi.\mathbf{n}ds = 1 \text{ and } \boldsymbol{\lambda}_\phi.\mathbf{n}\big|_{\Gamma_b} = 0$$
$$\text{i.e. } \boldsymbol{\lambda}_\phi \in H_{\Gamma_b}(\mathrm{div}0,\Omega) \qquad [3.273]$$

The surface Δ was introduced in this expression, as in equation [3.23]. It should be noted that by sliding Δ along the boundary Γ_b (see Figure 3.16), this surface may merge with the boundaries Γ_{h1} or Γ_{h2}. The boundary conditions imposed by expression [3.270] can thus be verified. Considering these properties and the fact that the studied domain is contractible, the support field $\boldsymbol{\lambda}_\phi$ can be expressed using an associated potential $\boldsymbol{\chi}_\phi$. As the boundary Γ_b is not simply connected, the construction of potential $\boldsymbol{\chi}_\phi$ requires some precautions. In fact, a cut must be introduced along the boundary Γ_b (see section 3.2.1.2.2). Based on relations [3.273], the properties of potential $\boldsymbol{\chi}_\phi$ have the following form:

$$\boldsymbol{\lambda}_\phi = \mathbf{curl}\boldsymbol{\chi}_\phi \text{ with } \boldsymbol{\chi}_\phi \wedge \mathbf{n}\big|_{\Gamma_b} = 0 \text{ and } \oint_{\gamma_0} \boldsymbol{\chi}_\phi.\mathbf{dl} = 1$$
$$\text{i.e. } \boldsymbol{\chi}_\phi \in H^{\Delta}_{\Gamma_b}(\mathbf{curl}0, \Omega) \qquad [3.274]$$

In this expression, γ_0 represents a contour belonging to Γ_b and that cannot be contracted to a point by successive transformations (see Figure 3.16).

As the source term $\mathbf{B}_s$ is assumed to be known, the field $\mathbf{B}'$ remains to be defined. Equation [3.271] indicates that $\mathbf{B}' \in H_{\Gamma b}(\text{div}0, \Omega)$. As the domain is contractible, it is then possible to express $\mathbf{B}'$ using a magnetic vector potential, denoted by $\mathbf{A}$ (see equation [2.31]), such that:

$$\mathbf{B}' = \mathbf{curl}\mathrm{A} \text{ with } \mathbf{A} \in H_{\Gamma_b}(\mathbf{curl}, \Omega) \qquad [3.275]$$

The vector potential $\mathbf{A}$ is not unique. It is defined up to a gradient (see equation [2.46]). The uniqueness of the solution can be imposed by adding a gauge condition similar to equations [2.50] or [2.51].

Based on equations [3.269] and [3.275], we obtain:

$$\mathbf{B} = \mathbf{B}_s + \mathbf{curlA} \qquad [3.276]$$

This can be rewritten using equation [3.272] in the form:

$$\mathbf{B} = \mathbf{curlA} + \phi\boldsymbol{\lambda}_\phi \qquad [3.277]$$

Or still by introducing the field $\boldsymbol{\chi}_\phi$ [3.274]:

$$\mathbf{B} = \mathbf{curlA} + \phi\mathbf{curl}\boldsymbol{\chi}_\phi \qquad [3.278]$$

As for the magnetic field **H**, it can be written via the behavior law [1.26]:

$$\mathbf{H} = \mu^{-1}(\mathbf{curlA} + \phi\boldsymbol{\lambda}_{\phi}) \quad [3.279]$$

Replacing the magnetic field in equation [3.207] by its expression given in equation [3.279], the following can be written:

$$\mathbf{curl}(\mu^{-1}(\mathbf{curlA} + \phi\boldsymbol{\lambda}_{\phi})) = 0 \quad [3.280]$$

Or still, if equation [3.278] and the behavior law are used, then:

$$\mathbf{curl}(\mu^{-1}(\mathbf{curlA} + \phi\mathbf{curl}\boldsymbol{\chi}_{\phi})) = 0 \quad [3.281]$$

This is the equation of a magnetostatics problem with the magnetic vector potential formulation, when the source term is the magnetic flux imposed on the boundaries of the domain.

It should be noted that the introduction of the associated potential $\boldsymbol{\chi}_{\phi}$ is not required for the resolution of the vector potential formulation, as it is sufficient to know the support vector $\boldsymbol{\lambda}_{\phi}$ (see relation [3.277]).

3.5.3.2. *Imposed magnetomotive force*

The studied geometry is still the one presented in Figure 3.16, considering as the source term the magnetomotive force imposed between the boundaries Γ_{h1} and Γ_{h2} of the domain.

In this case, for the magnetic vector potential formulation, the developments are more complex. The situation is equivalent to that in section 3.5.2.2 and the magnetomotive force must appear in the equations to be solved. To this end, an energy balance is used. The latter allows us to express the magnetomotive force as a function of the vector potential **A** and the magnetic flux ϕ, which becomes an unknown of the problem.

Based on the expression of the magnetic energy [3.223], expressing the magnetic flux density as a function of **A** and ϕ (see equation [3.277]), we have:

$$W_{mag} = \frac{1}{2}\iiint_{\Omega} \mathbf{H}.\mathbf{curlA}\,d\tau + \frac{1}{2}\iiint_{\Omega} \phi\boldsymbol{\lambda}_{\phi}.\mathbf{H}\,d\tau \quad [3.282]$$

Using the properties of the vector operators, equation [2.27], the first integral of this expression can be rewritten in the form:

$$\frac{1}{2}\iiint_{\Omega}\mathbf{H.curlA}d\tau=\frac{1}{2}\iiint_{\Omega}\mathbf{curlH.A}d\tau+\frac{1}{2}\oiint_{\Gamma}(\mathbf{H}\wedge\mathbf{n}).\mathbf{A}d\tau \qquad [3.283]$$

Considering equation [3.207], the first term on the right-hand side is zero. The same is applied to the second term by decomposing the boundary integral into two contributions, corresponding to $\Gamma_h = \Gamma_{h1} \cup \Gamma_{h2}$ and Γ_b. Indeed, with $\mathbf{H} \in H_{\Gamma h1 \cup \Gamma h2}(\mathbf{curl}, \Omega)$ (see equation [3.210]), the integral on the boundaries Γ_{h1} and Γ_{h2} is equal to zero. Similarly, it can be readily shown that the integral on Γ_b is equal to zero, by applying the mixed product and considering that $\mathbf{A} \in H_{\Gamma b}(\mathbf{curl}, \Omega)$. Under these conditions, equation [3.282] can be rewritten in the following form:

$$W_{mag} = \frac{1}{2}\iiint_{\Omega}\phi\mathbf{H}.\boldsymbol{\lambda}_{\phi}d\tau \qquad [3.284]$$

Expressing the magnetic energy as a function of global quantities f_m and ϕ, we can write:

$$W_{mag} = \frac{1}{2}f_m\phi = \frac{1}{2}\iiint_{\Omega}\phi\mathbf{H}.\boldsymbol{\lambda}_{\phi}d\tau \qquad [3.285]$$

and by identification the expression of the magnetomotive force is obtained:

$$f_m = \iiint_{\Omega}\mathbf{H}.\boldsymbol{\lambda}_{\phi}d\tau \qquad [3.286]$$

It can also be expressed as a function of $\boldsymbol{\chi}_{\phi}$ (see equation [3.274]) as follows:

$$f_m = \iiint_{\Omega}\mathbf{H.curl}\chi_{\varphi}d\tau \qquad [3.287]$$

If the field $\mathbf{H}$ is replaced in equation [3.286] by its expression given by [3.279], we obtain:

$$f_m = \iiint_{\Omega}\mu^{-1}\mathbf{curlA}.\boldsymbol{\lambda}_{\phi}d\tau+\iiint_{\Omega}\mu^{-1}\phi.\boldsymbol{\lambda}_{\phi}.\boldsymbol{\lambda}_{\phi}d\tau \qquad [3.288]$$

Under these conditions, when the source term is the magnetomotive force, the system of equations to be solved, with the vector potential formulation, is defined by equations [3.281] and [3.288]. The unknowns are then the magnetic vector potential $\mathbf{A}$ and the magnetic flux ϕ.

3.5.3.3. *Imposed current density*

Let us consider again the example presented in Figure 3.17 with the associated equations [3.231] and [3.232] and Γ_b-type conditions over the boundary. It can be noted that the source term, represented by the current density $\mathbf{J}_0$, is associated with a curl. As already seen in section 3.2.2.1, this source term can be expressed by means of vector fields, denoted by $\boldsymbol{\lambda}_I$ and $\boldsymbol{\chi}_I$, defined, respectively, by equations [3.38] and [3.39]. It should be recalled that, in order to overcome the constraints related to disconnectedness of the inductor (subdomain Ω_s), the support field $\boldsymbol{\lambda}_I$ and the potential $\boldsymbol{\chi}_I$ are defined on the entire domain Ω. Under these conditions, equation [3.231] can be rewritten as:

$$\begin{aligned}&\mathbf{curlH} = \mathbf{J}_0 = I\boldsymbol{\lambda}_I = I\mathbf{curl}\boldsymbol{\chi}_I \\ &\text{with } \boldsymbol{\lambda}_I \in H_0(\text{div}0,\Omega) \text{ and } \boldsymbol{\chi}_I \in H_0(\mathbf{curl}0,\Omega)\end{aligned} \qquad [3.289]$$

On the contrary, equation [3.232] and the boundary conditions on Γ_b show that $\mathbf{B} \in H_0(\text{div}0, \Omega)$. As the studied domain is contractible (see section 3.5.3.1), the magnetic flux density can be defined using a magnetic vector potential $\mathbf{A}$, such that:

$$\mathbf{B} = \mathbf{curlA} \text{ with } \mathbf{A} \in H_0(\mathbf{curl},\Omega) \qquad [3.290]$$

As shown in section 2.5.2.2, the uniqueness of the solution makes is necessary to impose a gauge condition on the vector potential $\mathbf{A}$.

Having defined the vector potential, the formulation results from equation [3.289] by replacing, via the behavior law [1.26], $\mathbf{H}$ by the magnetic flux density and $\mathbf{J}_0$ by the source field $I\boldsymbol{\lambda}_I$:

$$\mathbf{curl}\mu^{-1}\mathbf{curl}\,\mathbf{A} = I\boldsymbol{\lambda}_I \qquad [3.291]$$

Or by introducing the associated potential $\boldsymbol{\chi}_I$:

$$\mathbf{curl}\mu^{-1}\mathbf{curlA} = I\mathbf{curl}\boldsymbol{\chi}_I \qquad [3.292]$$

3.5.3.4. *Permanent magnet*

In order to study the case of a permanent magnet as the source term, let us consider again the problem presented in Figure 3.18. The equations to be solved are given by relations [3.251] and [3.252] with the behavior laws [1.26] for subdomains Ω_0 and Ω_1 and equation [1.27] for the subdomain Ω_{PM}. The boundary conditions on the boundary of the domain are of Γ_b type. The magnetic vector potential $\mathbf{A}$ can be used to automatically verify equation [3.252]. To establish the equation to be solved, the magnetic flux density is introduced into equation [3.252], via the behavior laws, as a function of the vector potential, as follows:

$$\mathbf{curl}(\mu^{-1}(\mathbf{curlA} - \mathbf{B}_r)) = 0 \quad \mathbf{A} \in H_0(\mathbf{curl}, \Omega) \qquad [3.293]$$

In this equation, permeability μ is equal to μ_0 in Ω_0, μ_1 in Ω_1 and μ_{PM} in Ω_{PM}. Likewise, the remanent magnetic flux density $\mathbf{B}_r$ is defined throughout the domain Ω and will be zero everywhere, except for Ω_{PM}.

NOTE.– As already seen in section 3.5.2.4 and particularly in Figure 3.19, the normal component of $\mu\mathbf{H}_c$ is discontinuous on the boundary Γ_{An}. Applying the same reasoning, it can be shown that the tangential component of the term $\mu^{-1}\mathbf{B}_r$ is discontinuous on the boundaries Γ_{At} of the permanent magnet. Under these conditions, the separation of the term $\mathbf{curl}(\mu^{-1}(\mathbf{curlA} - \mathbf{B}_r))$ into two terms, by using the linearity of the curl operator, would be abusive. Section 4.4.5.2 will show that this constraint is lifted, because the weighted residual method is used.

3.5.3.5. *General case*

Similar to the scalar potential formulation, this section studies the case when three source terms are simultaneously imposed on the studied domain (see Figure 3.20). First, let us consider as a source term: the flux ϕ imposed on the boundaries Γ_{h1} and Γ_{h2}, an inductor with a current density $\mathbf{J}_0$ and a permanent magnet represented by the remanent magnetic flux density $\mathbf{B}_r$. Then, the flux ϕ is replaced by a magnetomotive force f_m.

3.5.3.5.1. Source terms: ϕ, I, B_r

Similar to our approach in section 3.5.3.1, the magnetic flux density $\mathbf{B}$ can be decomposed into several terms and, based on equation [3.276], the following can be written:

$$\mathbf{B} = \mathbf{B}_{s\phi} + \mathbf{curlA} \qquad [3.294]$$

where $\mathbf{B}_{s\phi}$ represents the source term due to the flux imposed on the boundaries Γ_{h1} and Γ_{h2}. If $\mathbf{B}_{s\phi}$ is replaced by its expression as a function of $\boldsymbol{\lambda}_\phi$, similar to relation [3.272], we can write:

$$\mathbf{B} = \mathbf{curlA} + \phi\boldsymbol{\lambda}_\phi \quad \text{with} \quad \boldsymbol{\lambda}_\phi \in H_{\Gamma_b}(\text{div}0, \Omega) \qquad [3.295]$$

Based on this expression, using the behavior laws [1.26] and [1.27], the magnetic field $\mathbf{H}$ is written as follows:

$$\mathbf{H} = \mu^{-1}(\mathbf{curlA} + \phi\boldsymbol{\lambda}_\phi - \mathbf{B}_r) \qquad [3.296]$$

As already seen in section 3.5.3.1, equation [3.274], the support field can be expressed using an associated potential $\boldsymbol{\chi}_\phi$. Under these conditions, equation [3.296] can be rewritten as follows:

$$\mathbf{H} = \mu^{-1}(\mathbf{curlA} + \phi\mathbf{curl}\boldsymbol{\chi}_\phi - \mathbf{B}_r) \qquad [3.297]$$

It should be recalled that precautions must be taken to determine $\boldsymbol{\chi}_\phi$, as the boundary Γ_b is not simply connected (see section 3.2.1.2.2).

Consider now equation [3.289], in which the field **H** is replaced by its expression given by equation [3.296]. Expressing the current density $\mathbf{J}_0$ as a function of the support field $\boldsymbol{\lambda}_I$, after rearrangement we obtain:

$$\mathbf{curl}\mu^{-1}(\mathbf{curlA} + \phi\boldsymbol{\lambda}_\phi - \mathbf{B}_r) = I\boldsymbol{\lambda}_I \qquad [3.298]$$

or based on equation [3.297] and introducing the associated potential $\boldsymbol{\chi}_I$ via [3.289], we obtain:

$$\mathbf{curl}\mu^{-1}(\mathbf{curlA} + \phi\mathbf{curl}\boldsymbol{\chi}_\phi - \mathbf{B}_r) = I\mathbf{curl}\boldsymbol{\chi}_I \qquad [3.299]$$

Finally, to obtain the uniqueness of the equation system, a gauge condition and also boundary conditions on Γ_b must be imposed on **A**.

3.5.3.5.2. Source terms: f_m, I and B_r

When the source terms are f_m, I and $\mathbf{B}_r$, the approach is similar to the one presented in section 3.5.2.2. Indeed, in this case, the magnetomotive force f_m generates a flux ϕ, which becomes an unknown of the problem, but formulas [3.298] or [3.299] are not modified. Since there is an additional unknown, an equation must be added. To this end, the expression of the magnetomotive force between surfaces Γ_{h1} and Γ_{h2} can be used. Depending on the magnetic field, this one is given by equation [3.286] or [3.287]. Gathering equation [3.286] and expression [3.297], we obtain:

$$f_m = \iiint_\Omega \mu^{-1}(\mathbf{curlA} + \phi\boldsymbol{\lambda}_\phi - \mathbf{B}_r).\boldsymbol{\lambda}_\phi d\tau \qquad [3.300]$$

Then, the system to be solved is composed of equations [3.298] and [3.300] whose unknowns are the vector potential **A** and the flux ϕ. To impose the uniqueness of the solution, a gauge condition on **A** and the boundary conditions on Γ_b must be added.

3.5.4. *Summarizing tables*

In the case of magnetostatics, Table 3.5 summarizes the equations to be solved for various types of source term with the scalar potential formulation. Similarly, Table 3.6 summarizes the vector potential formulation.

Magnetostatics		
Scalar potential φ formulation	Source term: magnetomotive force f_m on the boundary (see Figure 3.16)	
	Source field support two possibilities: $\boldsymbol{\beta}_s$ or α_s	Decomposition of the magnetic field: $\mathbf{H} = \mathbf{H}_s + \mathbf{H}'$
		$\mathbf{H}_s = f_m \boldsymbol{\beta}_s$, $\boldsymbol{\beta}_s \in H_{\Gamma_{h1} \cup \Gamma_{h2}}(\mathbf{curl}0, \Omega)$
		$\mathbf{H}_s = -f_m \mathbf{grad}\, \alpha_s$, $\alpha_s \in H_{\Gamma_{h1} \cup \Gamma_{h2}}(\mathbf{grad}, \Omega)$
	Properties of the unknown $\mathbf{H}'$ and introduction of potential φ	$\mathbf{H}' \in H_{\Gamma_{h1} \cup \Gamma_{h2}}(\mathbf{curl}0, \Omega)$
		$\mathbf{H}' = -\mathbf{grad}\varphi$, $\varphi \in H_{\Gamma_{h1} \cup \Gamma_{h2}}(\mathbf{grad}, \Omega)$
	Equation to be solved: two possible forms depending on $\boldsymbol{\beta}_s$ or α_s	$\mathrm{div}(\mu(f_m \boldsymbol{\beta}_s - \mathbf{grad}\varphi)) = 0$
		$\mathrm{div}(\mu(f_m \mathbf{grad}\alpha_s + \mathbf{grad}\varphi)) = 0$
	Source term: flux ϕ on the boundary (see Figure 3.16)	
	f_m becomes an unknown; an additional equation is needed	
	$\phi = \iiint_\Omega \mathbf{B}.\boldsymbol{\beta}_s \, d\tau$	
	Source term: current intensity I (see Figure 3.17)	
	Source field support two possibilities: $\boldsymbol{\lambda}_I$ or $\boldsymbol{\chi}_I$	Decomposition of the magnetic field: $\mathbf{H} = \mathbf{H}_s + \mathbf{H}'$
		$\mathbf{curl}\mathbf{H}_s = I\boldsymbol{\lambda}_I$, $\boldsymbol{\lambda}_I \in H_0(\mathrm{div}0, \Omega)$
		$\mathbf{H}_s = I\boldsymbol{\chi}_I$, $\boldsymbol{\chi}_I \in H_0(\mathbf{curl}, \Omega)$
	Properties of the unknown $\mathbf{H}'$ and introduction of scalar potential φ	$\mathbf{H}' \in H(\mathbf{curl}0, \Omega)$
		$\mathbf{H}' = -\mathbf{grad}\varphi$, $\varphi \in H(\mathbf{grad}, \Omega)$
	Equation to be solved as a function of $\boldsymbol{\chi}_I$	$\mathrm{div}(\mu(I\boldsymbol{\chi}_I - \mathbf{grad}\varphi)) = 0$
	Source term: permanent magnet (see Figure 3.18)	
	$\mathbf{B} = \mu(\mathbf{H} - \mathbf{H}_c)$ with $\mathbf{H} = -\mathbf{grad}\varphi$, $\varphi \in H(\mathbf{grad}, \Omega)$	
	$\mathrm{div}(\mu(\mathbf{grad}\varphi + \mathbf{H}_c)) = 0$	

Table 3.5. *Summary of equations to be solved in magnetostatics with the scalar potential formulation for various source terms*

Magnetostatics		
Vector potential A formulation	Source term: flux ϕ on the boundary (see Figure 3.16)	
	Source field support two possibilities: $\boldsymbol{\lambda}_\phi$ or $\boldsymbol{\chi}_\phi$ with a cut on not simply connected Γ_b	Decomposition of the magnetic field: $\mathbf{B} = \mathbf{B}_s + \mathbf{B}'$
		$\mathbf{B}_s = \phi\boldsymbol{\lambda}_\phi,\ \boldsymbol{\lambda}_\phi \in \mathrm{H}_{\Gamma_b}(\mathbf{curl}0, \Omega)$
		$\mathbf{B}_s = \phi\mathbf{curl}\boldsymbol{\chi}_\phi,\ \boldsymbol{\chi}_\phi \in \mathrm{H}^{\Delta}_{\Gamma_b}(\mathbf{curl}, \Omega)$
	Properties of the unknown **B'** and introduction of the vector potential **A**	$\mathbf{B}' \in \mathrm{H}_{\Gamma_b}(\mathrm{div}0, \Omega)$
		$\mathbf{B}' = \mathbf{curl}\mathbf{A},\ \mathbf{A} \in \mathrm{H}_{\Gamma_b}(\mathbf{curl}, \Omega)$
	Equation to be solved: two possible forms depending on $\boldsymbol{\lambda}_\phi$ or $\boldsymbol{\chi}_\phi$	$\mathbf{curl}(\mu^{-1}(\mathbf{curl}\mathbf{A} + \phi\boldsymbol{\lambda}_\phi)) = 0$
		$\mathbf{curl}(\mu^{-1}(\mathbf{curl}\mathbf{A} + \phi\mathbf{curl}\boldsymbol{\chi}_\phi)) = 0$
	Source term: magnetomotive force f_m on the boundary (see Figure 3.16)	
	ϕ becomes unknown; an additional equation is needed	
	$f_m = \iiint_\Omega \mathbf{H}.\boldsymbol{\lambda}_\phi d\tau$	
	Source term: current intensity I (see Figure 3.17)	
	Source field support two possibilities: $\boldsymbol{\lambda}_I$ or $\boldsymbol{\chi}_I$	$\mathbf{curl}\mathbf{H} = \mathrm{I}\boldsymbol{\lambda}_\mathrm{I},\ \boldsymbol{\lambda}_\mathrm{I} \in \mathrm{H}_0(\mathrm{div}0, \Omega)$
		$\mathbf{curl}\mathbf{H} = \mathrm{I}\mathbf{curl}\boldsymbol{\chi}_\mathrm{I},\ \boldsymbol{\chi}_\mathrm{I} \in \mathrm{H}_0(\mathbf{curl}, \Omega)$
	Properties of the unknown **B** and introduction of the vector potential **A**	$\mathbf{B} \in \mathrm{H}_0(\mathrm{div}0, \Omega)$
		$\mathbf{B} = \mathbf{curl}\mathbf{A},\ \mathbf{A} \in \mathrm{H}_0(\mathbf{curl}, \Omega)$
	Equation to be solved depending on $\boldsymbol{\chi}_I$	$\mathbf{curl}\mu^{-1}\mathbf{curl}\mathrm{A} = \mathrm{I}\mathbf{curl}\boldsymbol{\chi}_\mathrm{I}$
	Source term: permanent magnet (see Figure 3.18)	
	$\mathbf{H} = \mu^{-1}(\mathbf{B} - \mathbf{B}_r)$ with $\mathbf{B} = \mathbf{curl}\mathbf{A},\ \mathbf{A} \in \mathrm{H}_0(\mathbf{curl}, \Omega)$	
	$\mathbf{curl}\left(\mu^{-1}\left(\mathbf{curl}\mathbf{A} - \mu^{1}\mathbf{B}_r\right)\right) = 0$	

Table 3.6. *Summary of the equations to be solved in magnetostatics with the vector potential formulation for various source terms*

3.5.5. *Tonti diagram*

Similar to the approach taken for electrostatics and electrokinetics for obtaining the Tonti diagram in magnetostatics, the fields **H**, $\mathbf{J}_0$ and **B** are placed to verify equations [3.207] and [3.208] (see Figure 3.21). As can be noted in this figure, quite naturally, the current density $\mathbf{J}_0$ is divergence free. The source terms $\mathbf{H}_s$ and $\mathbf{B}_s$ as well as the scalar potential φ and the vector potential **A** can also be positioned.

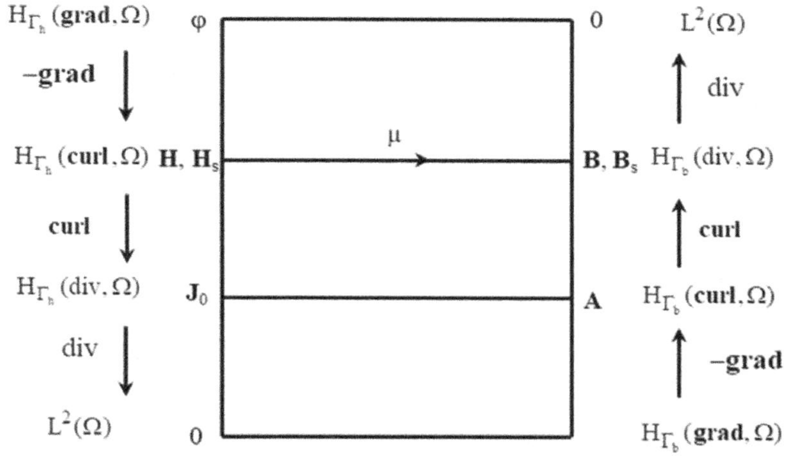

Figure 3.21. *Tonti diagram for magnetostatics*

3.6. Magnetodynamics

A magnetodynamics problem, as shown in Figure 1.12, leads to the study of electromagnetic phenomena at industrial frequencies. To simplify the notations, the time dependency of all electric and magnetic quantities is not explicit but rather implicit, similar to space dependency. Nevertheless, concerning the source terms e(t), ϕ(t), f_m(t) and I(t), this time dependency will be recalled at the beginning of sections 3.6.1 and 3.6.2.

Under these conditions, in a domain Ω, the magnetodynamic equations are recalled as follows:

$$\mathbf{curlE} = -\frac{\partial \mathbf{B}}{\partial t} \qquad [3.301]$$

$$\mathbf{curlH} = \mathbf{J} \qquad [3.302]$$

Applying the divergence operator to equation [3.301], we deduce:

$$\text{div}\mathbf{B} = 0 \quad [3.303]$$

Likewise, applying the divergence operator to equation [3.302], it can be readily verified that:

$$\text{div}\mathbf{J} = 0 \quad [3.304]$$

The behavior laws [1.20] and [1.26], as well as the homogeneous boundary conditions, are added to these equations. As indicated in section 1.3.2, there is a link between the fields **E** and **B**, and also between **H** and **J**, which can be written as follows:

$$\mathbf{n} \wedge \mathbf{E}\big|_{\Gamma_e} = 0 \;\Rightarrow\; \mathbf{B.n}\big|_{\Gamma_e} = 0 \quad [3.305]$$

$$\mathbf{n} \wedge \mathbf{H}\big|_{\Gamma_h} = 0 \Rightarrow \; \mathbf{J_n}\big|_{\Gamma_h} = 0 \quad [3.306]$$

It is important to note that reciprocity does not always apply to these two equations, and this depends on the topology of the boundaries Γ_e or Γ_h.

This study first considers the problem presented in Figure 3.22. Given a domain Ω of boundary Γ, composed of two subdomains denoted by Ω_c and Ω_0. The subdomain Ω_c, of boundary Γ_c, is a conductor whose conductivity is denoted by σ. It holds two gates Γ_{n1} and Γ_{n2}, in contact with an external source. Electric quantities e(t) or I(t), or magnetic quantities $f_m(t)$ or $\phi(t)$, can be imposed on these two gates. The remaining boundary of the conducting subdomain Ω_c, denoted by Γ_j is in contact with the subdomain Ω_0 and represents a wall for the current density **J**. The subdomain Ω_0 is a non-conducting material. Nevertheless, the conductivity σ is defined on the entire domain as follows:

$$\sigma > 0 \text{ in } \Omega_c \text{ and } \sigma = 0 \text{ in } \Omega_0 \quad [3.307]$$

The conductivity is not necessarily constant on the conducting domain Ω_c and can vary depending on the position.

The subdomain Ω_0 (see Figure 3.22) is not simply connected. The part of its boundary, in contact with the exterior, is a wall for the magnetic flux density and is denoted by Γ_b. Finally, magnetic permeability, which depends on space, will be denoted by μ throughout the domain.

Equation [3.301] is valid throughout the domain Ω. In fact, there is an electric field related to the variations in time of the magnetic flux density. However, in the subdomain Ω_0, the conductivity being zero, the current density will be zero irrespective of the value of the electric field **E**. Under these conditions, the field **E** does not need to satisfy equation [3.301] on Ω_0. Even though the magnetic flux density **B** is defined uniquely on Ω_0, this is not applicable to the electric field, which is defined up to a gradient. On the contrary, it is perfectly defined in the domain Ω_c. The equations of magnetodynamics can therefore be solved on the domain Ω_c and those of magnetostatics in the domain Ω_0. The next section shows that the coupling of these two problems is quite natural for potential formulations.

Therefore, various potential formulations (Bouillault and Ren 2008; Alonso Rodriguez and Valli 2010) will be implemented when electric or magnetic quantities are imposed on gates Γ_{n1} and Γ_{n2}.

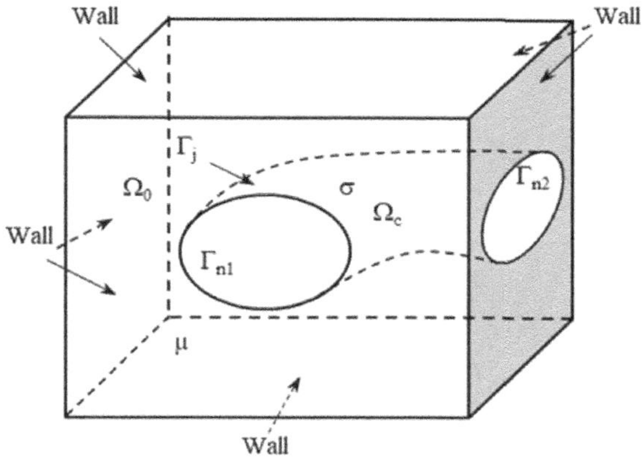

Figure 3.22. *Basic geometry for the magnetodynamics study*

Depending on the cases studied, boundary topology-related problems will be highlighted. Consider a divergence free vector field respecting, on a not simply connected part of the boundary, wall-type conditions. As shown (see section 3.2.1.2.2), to define a "χ" type source potential, a cut can be introduced. Sections 3.6.1 and 3.6.2 will show that for the same topology of the studied domain, depending on the type of electric or magnetic source, introducing a cut may or may not be necessary.

The studied example considers a single conductor in a given environment. Nevertheless, the proposed approach can be generalized to multiple conductors using the linearity of differential operators.

3.6.1. *Imposed electric quantities*

This section again uses the example in Figure 3.22 by imposing electric quantities e(t) or I(t) on the two conductor boundaries in contact with an external source, which can be a voltage or current source. These two boundaries are then gates for the electric field with boundary conditions denoted by Γ_{e1} and Γ_{e2} (see Figure 3.23). The other boundary conditions remain unchanged. The following can then be written for the boundary Γ of the domain Ω:

$$\Gamma = \Gamma_{e1} \cup \Gamma_{e2} \cup \Gamma_b \qquad [3.308]$$

and for the boundary Γ_c of the conductor Ω_c:

$$\Gamma_c = \Gamma_{e1} \cup \Gamma_{e2} \cup \Gamma_j \qquad [3.309]$$

As already noted, inside the conducting domain Ω_c, the magnetodynamic equations are solved, while the subdomain Ω_0 is governed by magnetostatic equations.

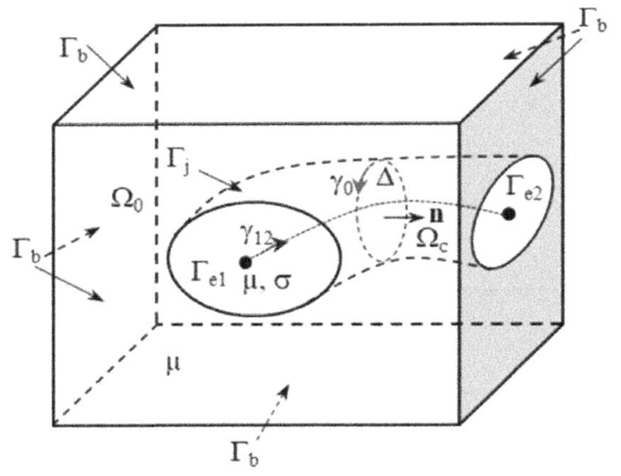

Figure 3.23. *Geometry studied in magnetodynamics: electric quantities imposed on the boundary*

For the problem at hand, the fields **E** and **J** are defined in the domain Ω_c and, considering equations [3.301], [3.304] and the previously introduced boundary conditions, the function spaces to which they belong can be written as follows:

$$\mathbf{E} \in H_{\Gamma_{e1} \cup \Gamma_{e2}}(\mathbf{curl}, \Omega_c), \ \mathbf{J} \in H_{\Gamma_j}(\mathrm{div}0, \Omega_c) \qquad [3.310]$$

It is possible to extend the domain of definition of the current density to the entire domain Ω by extending **J** over Ω_0 and posing $\mathbf{J} = 0$ in Ω_0. It can be noted that $\text{div}\mathbf{J} = 0$ on Ω_0. The extension of **J** to the entire domain Ω raises no problem for the continuity of the normal component since $\mathbf{J}.\mathbf{n}|_{\Gamma j} = 0$. On the contrary, as will be noted in the following, it allows for the definition of a source term without dealing with the non-connected domain and the introduction of cuts. The function space of the current density, defined by relation [3.310], is then written as:

$$\mathbf{J} \in \mathrm{H}(\mathrm{div}0, \Omega) \qquad [3.311]$$

As for the fields **H** and **B**, they are defined throughout the domain Ω. The properties of the magnetic field **H** are governed, at the beginning, by equations [3.302] in Ω_c and [3.207] in Ω_0. Nevertheless, due to the extension of current density **J** to the entire domain Ω, the field **H** verifies equation $\mathbf{curl}\mathbf{H} = \mathbf{J}$ on Ω. As for the magnetic flux density **B**, it is defined, in the entire domain Ω, by equation [3.303]. Taking into account the boundary conditions on the boundary Γ (see equation [3.308]), it can be noted that, considering equation [3.305], the normal component of the magnetic flux density is zero.

The function spaces of the fields **H** and **B** can be introduced as follows:

$$\mathbf{H} \in \mathrm{H}(\mathbf{curl}, \Omega), \;\; \mathbf{B} \in \mathrm{H}_0(\mathrm{div}0, \Omega) \qquad [3.312]$$

In order to solve the magnetodynamic equations, two potential formulations can be used. The first one, known as "electric formulation", is based on the magnetic vector potential **A** and the electric scalar potential V. The second, known as "magnetic formulation", uses the electric vector potential **T** and the magnetic scalar potential φ. For these two formulations, it can be seen that, considering the choice of potentials, the coupling with magnetostatics can be readily made.

3.6.1.1. *Electric formulation A-V*

The development of the **A**-V formulation, when an electromotive force e(t) is imposed, is quite natural. On the contrary, when the current intensity I(t) is imposed, an additional equation is required.

3.6.1.1.1. Imposed electromotive force

For this formulation, in order to introduce the source term e(t) into local form, the approach presented in section 3.4.1.1.1 is used. The electric field is decomposed (see equation [3.153]) in the form of a source field $\mathbf{E}_s$ and the field **E'** that becomes the unknown of the problem. The properties of the source field are given in equation [3.154] with $\mathbf{E}_s \in \mathrm{H}_{\Gamma e1 \cup \Gamma e2}(\mathbf{curl}0, \Omega_c)$. It should be recalled that $\mathbf{E}_s$ can be expressed using a support field $\boldsymbol{\beta}_e$ (see equation [3.157]) or an associated scalar potential α_e (see equation [3.158]). For our example, $\boldsymbol{\beta}_e$ and α_e are defined in the function spaces:

$$\boldsymbol{\beta}_e \in H_{\Gamma_{e1}\cup\Gamma_{e2}}(\mathbf{curl}0, \Omega_c) \text{ and } \alpha_e \in H(\mathbf{grad}, \Omega_c) \qquad [3.313]$$

Introducing the fields $\mathbf{E}'$ and $\mathbf{E}_s$ into equation [3.301] and, considering the properties of $\mathbf{E}_s$, the following succession of equations can be written as:

$$\mathbf{curlE} = \mathbf{curl}(\mathbf{E}' + \mathbf{E}_s) = \mathbf{curlE}' = -\frac{\partial \mathbf{B}}{\partial t} \qquad [3.314]$$

Under these conditions, the properties of the field $\mathbf{E}'$ are stated as follows:

$$\mathbf{curlE}' = -\frac{\partial \mathbf{B}}{\partial t}, \quad \int_{\gamma_{12}} \mathbf{E}'.\mathbf{dl} = 0, \quad \mathbf{E}' \wedge \mathbf{n}\big|_{\Gamma_{e1}\cup\Gamma_{e2}} = 0$$
$$\text{i.e. } \mathbf{E}' \in H_{\Gamma_{e1}\cup\Gamma_{e2}}(\mathbf{curl}, \Omega_c) \qquad [3.315]$$

Concerning the magnetic flux density, as shown in relation [3.312], it belongs to the function space $H_0(\text{div}0, \Omega)$. Therefore, it can be expressed using the magnetic vector potential $\mathbf{A}$ defined on the entire domain Ω, which is contractible, as follows:

$$\mathbf{B} = \mathbf{curlA} \text{ with } \mathbf{A} \in H_0(\mathbf{curl}, \Omega) \qquad [3.316]$$

Here we again find the potential $\mathbf{A}$, which was introduced in magnetostatics (see section 3.5.3). Using the behavior law [1.26], the magnetic field is then written as:

$$\mathbf{H} = \mu^{-1}\mathbf{curlA} \qquad [3.317]$$

On the domain Ω_c, if the magnetic flux density is replaced in the first equation of relation [3.315] by its expression given in equation [3.316], the following is deduced:

$$\mathbf{curl}(\mathbf{E}' + \frac{\partial \mathbf{A}}{\partial t}) = 0 \qquad [3.318]$$

The field $(\mathbf{E}' + \partial\mathbf{A}/\partial t) \in H_{\Gamma e1\cup\Gamma e2}(\mathbf{curl}0, \Omega_c)$ and, Ω_c being contractible, an electric scalar potential can be defined (see Figure 2.5) such that:

$$\mathbf{E}' = -(\frac{\partial \mathbf{A}}{\partial t} + \mathbf{grad}V) \text{ with } V \in H_{\Gamma_{e1}\cup\Gamma_{e2}}(\mathbf{grad}, \Omega_c) \qquad [3.319]$$

NOTE.– It can be noted that since the source term is supported by the source fields $\boldsymbol{\beta}_e$ or α_e (see equation [3.313]), the boundary conditions of the potential V, on the boundaries Γ_{e1} and Γ_{e2}, are zero.

Based on equation [3.319], by adding a source term $\mathbf{E}_s$ (see equation [3.153]) expressed by means of the potential α_e (see equation [3.158]), the following expression of the electric field $\mathbf{E}$ is obtained:

$$\mathbf{E} = -(\frac{\partial \mathbf{A}}{\partial t} + \mathbf{grad}\,V + e\,\mathbf{grad}\,\alpha_e) \qquad [3.320]$$

In this equation, besides the source term, the electric field is expressed using the **A**-V pair of potentials, hence its name of electric formulation. Moreover, equation [3.320] automatically verifies relations [3.301] and [3.303]. Then, using the behavior law [1.20], the expression of the current density in Ω_c can be written in the following form:

$$\mathbf{J} = -\sigma(\frac{\partial \mathbf{A}}{\partial t} + \mathbf{grad}\,V + e\mathbf{grad}\,\alpha_e) \qquad [3.321]$$

If in equation [3.302] the magnetic field is replaced by its expression given by equation [3.317] and the current density by equation [3.321], we then obtain:

$$\mathbf{curl}\mu^{-1}(\mathbf{curl A}) + \sigma(\frac{\partial \mathbf{A}}{\partial t} + \mathbf{grad}V) = -e\sigma\mathbf{grad}\alpha_e \qquad [3.322]$$

Introducing instead the support field $\boldsymbol{\beta}_e$ (see equation [3.158]), we have:

$$\mathbf{curl}\mu^{-1}(\mathbf{curl}A) + \sigma(\frac{\partial \mathbf{A}}{\partial t} + \mathbf{grad}V) = e\sigma\boldsymbol{\beta}_e \qquad [3.323]$$

It is recommended to also verify equation [3.304], which is performed by imposing the divergence operator to relation [3.321].

Then, we can write:

$$\text{div}(\sigma(\frac{\partial \mathbf{A}}{\partial t} + \mathbf{grad}V + e\mathbf{grad}\alpha_e)) = 0 \qquad [3.324]$$

Similarly, with the support field $\boldsymbol{\beta}_e$:

$$\text{div}(\sigma(\frac{\partial \mathbf{A}}{\partial t} + \mathbf{grad}V - e\,\boldsymbol{\beta}_e)) = 0 \qquad [3.325]$$

The systems of equations [3.322] and [3.324] or [3.323] and [3.325] correspond to the electric formulation of a magnetodynamics problem with an electromotive force as the source term. It should be recalled that the source field $\mathbf{E}_s$ is defined only on the conducting domain Ω_c. The same is valid for the support field $\boldsymbol{\beta}_e$ and the associated potential α_e.

To extend the equations to be solved to the complete domain Ω, the coupling with the magnetostatics formulation is naturally made. Indeed, the vector potential $\mathbf{A}$ is defined on the entire domain (see equation [3.316]). Therefore, it also appears as an unknown of the magnetostatics problem, on the domain Ω_0. Moreover, since the fields $\mathbf{B}$ and $\mathbf{H}$ depend only on the vector potential (see equations [3.316] and [3.317]), the normal component of $\mathbf{B}$ and the tangential component of $\mathbf{H}$ will be naturally continuous on the boundary Γ_j between Ω_c and Ω_0. Furthermore, it can be verified that the vector potential magnetostatic formulation in the domain Ω_0 can be deduced from the magnetodynamic formulation. Indeed, as the conductivity σ is zero in Ω_0, equation [3.325] is naturally verified and relation [3.323] becomes:

$$\mathbf{curl}(\mu^{-1}\mathbf{curl}\mathbf{A}) = 0 \qquad [3.326]$$

It can be noted that the above equation relates only to $\mathbf{A}$ and that in the non-conducting domain, where the electric field $\mathbf{E}$ is not defined, the scalar potential V is not defined either, and therefore there is no need to determine it. The scalar potential will only be calculated on Ω_c, though the latter may be considered as existing throughout the domain.

As conductivity σ is equal to zero on Ω_0 (see equation [3.307]), and the vector potential $\mathbf{A}$ is defined throughout the domain Ω, equations [3.323] and [3.326] can be regrouped. The system of equations to be solved then has the following form:

$$\mathbf{curl}\mu^{-1}(\mathbf{curl}\mathbf{A}) + \sigma(\frac{\partial \mathbf{A}}{\partial t} + \mathbf{grad}V) - \sigma e \boldsymbol{\beta}_e = 0 \text{ on } \Omega \qquad [3.327]$$

$$\mathrm{div}(\sigma(\frac{\partial \mathbf{A}}{\partial t} + \mathbf{grad}V - e\boldsymbol{\beta}_e)) = 0 \text{ on } \Omega_c \qquad [3.328]$$

NOTE.– It should be noted that equation [3.328] is obtained by applying the divergence operator to equation [3.327]. This result is quite expected, considering that equations [3.327] and [3.328] are built, respectively, from equations [1.5] and [1.6] and that equation [1.6] is obtained by applying the divergence operator to equation [1.5]. However, to simplify the developments, the two expressions will be used in the following.

It should be recalled that the scalar potential V should be determined only on the domain Ω_c and the vector potential **A** on the entire domain Ω.

3.6.1.1.2. Imposed current intensity

The current intensity I(t) is now imposed on the boundaries Γ_{e1} and Γ_{e2}. In order to solve the problem in the domain Ω, we can use the system of equations [3.327] and [3.328] where the unknowns are the potentials **A** and V and the source term "e". As the current I is imposed, the electromotive force e becomes an additional unknown of the problem. A new equation should then be introduced, and this is obtained by expressing I as a function of **A**-V potentials and of the electromotive force. To this end, a power balance is written.

In magnetodynamics, the instantaneous power "p" is written as:

$$p = \iiint_\Omega \mathbf{E}.\mathbf{J}d\tau + \iiint_\Omega \frac{\partial \mathbf{B}}{\partial t}.\mathbf{H}d\tau \qquad [3.329]$$

Expressing **E** according to [3.320] and **B** as a function of the potential **A**, we have:

$$\begin{aligned} p = &-\iiint_\Omega \frac{\partial \mathbf{A}}{\partial t}.\mathbf{J}d\tau - \iiint_\Omega \mathbf{grad}V.\mathbf{J}d\tau \\ &- \iiint_\Omega e\,\mathbf{grad}\alpha_e.\mathbf{J}d\tau + \iiint_\Omega \frac{\partial(\mathbf{curlA})}{\partial t}.\mathbf{H}d\tau \end{aligned} \qquad [3.330]$$

Let us now consider the first term of the integral on the right and replace the current density **J** by its expression as a function of **H** (see equation [3.302]). Using relation [2.27], for vector operators, we can write:

$$-\iiint_\Omega \frac{\partial \mathbf{A}}{\partial t}.\mathbf{curlH}d\tau = -\iiint_\Omega \left(\frac{\partial \mathbf{curlA}}{\partial t}\right).\mathbf{H}d\tau - \oiint_\Gamma \left(\mathbf{n} \wedge \frac{\partial \mathbf{A}}{\partial t}\right).\mathbf{H}dS \qquad [3.331]$$

Concerning the surface integral of the above equation, given that for the example mentioned $\mathbf{A} \in H_0(\mathbf{curl}, \Omega)$, it is equal to zero. Based on this result, equation [3.331] can be rewritten as follows:

$$-\iiint_\Omega \frac{\partial \mathbf{A}}{\partial t}.\mathbf{J}d\tau = -\iiint_\Omega \left(\frac{\partial \mathbf{curlA}}{\partial t}\right).\mathbf{H}d\tau \qquad [3.332]$$

Taking this result into account, the first and the last term of equation [3.330] cancel each other out. On the contrary, it can be shown that the second integral, on the right-hand side of equation [3.330], is also equal to zero. Indeed, it can be rewritten using formula [2.23] of vector operators. Moreover, as the current density **J** is zero in the subdomain Ω_0 (see equations [3.310] and [3.311]), the volume integral is limited to the subdomain Ω_c. Then, we have:

$$\iiint_{\Omega_c} \mathbf{grad}V.\mathbf{J}d\tau = -\iiint_{\Omega_c} V\mathrm{div}\mathbf{J}d\tau + \oiint_{\Gamma_c} V\mathbf{J}.\mathbf{n}dS = 0 \qquad [3.333]$$

Since the current density is divergence free, the first term on the right-hand side is equal to zero. As for the term related to the surface integral on Γ_c, it can be decomposed into three parts, as shown in equation [3.309], namely Γ_{e1}, Γ_{e2} and Γ_j. On the one hand, on the boundaries Γ_{e1} and Γ_{e2}, the scalar potential V is imposed to zero due to the introduction of the source term α_e (see the function space of potential V, equation [3.319]). On the other hand, the normal component of the current density is zero on the boundary Γ_j.

Let us again consider the expression of power, equation [3.330]. Considering equations [3.332] and [3.333], the following can be written as:

$$p = -\iiint_{\Omega} e\,\mathbf{grad}\,\alpha_e.\mathbf{J}d\tau \qquad [3.334]$$

Given that the current density **J** is zero in the non-conducting subdomain Ω_0, the integral over volume Ω can be contracted to an integration over Ω_c. The power can also be expressed using the global electric quantities "e" and "I". This leads to the following succession of equations:

$$p = eI = -\iiint_{\Omega_c} e\,\mathbf{grad}\alpha_e.\mathbf{J}d\tau \qquad [3.335]$$

By identification, the current I can be expressed as follows:

$$I = -\iiint_{\Omega_c} \mathbf{grad}\alpha_e.\mathbf{J}d\tau \qquad [3.336]$$

Or, by introducing the support vector $\boldsymbol{\beta}_e$ (see equation [3.158]), we have:

$$I = \iiint_{\Omega_c} \boldsymbol{\beta}_e.\mathbf{J}d\tau \qquad [3.337]$$

If the current density is replaced by its expression given in equation [3.321], we can write:

$$I = -\iiint_{\Omega_c} \boldsymbol{\beta}_e.\sigma(e\mathbf{grad}\alpha_e + \frac{\partial \mathbf{A}}{\partial t} + \mathbf{grad}V)d\tau \qquad [3.338]$$

Using equation [3.158], this relation can be rewritten as follows:

$$I = -\iiint_{\Omega_c} \boldsymbol{\beta}_e.\sigma(-e\boldsymbol{\beta}_e + \frac{\partial \mathbf{A}}{\partial t} + \mathbf{grad}\,V)d\tau \qquad [3.339]$$

In this configuration, the objective is to solve the system composed of equations [3.327] and [3.328] coupled with equation [3.338] or [3.339]. Furthermore, it should be recalled that the gauge condition on **A** should be added. Concerning the electric scalar potential V, the gauge condition is naturally provided, as $V = 0$ is imposed on the boundaries Γ_{e1} and Γ_{e2}.

3.6.1.2. *Magnetic formulation T-φ*

Let us now focus on the magnetic formulation in terms of the electric vector potential **T** and the magnetic scalar potential φ. For this formulation, when the current intensity I(t) is imposed, the source terms appear naturally in the developments. On the contrary, when an electromotive force e(t) is imposed, an additional equation is required, similar to the **A**-V formulation when current intensity is imposed.

3.6.1.2.1. Imposed current density flux

Let us consider the problem represented in Figure 3.23 having as a source term the flux I(t) of the current density. Moreover, let us recall the boundary conditions for the domain Ω_c, which can be stated as follows:

$$\Gamma_c = \Gamma_{e1} \cup \Gamma_{e2} \cup \Gamma_j \text{ with } \mathbf{E} \wedge \mathbf{n}\big|_{\Gamma_{ek}} = 0,\ k \in \{1,2\} \text{ and } \mathbf{J}.\mathbf{n}\big|_{\Gamma_j} = 0 \qquad [3.340]$$

As previously mentioned (see equation [3.311]), the current density is extended to the entire domain with $\mathbf{J} = 0$ in Ω_0, while maintaining $\mathbf{J}.\mathbf{n}|_{\Gamma j} = 0$.

Finally, on the boundary Γ of the domain Ω, we have:

$$\Gamma = \Gamma_b \cup \Gamma_{e1} \cup \Gamma_{e2} \text{ i.e.: } \mathbf{B}.\mathbf{n}\big|_{\Gamma} = 0 \qquad [3.341]$$

Similar to electrokinetics (see section 3.4.1.2.1), in the presence of an imposed current I, a source current density $\mathbf{J}_s$ is defined, extended to the entire domain Ω (with $\mathbf{J}_s = 0$ in Ω_0 and $\mathbf{J}_s.\mathbf{n}|_{\Gamma j} = 0$), such that:

$$\text{div}\mathbf{J}_S = 0,\ \iint_{\Gamma_{ek}} \mathbf{J}_S.\mathbf{n}ds = \pm I, \text{ with } k \in \{1,2\} \text{ i.e. } \mathbf{J}_S \in H(\text{div}0,\Omega) \qquad [3.342]$$

In this expression, $\mathbf{n}$ represents the outgoing normal on the boundaries. On the contrary, as noted after equation [3.19], for a vector field with conservative flux, in this case the current density $\mathbf{J}_s$, the integral over Γ_{e1} or Γ_{e2} can be replaced by any surface Δ (see Figure 3.23) whose contour γ_0 belongs to the boundary Γ_j surrounding the domain Ω_c.

Introducing the source term $\mathbf{J}_s$ (see equation [3.172]), the current density in the domain Ω is written as:

$$\mathbf{J} = \mathbf{J}_S + \mathbf{J}' \qquad [3.343]$$

where $\mathbf{J}'$ represents an unknown of the problem. Considering the properties of $\mathbf{J}$ and $\mathbf{J}_s$, the current density $\mathbf{J}'$ is extended to the entire domain Ω. Then, the current density $\mathbf{J}'$ is defined by:

$$\mathrm{div}\mathbf{J}' = 0, \quad \iint_\Delta \mathbf{J}'.\mathbf{n}ds = 0 \quad \text{with} \quad \mathbf{J}' \in H(\mathrm{div}0, \Omega) \qquad [3.344]$$

Like the fields $\mathbf{J}$ and $\mathbf{J}_s$, $\mathbf{J}' = 0$ in Ω_0 and $\mathbf{J}'.\mathbf{n}|_{\Gamma j} = 0$.

The source current $\mathbf{J}_s$ can be written, using a $\boldsymbol{\lambda}_I$ support vector field, as shown in the case of electrokinetics by equations [3.174] and [3.175]. However, $\boldsymbol{\lambda}_I$ extends to the entire domain by considering it equal to zero on Ω_0 similarly to the current density $\mathbf{J}_s$ (see equation [3.344]). Then, we have:

$$\begin{aligned} &\mathbf{J}_S = I\boldsymbol{\lambda}_I, \ \boldsymbol{\lambda}_I = 0 \text{ in } \Omega_0, \ \mathrm{div}\boldsymbol{\lambda}_I = 0, \ \iint_\Delta \boldsymbol{\lambda}_I.\mathbf{n}ds = 1 \\ &\text{and } \ \boldsymbol{\lambda}_I.\mathbf{n}\big|_{\Gamma_j} = 0 \ \text{ i.e. } \ \boldsymbol{\lambda}_I \in H(\mathrm{div}0, \Omega) \end{aligned} \qquad [3.345]$$

Let us recall that surface Δ, whose contour is denoted by γ_0, lies on the boundary Γ_j (see Figure 3.23). It is possible to consider a surface Δ, and therefore, γ_0 its contour in Ω_0, in contact with the boundary Γ_b. This does not change the properties of $\boldsymbol{\lambda}_I$ since, by continuous transformation, γ_0 can be back to a contour surrounding Γ_j.

Since the domain Ω is simply connected, the support field $\boldsymbol{\lambda}_I$ can be expressed using an associated vector potential $\boldsymbol{\chi}_I$ such that:

$$\boldsymbol{\lambda}_I = \mathbf{curl}\boldsymbol{\chi}_I, \quad \text{with} \quad \boldsymbol{\chi}_I \in H(\mathbf{curl}, \Omega) \qquad [3.346]$$

NOTE.– The current density **J** and therefore **J'**, $\mathbf{J}_s$ and the source field $\boldsymbol{\lambda}_I$ are extended to the entire domain Ω, which is simply connected. Then, there is no condition on normal components of **J**, $\mathbf{J}_s$, **J'** and $\boldsymbol{\lambda}_I$, on the boundary Γ_b. A source potential $\boldsymbol{\chi}_I$ can then be introduced without introducing any cut. This would not have been the case if the domain of definition of **J** and its associated quantities had been restricted Ω_c (see equations [3.310] and [3.311]). In this case, the domain of definition for $\mathbf{J}_s$, and therefore for $\boldsymbol{\lambda}_I$, would have been restrained to Ω_c, with a not simply connected boundary of Γ_j type on which $\boldsymbol{\lambda}_I.\mathbf{n} = 0$ should be imposed. But such a constraint, in the case of a not simply connected boundary (see section 3.2.1.2.2), a cut should be introduced to define correctly a $\boldsymbol{\chi}_I$ type potential. Similarly, the domain Ω_0 being not simply connected, a cut should have also been introduced for the latter, to take into account the current I flowing through Ω_c.

Having defined the current density $\mathbf{J}_s$, the objective is to determine the expression of **J'**. As shown in relation [3.344], $\mathbf{J'} \in$ H(div0, Ω) and can therefore be defined using an electric vector potential **T**. Then, we have:

$$\mathbf{J}' = \mathbf{curlT} \text{ i.e. } \mathbf{T} \in \mathrm{H}(\mathbf{curl}, \Omega) \qquad [3.347]$$

Since the current density **J'** is equal to zero on Ω_0, the potential **T** equal to zero can be imposed on Ω_0. This is quite compatible with the continuity of the tangential component of **T** on the boundary Γ_j and with $\mathbf{J'} = \mathbf{curlT} = 0$ on Ω_0. In this case, the potential **T** remains unknown only on Ω_c. It should be recalled (see equation [3.344]) that the flux of the current density **J'** through a section of the conductor is equal to zero. Though Γ_j is not simply connected, this property allows us to impose homogeneous boundary conditions on the tangential component of **T** on the entire surface Γ_j. Under these conditions, based on equation [3.347], a restriction of **T** on Ω_c can be defined such that:

$$\mathbf{J}' = \mathbf{curlT} \text{ with } \mathbf{T} \wedge \mathbf{n}\big|_{\Gamma_j} = 0 \text{ i.e. } \mathbf{T} \in \mathrm{H}_{\Gamma_j}(\mathbf{curl}, \Omega_c) \qquad [3.348]$$

Considering equations [3.343], [3.345] and [3.347], the current density **J** can then be written as follows:

$$\mathbf{J} = \mathrm{I}\boldsymbol{\lambda}_I + \mathbf{curlT} \qquad [3.349]$$

Or using the associated potential $\boldsymbol{\chi}_I$:

$$\mathbf{J} = \mathrm{I}\mathbf{curl}\boldsymbol{\chi}_I + \mathbf{curlT} \qquad [3.350]$$

Using the behavior law [1.20], the expression of the electric field on Ω_c is:

$$\mathbf{E} = \sigma^{-1}(\mathbf{I}\mathbf{curl}\chi_I + \mathbf{curl}\mathbf{T}) \quad [3.351]$$

Using equations [3.302] and [3.350], the following relation can be written on the entire domain Ω (**T** being defined on Ω_0) (see equation [3.347]):

$$\mathbf{curl}\mathbf{H} = \mathbf{I}\mathbf{curl}\chi_I + \mathbf{curl}\mathbf{T} \quad [3.352]$$

which can be written as:

$$\mathbf{curl}(\mathbf{H} - \mathbf{T} - \mathrm{I}\chi_I) = 0 \quad [3.353]$$

Since the domain Ω is simply connected, the magnetic scalar potential φ (see equation [2.32]) can be introduced as follows:

$$\mathbf{H} = \mathrm{I}\chi_I + \mathbf{T} - \mathbf{grad}\varphi, \text{ with } \varphi \in H(\mathbf{grad}, \Omega) \quad [3.354]$$

Besides the source term $\mathbf{I}\boldsymbol{\chi}_I$, the magnetic field is expressed using the **T**-φ pair of potentials, hence the name magnetic formulation. Finally, for the scalar potential φ to be uniquely defined, its value must be fixed at a point of the domain Ω.

Based on relation [3.354], the magnetic flux density can be written, via the magnetic behavior law [1.26], as follows:

$$\mathbf{B} = \mu(\mathrm{I}\chi_I + \mathbf{T} - \mathbf{grad}\varphi) \quad [3.355]$$

At this stage of our developments, equations [3.350] and [3.354] verify, respectively, equations [3.302] and [3.304]. Equations [3.301] and [3.303] should also be verified. As for the first equation, replacing the electric field by its expression given in equation [3.351] and the magnetic flux density by equation [3.355], the following relation is obtained on Ω_c:

$$\mathbf{curl}(\sigma^{-1}(\mathbf{I}\mathbf{curl}\chi_I + \mathbf{curl}\mathbf{T})) = -\frac{\partial}{\partial t}(\mu(\mathrm{I}\chi_I + \mathbf{T} - \mathbf{grad}\varphi)) \quad [3.356]$$

Gathering the source term on the right-hand side, the following can be written:

$$\begin{aligned}&\mathbf{curl}(\sigma^{-1}\mathbf{curl}\mathbf{T}) + \frac{\partial}{\partial t}(\mu(\mathbf{T} - \mathbf{grad}\varphi)) = \\ &-\mathbf{curl}(\sigma^{-1}\mathbf{I}\mathbf{curl}\chi_I) - \frac{\partial}{\partial t}(\mu\ (\mathrm{I}\chi_I)\end{aligned} \quad [3.357]$$

For the second equation [3.303], the magnetic flux density is replaced by expression [3.355], hence:

$$\mathrm{div}(\mu(\mathrm{I}\boldsymbol{\chi}_\mathrm{I} + \mathbf{T} - \mathbf{grad}\varphi)) = 0 \qquad [3.358]$$

To obtain the solution to the problem on the domain Ω_c, the system consisting of equations [3.357] and [3.358] must be solved, in which the unknowns are the vector potential **T** and the scalar potential φ, coupled with magnetostatic equations.

The potentials $\boldsymbol{\chi}_\mathrm{I}$ and φ are defined on the entire domain (see equations [3.346] and [3.354]). In the subdomain Ω_0, since the vector potential **T** is equal to zero, the magnetic field **H** is written as follows:

$$\mathbf{H} = \mathrm{I}\boldsymbol{\chi}_\mathrm{I} - \mathbf{grad}\varphi \qquad [3.359]$$

Similarly, based on relation [3.355], the expression of the magnetic flux density is:

$$\mathbf{B} = \mu(\mathrm{I}\boldsymbol{\chi}_\mathrm{I} - \mathbf{grad}\varphi) \qquad [3.360]$$

For the magnetostatics part, based on equations [3.303] and [3.360], the equation to be solved has the form:

$$\mathrm{div}(\mu(\mathrm{I}\boldsymbol{\chi}_\mathrm{I} - \mathbf{grad}\varphi)) = 0 \qquad [3.361]$$

Finally, the coupling between magnetodynamics and magnetostatics is quite natural. Indeed, the vector potential **T** is zero on Ω_0 and φ is defined on the entire domain Ω. At the interface between Ω_c and Ω_0, namely the boundary Γ_j, the condition $\mathbf{J}.\mathbf{n} = 0$ is imposed by means of the properties of the vector potential **T** and of the support field $\boldsymbol{\lambda}_\mathrm{I}$ (see equation [3.349]). Similarly, still on Γ_j, the continuity of the tangential component of the magnetic field **H** is ensured via the continuity of the tangential component of $\boldsymbol{\chi}_\mathrm{I}$ and of $\mathbf{grad}\varphi$. As already noted, the electric field **E** was only defined in the domain Ω_c.

The system to be solved is given by equations [3.356] and [3.358] in Ω_c and [3.361] in Ω_0. However, it can be noted that the vector potential **T** is equal to zero on Ω_0 (see equation [3.348]) and the scalar potential φ is defined (see equation [3.354]) on the entire domain. It is therefore possible to gather relations [3.358] and [3.361]. In this case, the system to be solved can be written as follows:

$$\mathbf{curl}(\sigma^{-1}(\mathrm{I}\mathbf{curl}\chi_\mathrm{I} + \mathbf{curlT})) + \frac{\partial}{\partial t}(\mu(\mathrm{I}\chi_\mathrm{I} + \mathbf{T} - \mathbf{grad}\varphi)) = 0 \text{ on } \Omega_c \qquad [3.362]$$

$$\mathrm{div}(\mu(\mathrm{I}\chi_\mathrm{I} + \mathbf{T} - \mathbf{grad}\varphi)) = 0 \text{ on } \Omega \qquad [3.363]$$

3.6.1.2.2. Imposed electromotive force

The source term is now the electromotive force. With the **T**-φ electric formulation, the equations to be solved are given by expressions [3.362] and [3.363]. However, in this system, current intensity I becomes an unknown. To obtain a full equation system, a new equation is added, in which the electromotive force is expressed as a function of **T**-φ potentials and of the current intensity I. To this end, the power conservation equation [3.329] is again used in the following form:

$$p = e\mathrm{I} = \iiint_\Omega \mathbf{E.J}d\tau + \iiint_\Omega \frac{\partial \mathbf{B}}{\partial t}.\mathbf{H}d\tau \qquad [3.364]$$

In this equation, **J** and **H** are replaced by, respectively, expressions [3.350] and [3.354]. Then, we obtain:

$$e\mathrm{I} = \iiint_\Omega \mathbf{E.(curlT + Icurl\chi_I)}d\tau + \iiint_\Omega \frac{\partial \mathbf{B}}{\partial t}.(\mathrm{I}\chi_\mathrm{I} + \mathbf{T} - \mathrm{grad}\varphi)d\tau \qquad [3.365]$$

Using the formulas of vector operators (see equation [2.27]), the first term of the first integral on the right-hand side can be written as follows:

$$\iiint_\Omega \mathbf{E.curlT}\,d\tau = \iiint_\Omega \mathbf{curlE.T}d\tau + \oiint_\Gamma (\mathbf{n} \wedge \mathbf{T}).\mathbf{E}dS \qquad [3.366]$$

The surface integral of this equation, on the boundary Γ of the domain, is equal to zero. Indeed, Γ is the union of three boundaries (see relation [3.308]) or $\Gamma = \Gamma_{e1} \cup \Gamma_{e2} \cup \Gamma_b$. If the integral is decomposed into three terms, considering the properties of the electric field, the integrals over Γ_{e1} and Γ_{e2} are equal to zero. The same is valid for the integral on Γ_b as the vector potential **T** is zero in the subdomain Ω_0 and therefore on the boundary. Considering this result, we can write:

$$\iiint_\Omega \mathbf{E.curlT}d\tau = \iiint_\Omega \mathbf{curlE.T}d\tau \qquad [3.367]$$

Consider now the last term of the second integral of equation [3.365]. Using the formulas of vector operators (see equation [2.23]), we can write:

$$\iiint_{\Omega} \frac{\partial \mathbf{B}}{\partial t}.\mathbf{grad}\varphi d\tau = -\iiint_{\Omega} \varphi \mathrm{div}(\frac{\partial \mathbf{B}}{\partial t})d\tau + \oiint_{\Gamma} \varphi \frac{\partial \mathbf{B}}{\partial t}.\mathbf{n}dS = 0 \qquad [3.368]$$

This equation is also zero as the magnetic flux density **B** belongs to $H_0(\mathrm{div}0, \Omega)$. Under these conditions, considering relations [3.367] and [3.368], equation [3.365] is written as:

$$eI = \iiint_{\Omega} (\mathbf{T}.\mathbf{curlE} + I\mathbf{E}.\mathbf{curl}\chi_I)d\tau + \iiint_{\Omega} \frac{\partial \mathbf{B}}{\partial t}.(I\chi_I + \mathbf{T})d\tau \qquad [3.369]$$

This equation can be simplified by replacing "**curlE**" by its expression given in equation [3.301]. After simplification, we obtain:

$$eI = \iiint_{\Omega} I\mathbf{E}.\mathbf{curl}\chi_I d\tau + \iiint_{\Omega} \frac{\partial \mathbf{B}}{\partial t}.I\chi_I d\tau \qquad [3.370]$$

The expression of the electromotive force "e" is deduced by identification as follows:

$$e = \iiint_{\Omega} \mathbf{E}.\mathbf{curl}\chi_I d\tau + \iiint_{\Omega} \frac{\partial \mathbf{B}}{\partial t}.\chi_I d\tau \qquad [3.371]$$

This relation can also be written using the properties of the source vector field $\boldsymbol{\lambda}_I$ (see equation [3.345]) as follows:

$$e = \iiint_{\Omega} \mathbf{E}.\boldsymbol{\lambda}_I d\tau + \iiint_{\Omega} \frac{\partial \mathbf{B}}{\partial t}.\chi_I d\tau \qquad [3.372]$$

If **E** and **B** are, respectively, replaced by equations [3.351] and [3.355], we obtain:

$$e = \iiint_{\Omega} \sigma^{-1}(\mathbf{curlT} + I\mathbf{curl}\chi_I).\boldsymbol{\lambda}_I)d\tau + \iiint_{\Omega} \frac{\partial}{\partial t}\mu(\mathbf{T} - \mathbf{grad}\varphi + I\chi_I).\chi_I d\tau \qquad [3.373]$$

Using relations [3.371]–[3.373], the electromotive force "e" can be expressed as a function of **T**-φ potentials and the current intensity "I".

To solve a magnetodynamics problem, with the **T**-φ magnetic formulation, when the source term is the electromotive force, we have to solve the system consisting of equations [3.362] and [3.363], in association with expression [3.373].

3.6.2. *Imposed magnetic quantities*

This section considers the same domain Ω, consisting of a conducting subdomain Ω_c, where two parts of its boundary Γ_c are in contact with the external domain.

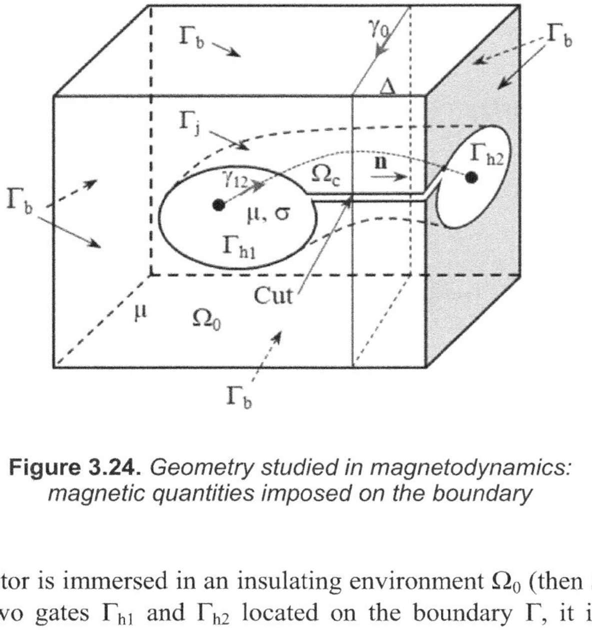

Figure 3.24. *Geometry studied in magnetodynamics: magnetic quantities imposed on the boundary*

The conductor is immersed in an insulating environment Ω_0 (then $\Omega = \Omega_0 \cup \Omega_c$). Through the two gates Γ_{h1} and Γ_{h2} located on the boundary Γ, it is supplied by magnetic source terms that could be a magnetomotive force $f_m(t)$ or a flux φ(t) (see Figure 3.24). On the external boundary of Ω_0, Γ_b boundary conditions are imposed. It can be noted that the external surface Γ_b of Ω (on which **B**.**n** = 0 is imposed) is not simply connected. This is due to the presence of surfaces Γ_{h1} and Γ_{h2} on which $\mathbf{H} \wedge \mathbf{n} = 0$ is imposed. Then, for the boundary Γ of the domain Ω, the following boundary conditions apply:

$$\Gamma = \Gamma_{h1} \cup \Gamma_{h2} \cup \Gamma_b \text{ with } \mathbf{H} \wedge \mathbf{n}\big|_{\Gamma_{hk}} = 0,\ k \in \{1,2\} \text{ and } \mathbf{B}.\mathbf{n}\big|_{\Gamma_b} = 0 \qquad [3.374]$$

The boundary Γ_c of the conducting domain represents a wall for the current density. Indeed, the boundary conditions on the field **H**, imposed on Γ_{h1} and Γ_{h2}, imply that $\mathbf{J.n} = 0$ (see equation [3.306]). Moreover, since the domain Ω, in which the conductor is immersed, is an insulator, the remaining part of the boundary, denoted by Γ_j, is a wall for the current density. Then:

$$\Gamma_c = \Gamma_{h1} \cup \Gamma_{h2} \cup \Gamma_j \qquad [3.375]$$

The magnetic permeability, which depends on the space, will be denoted by μ throughout the domain. We recall that, in the subdomain Ω_c, the conductivity is $\sigma \neq 0$. On the contrary, its values are zero in Ω_0 (see equation [3.307]).

Similar to section 3.6.1.1, inside the conducting domain Ω_c, the magnetodynamic equations are solved, while in Ω_0, those of magnetostatics are solved.

In this new configuration, the fields **E** and **J** are defined only in the domain Ω_c and, taking into account equations [3.301], [3.304] and the boundary conditions, the associated function spaces are written as:

$$\mathbf{E} \in \mathrm{H}(\mathbf{curl}, \Omega_c), \ \mathbf{J} \in \mathrm{H}_0(\mathrm{div}0, \Omega_c) \qquad [3.376]$$

The fields **H** and **B** are defined in the entire domain Ω. For the magnetic field, equation [3.302] must be solved in Ω_c and equation [3.207] in Ω_0. As for the magnetic flux density, it is governed by equation [3.303] on the entire domain. Based on their properties, fields **H** and **B** are defined, respectively, in the function spaces:

$$\mathbf{H} \in \mathrm{H}_{\Gamma_{h1} \cup \Gamma_{h2}}(\mathbf{curl}, \Omega), \quad \mathbf{B} \in \mathrm{H}_{\Gamma_b}(\mathrm{div}0, \Omega) \qquad [3.377]$$

Applying the same approach as in section 3.6.1, the electric and magnetic formulations will be introduced for the magnetic source terms $f_m(t)$ and $\phi(t)$. The equations of magnetodynamics will be developed in the conducting domain and those of magnetostatics in Ω_0. It will be noted that, depending on the chosen potentials, the coupling is naturally achieved at the interface between these two domains.

3.6.2.1. *Electric formulation A-V*

3.6.2.1.1. Imposed magnetic flux

In order to take into account the source term $\phi(t)$, a similar approach to the one in section 3.5.3.1 will be used. The magnetic flux density is decomposed into two terms, $\mathbf{B}_s$ and $\mathbf{B}'$, as shown by equation [3.269], as follows:

$$\mathbf{B} = \mathbf{B}_S + \mathbf{B}' \tag{3.378}$$

where $\mathbf{B}_s$ represents a source field, the image of the flux ϕ imposed on the boundaries Γ_{h1} and Γ_{h2}. The properties of this field (see equation [3.21]) are given by the following relations:

$$\mathrm{div}\mathbf{B}_S = 0,\ \iint_{\Gamma_{hk}} \mathbf{B}_S.\mathbf{n}ds = \pm\phi \text{ with } k \in \{1,2\},\ \mathbf{B}_S.\mathbf{n}\Big|_{\Gamma_b} = 0$$
$$\text{i.e. } \mathbf{B}_S \in H_{\Gamma_b}(\mathrm{div}0,\Omega) \tag{3.379}$$

The source term $\mathbf{B}_s$ can be expressed on the entire domain Ω using a support field $\boldsymbol{\lambda}_\phi$ (see equation [3.272]). Based on the field $\mathbf{B}_s$, $\boldsymbol{\lambda}_\phi$ must verify the following properties:

$$\mathrm{div}\boldsymbol{\lambda}_\phi = 0 \text{ with } \iint_{\Delta} \boldsymbol{\lambda}_\phi.\mathbf{n}ds = 1 \text{ and } \boldsymbol{\lambda}_\phi.\mathbf{n}\Big|_{\Gamma_b} = 0$$
$$\text{i.e., } \boldsymbol{\lambda}_\phi \in H_{\Gamma_b}(\mathrm{div}0,\Omega) \tag{3.380}$$

NOTE.– The flux of the vector field $\boldsymbol{\lambda}_\phi$ is conservative, and therefore the conditions of the note following equation [3.19] are met. The surface Δ (see Figure 3.24) represents an arbitrary section of the domain Ω whose contour γ_0 must rely on the boundary Γ_b. If Δ is reduced, by deformation on the surface Γ_b, the latter merges in the end with the boundaries Γ_{h1} or Γ_{h2} and it can no longer be contracted to a point on Γ_b. This highlights the fact that the surface Γ_b is not simply connected.

Considering its properties, the support field $\boldsymbol{\lambda}_\phi$ can be expressed by means of a potential $\boldsymbol{\chi}_\phi$ (see equation [3.274]). As the boundary Γ_b is not simply connected, imposing the constraints on $\boldsymbol{\chi}_\phi$ requires the introduction of a cut, as shown in section 3.2.1.2.2. Figure 3.24 shows an example of a possible cut. Under these conditions, the properties of potential $\boldsymbol{\chi}_\phi$ can be written as follows:

$$\boldsymbol{\lambda}_\phi = \mathbf{curl}\boldsymbol{\chi}_\phi,\ \int_{\gamma_0} \boldsymbol{\chi}_\phi.\mathbf{dl} = \pm 1 \text{ with } \boldsymbol{\chi}_\phi \in H_{\Gamma_b}^{\Delta}(\mathbf{curl},\Omega) \tag{3.381}$$

The function space to which χ_ϕ belongs is built according to equation [3.28].

Having defined the source field $\mathbf{B}_s$, let us now focus on the magnetic flux density $\mathbf{B}'$ that can be expressed, as shown by equation [3.275], as a function of the vector potential $\mathbf{A}$, as follows:

$$\mathbf{B}' = \mathbf{curlA}, \text{ with } \mathbf{A} \in H_{\Gamma_b}(\mathbf{curl}, \Omega) \qquad [3.382]$$

If $\mathbf{B}'$ and $\mathbf{B}_s$ are replaced in equation [3.378] by $\mathbf{curlA}$ and $\phi\boldsymbol{\lambda}_\phi$ (or $\phi\mathbf{curl}\boldsymbol{\chi}_\phi$), respectively, we obtain:

$$\mathbf{B} = \mathbf{curlA} + \phi\boldsymbol{\lambda}_\phi = \mathbf{curlA} + \phi\mathbf{curl}\boldsymbol{\chi}_\phi \qquad [3.383]$$

The behavior law of magnetic materials [1.26] can be used to express the field $\mathbf{H}$ as follows:

$$\mathbf{H} = \mu^{-1}(\mathbf{curlA} + \phi\boldsymbol{\lambda}_\phi) = \mu^{-1}(\mathbf{curlA} + \phi\mathbf{curl}\boldsymbol{\chi}_\phi) \qquad [3.384]$$

To obtain the $\mathbf{A}$-V formulation, equation [3.383] is introduced into equation [3.301], which leads, on the domain Ω_c, to the following relation:

$$\mathbf{curlE} = -\frac{\partial}{\partial t}(\mathbf{curlA} + \phi\mathbf{curl}\boldsymbol{\chi}_\phi) \qquad [3.385]$$

This equation can be rewritten as:

$$\mathbf{curl}(\mathbf{E} + \frac{\partial \mathbf{A}}{\partial t} + \frac{\partial \phi\boldsymbol{\chi}_\phi}{\partial t}) = 0 \qquad [3.386]$$

Considering equation [3.386] and, since the domain Ω_c is simply connected, the electric scalar potential V can be defined such that:

$$\mathbf{E} + \frac{\partial \mathbf{A}}{\partial t} + \frac{\partial \phi\boldsymbol{\chi}_\phi}{\partial t} = -\mathbf{grad}V \text{ with } V \in H(\mathbf{grad}, \Omega_c) \qquad [3.387]$$

Under these conditions, the electric field $\mathbf{E}$ has the form:

$$\mathbf{E} = -(\frac{\partial \mathbf{A}}{\partial t} + \frac{\partial \phi\boldsymbol{\chi}_\phi}{\partial t} + \mathbf{grad}\,V) \qquad [3.388]$$

The current density is then written as:

$$\mathbf{J} = -\sigma(\frac{\partial \mathbf{A}}{\partial t} + \frac{\partial \phi \chi_\phi}{\partial t} + \mathbf{grad}\,V) \qquad [3.389]$$

If in equation [3.302], the magnetic field is replaced by equation [3.384] and the current density by equation [3.389], we obtain:

$$\mathbf{curl}\,(\mu^{-1}\mathbf{curl}(\mathbf{A} + \phi\chi_\phi)) + \sigma(\frac{\partial \mathbf{A}}{\partial t} + \mathbf{grad}V + \frac{\partial \phi \chi_\phi}{\partial t}) = 0 \qquad [3.390]$$

The system is completed considering that equation [3.304] is also verified, hence:

$$\mathrm{div}(\sigma(\frac{\partial \mathbf{A}}{\partial t} + \frac{\partial \phi \chi_\phi}{\partial t} + \mathbf{grad}\,V)) = 0 \qquad [3.391]$$

In conclusion, with the **A**-V formulation, when the source term is the magnetic flux ϕ, the system consisting of equations [3.390] and [3.391] must be solved in the domain Ω_c. To these equations should be added the equations related to the magnetostatics formulation in Ω_0 and the continuity conditions of fields **H** and **B** should be imposed at the interface Γ_j.

The equations to be solved in the domain Ω_0 are [3.207] and [3.303]. The choice of the magnetic vector potential (see equation [3.383]) allows for the verification of property [3.303] throughout the domain Ω. Similarly, the magnetic field, via equation [3.384], is defined in Ω. Under these conditions, if the magnetic field, expressed by relation [3.384], is replaced in equation [3.207], the equation to be solved in magnetostatics is:

$$\mathbf{curl}\,\mu^{-1}(\mathbf{curl}(\mathbf{A} + \phi\chi_\phi) = 0 \qquad [3.392]$$

The electric field is defined only in the domain Ω_c and no constraint is imposed on its tangential component. On the contrary, the normal component of the current density must be equal to zero on the boundary of the conductor. This property is imposed on the boundaries Γ_{h1} and Γ_{h2} via the magnetic field (see equation [3.306]). For the boundary Γ_j, it is imposed via equation [3.391]. The fields **H** and **B** are defined on the entire domain and no constraint is imposed at the interface between Ω_0 and Ω_c.

Under these conditions, for the study of the problem represented in Figure 3.24, equations [3.390] and [3.391] in the domain Ω_c, and in addition equation [3.392], must be solved in Ω_0. The vector potential **A** is defined on the entire domain Ω [3.382]. Moreover, conductivity σ is equal to zero in the subdomain Ω_0 (see equation [3.307]).

Under these conditions, relations [3.390] and [3.392] can be gathered. Then, the system of equations has the following form:

$$\mathbf{curl}(\mu^{-1}\mathbf{curl}(\mathbf{A}+\phi\chi_\phi)+\sigma(\frac{\partial \mathbf{A}}{\partial t}+\mathbf{grad}V+\frac{\partial \phi\chi_\phi}{\partial t})=0 \text{ on } \Omega \quad [3.393]$$

$$\mathrm{div}(\sigma(\frac{\partial \mathbf{A}}{\partial t}+\frac{\partial \phi\chi_\phi}{\partial t}+\mathbf{grad}V))=0 \text{ on } \Omega_c \quad [3.394]$$

NOTE.– It can be noted that equation [3.394] is deduced from equation [3.393] when the divergence operator is applied. In fact, a configuration equivalent to the one analyzed at the end of section 3.6.1.1.1 (see note after equation [3.328]) is obtained.

3.6.2.1.2. Imposed magnetomotive force

For the studied geometry (see Figure 3.24), the magnetomotive force f_m is imposed. The flux ϕ then becomes an unknown of the problem. The system consisting of equations [3.393] and [3.394] should therefore be completed with an additional equation that is obtained from a power balance. The objective is to express the magnetomotive force as a function of **A**-V potentials and of the magnetic flux ϕ.

The power "p" (see equation [3.329]) is written as follows:

$$p = \iiint_{\Omega_c} \mathbf{E}.\mathbf{J}d\tau + \iiint_{\Omega} \frac{\partial \mathbf{B}}{\partial t}.\mathbf{H}d\tau \quad [3.395]$$

The first term on the right-hand side is integrated over Ω_c, taking into account the function space to which the current density **J** belongs (see equation [3.376]). **E** is expressed using relation [3.388] and **B** as a function of **A** and $\boldsymbol{\lambda}_\phi$ (see equation [3.383]).

Then, the expression of power is:

$$p = -\iiint_{\Omega_c} \left(\frac{\partial \mathbf{A}}{\partial t} + \frac{\partial \phi \chi_\phi}{\partial t}\right).\mathbf{J}d\tau - \iiint_{\Omega_c} \mathbf{grad}V.\mathbf{J}d\tau + \iiint_{\Omega} \left(\frac{\partial(\mathbf{curlA})}{\partial t} + \frac{\partial(\phi \lambda_\phi)}{\partial t}\right).\mathbf{H}d\tau \qquad [3.396]$$

Let us now consider the first term of the integral on the right-hand side. Considering equation [3.302] and the properties of the vector operators (see equation [2.27]), this integral can be rewritten on the domain Ω (since $\mathbf{curlH} = \mathbf{J}$ on Ω_c and $\mathbf{curlH} = 0$ on Ω_0) in the form:

$$-\iiint_{\Omega} \frac{\partial \mathbf{A}}{\partial t}.\mathbf{curlH}d\tau = -\iiint_{\Omega} \left(\frac{\partial \mathbf{curlA}}{\partial t}\right).\mathbf{H}d\tau - \oiint_{\Gamma} \left(\mathbf{n} \wedge \frac{\partial \mathbf{A}}{\partial t}\right).\mathbf{H}dS \qquad [3.397]$$

Concerning the surface integral of equation [3.397], given that $\mathbf{A} \in H_{\Gamma b}(\mathbf{curl}, \Omega)$ and $\mathbf{H} \in H_{\Gamma h1 \cup \Gamma h2}(\mathbf{curl}, \Omega)$, it is equal to zero. In fact, by decomposing the boundary Γ and using the properties of the mixed product, we have:

$$\oiint_{\Gamma} \left(\mathbf{n} \wedge \frac{\partial \mathbf{A}}{\partial t}\right).\mathbf{H}dS = \iint_{\Gamma_b} \left(\mathbf{n} \wedge \frac{\partial \mathbf{A}}{\partial t}\right).\mathbf{H}dS + \iint_{\Gamma_{h_1} \cup \Gamma_{h2}} (\mathbf{H} \wedge \mathbf{n}).\frac{\partial \mathbf{A}}{\partial t}dS = 0 \qquad [3.398]$$

Based on this result, equation [3.397] can be rewritten as:

$$-\iiint_{\Omega} \frac{\partial \mathbf{A}}{\partial t}.\mathbf{curlH}d\tau = -\iiint_{\Omega} \left(\frac{\partial \mathbf{curlA}}{\partial t}\right).\mathbf{H}d\tau \qquad [3.399]$$

On the contrary, the second integral on the right-hand side of equation [3.396] is also zero. In fact, it can be rewritten, again using the formulas of vector operators [2.23], in the following form:

$$\iiint_{\Omega_c} \mathbf{grad}V.\mathbf{J}d\tau = -\iiint_{\Omega_c} V\mathrm{div}\mathbf{J}d\tau + \oiint_{\Gamma_c} V\mathbf{J}.\mathbf{n}dS = 0 \qquad [3.400]$$

Since the current density is divergence free, the first integral on the right-hand side is equal to zero. The same is true for the second integral, as on $\Gamma_c = \Gamma_j \cup \Gamma_{h1} \cup \Gamma_{h2}$ (see equation [3.376]), we have $\mathbf{J}.\mathbf{n} = 0$.

Consider now the expression of power, defined by equation [3.396]. Taking into account the results given by equations [3.399] and [3.400] and after rearrangement, we can write:

$$\begin{aligned} p = & -\iiint_{\Omega} \left(\frac{\partial \mathbf{curlA}}{\partial t}\right).\mathbf{H}d\tau - \frac{d\phi}{dt}\iiint_{\Omega} \boldsymbol{\chi}_{\phi}.\mathbf{J}d\tau \\ & + \iiint_{\Omega} \frac{\partial(\mathbf{curlA})}{\partial t}.\mathbf{H}d\tau + \frac{d\phi}{dt}\iiint_{\Omega} \boldsymbol{\lambda}_{\phi}.\mathbf{J}d\tau \end{aligned} \quad [3.401]$$

After simplification and by expressing the power "p" as a function of global quantities, we obtain:

$$p = f_m \frac{d\phi}{dt} = -\frac{d\phi}{dt}\iiint_{\Omega_c} \boldsymbol{\chi}_{\phi}.\mathbf{J}.d\tau + \frac{d\phi}{dt}\iiint_{\Omega_c} \boldsymbol{\lambda}_{\phi}.\mathbf{H}d\tau \quad [3.402]$$

By identification, the expression of the magnetomotive force is deduced as follows:

$$f_m = \iiint_{\Omega} \boldsymbol{\lambda}_{\phi}.\mathbf{H}d\tau - \iiint_{\Omega} \boldsymbol{\chi}_{\phi}.\mathbf{J}d\tau \quad [3.403]$$

NOTE.– The above expression can be rewritten by again using the vector operators. Indeed, let us consider the first integral term on the right-hand side, in which the support field $\boldsymbol{\lambda}_{\phi}$ is replaced by its potential $\boldsymbol{\chi}_{\phi}$, defined by relation [3.381]. Applying the formula [2.27], we obtain:

$$\iiint_{\Omega} \mathbf{curl}\boldsymbol{\chi}_{\phi}.\mathbf{H}\, d\tau = \iiint_{\Omega} \boldsymbol{\chi}_{\phi}.\mathbf{curlH}\, d\tau - \oiint_{\Gamma} (\mathbf{n} \wedge \mathbf{H}).\boldsymbol{\chi}_{\phi} dS \quad [3.404]$$

Let us now consider the surface integral, which can be decomposed as follows:

$$\oiint_{\Gamma} (\mathbf{n} \wedge \mathbf{H}).\boldsymbol{\chi}_{\phi} dS = \iint_{\Gamma_{h1} \cup \Gamma_{h2}} (\mathbf{n} \wedge \mathbf{H}).\boldsymbol{\chi}_{\phi} dS + \iint_{\Gamma_b} (\mathbf{n} \wedge \mathbf{H}).\boldsymbol{\chi}_{\phi} dS \quad [3.405]$$

Considering the first integral on the right-hand side, on the boundary $\Gamma_{h1} \cup \Gamma_{h2}$, it appears as the tangential component of **H**. As this tangential component is equal to zero on Γ_{h1} and Γ_{h2}, the integral term is also equal to zero. For the second integral on the right, we again refer to the domain of definition of potential $\boldsymbol{\chi}_{\phi}$ (see equations [3.381] and [3.31]). On Γ_b, $\boldsymbol{\chi}_{\phi}$ is decomposed into two terms $\boldsymbol{\chi}'_{\phi}$ and $\boldsymbol{\chi}_{c\phi}$. For $\boldsymbol{\chi}'_{\phi}$, its tangential component on Γ_b is equal to zero. As for the second term $\boldsymbol{\chi}_{c\phi}$, it is tangential to Γ_b and perpendicular to the cut (see Figure 3.2b). Then, equation [3.405] is written as:

$$\oiint_{\Gamma}(\mathbf{n} \wedge \mathbf{H}).\chi_{\phi}dS = \iint_{\Gamma_b}(\chi_{c\phi} \wedge \mathbf{n}).\mathbf{H}dS \qquad [3.406]$$

Grouping equations [3.403], [3.404] and [3.405] and considering that **curlH** = **J**, after simplification, we have:

$$\begin{aligned} f_m &= \iiint_{\Omega}\chi_{\phi}.\mathbf{curlH}\, d\tau - \oiint_{\Gamma}(\mathbf{n} \wedge \mathbf{H}).\chi_{\phi}dS \\ &- \iiint_{\Omega}\chi_{\phi}.\mathbf{J}d\tau = -\iint_{\Gamma_b}(\chi_{c\phi} \wedge \mathbf{n}).\mathbf{H}dS \end{aligned} \qquad [3.407]$$

The vector product $\chi_{c\phi}\wedge\mathbf{n}$ represents a field collinear with the cut. Under these conditions, the last integral term corresponds to the circulation of field **H** along the cut between gates Γ_{h1} and Γ_{h2}. This validates equation [3.403] as the definition of the magnetomotive force f_m.

Let us consider equation [3.403] again and replace, on the one hand, the field **H**, via the behavior law [1.26], with the magnetic flux density **B** expressed by equation [3.383] and, on the other hand, the current density **J** with its expression [3.389]. Then, we obtain:

$$\begin{aligned} f_m &= \iiint_{\Omega}\mu^{-1}(\mathbf{curlA} + \phi\boldsymbol{\lambda}_{\phi}).\boldsymbol{\lambda}_{\phi}d\tau \\ &+ \iiint_{\Omega}\sigma(\frac{\partial}{\partial t}(\mathbf{A} + \phi\chi_{\phi}) + \mathbf{grad}V).\chi_{\phi}d\tau \end{aligned} \qquad [3.408]$$

For this formulation, the system of equations to be solved has the form of equations [3.393], [3.394] and [3.408], where the unknowns are the potentials **A** and V and the flux ϕ.

3.6.2.2. *Magnetic formulation T-φ*

For the magnetic quantities imposed on the boundary (see Figure 3.24), the **T**-φ formulation will be developed. With this formulation, it is quite natural to introduce the magnetomotive force $f_m(t)$ as a source term. On the contrary, an additional equation should be considered in order to impose the magnetic flux $\phi(t)$.

3.6.2.2.1. Imposed magnetomotive force

For the electric formulation, having $f_m(t)$ as the source term, let us first use the property of the current density **J** defined by equation [3.304]. Moreover, as noted above, the boundary condition of **J** is wall-type, throughout the boundary of the domain Ω_c (see equation [3.376]). With $\mathbf{J} \in H_0(\text{div}0, \Omega_c)$ and since the domain Ω_c is contractible, an electric vector potential **T** can be introduced (see equation [2.31]), such that:

$$\mathbf{J} = \mathbf{curl T} \text{ and } \mathbf{T} \wedge \mathbf{n}\big|_{\Gamma_c} = 0 \text{ i.e.: } \mathbf{T} \in H_0(\mathbf{curl}, \Omega_c) \qquad [3.409]$$

It can be noted that $\mathbf{T} \wedge \mathbf{n} = 0$ can be directly imposed on Γ_j even though the latter is not simply connected, because, unlike the case of section 3.6.1.2.1, the flux of the current density flowing through a surface Δ (see Figure 3.24) is zero. The circulation of **T** on the contour surrounding the domain Ω_c is therefore equal to zero this time, unlike the case in which electric quantities are imposed. Indeed, the addition of a support potential is not required in order to take into account the current I, as in the case of section 3.6.1.2. Moreover, the domain of the definition of **J**, and therefore that of **T**, can be extended, on the entire domain Ω taking $\mathbf{J} = 0$ and $\mathbf{T} = 0$ on Ω_0. The conditions [3.409] can then be rewritten in the following form:

$$\mathbf{J} = \mathbf{curl T}, \ \mathbf{T} \wedge \mathbf{n}\big|_{\Gamma_c} = 0, \ \mathbf{T} = 0 \text{ on } \Omega_0 \text{ i.e. } \mathbf{T} \in H_0(\mathbf{curl}, \Omega) \qquad [3.410]$$

Since the source term is the magnetomotive force, the developments used for its introduction into the formulation are similar to those presented in section 3.5.2.1. The magnetic field **H** is then decomposed into two terms:

$$\mathbf{H} = \mathbf{H}_s + \mathbf{H'} \qquad [3.411]$$

Considering the properties of **H** (see equation [3.377]), the field **H'** belongs to the function space $H_{\Gamma h1 \cup \Gamma h2}(\mathbf{curl}, \Omega)$. Concerning the source field $\mathbf{H}_s$, it makes it possible to take into account the constraints imposed to the magnetic field on the boundaries Γ_{h1} and Γ_{h2}. It is therefore defined in the function space $H_{\Gamma h1 \cup \Gamma h2}(\mathbf{curl}0, \Omega)$ and can be represented by a support vector field $\boldsymbol{\beta}_s$ (see equation [3.213]) such that:

$$\mathbf{H}_s = f_m \boldsymbol{\beta}_s \text{ with } \boldsymbol{\beta}_s \in H_{\Gamma_{h1} \cup \Gamma_{h2}}(\mathbf{curl}0, \Omega) \qquad [3.412]$$

The properties of field $\boldsymbol{\beta}_s$ are identical to those defined in section 3.5.2.1 (see equation [3.214]), hence:

$$\mathbf{curl}\boldsymbol{\beta}_s = 0, \ \boldsymbol{\beta}_s \wedge \mathbf{n}\big|_{\Gamma_{h1} \cup \Gamma_{h2}} = 0 \text{ and } \int_{\gamma 12} \boldsymbol{\beta}_s . \mathbf{dl} = 1 \qquad [3.413]$$

where γ_{12} represents an arbitrary path, in the domain Ω, linking the gates Γ_{h1} and Γ_{h2}. A possible path γ_{12} is represented in Figure 3.24. Based on the function space in which the support field $\boldsymbol{\beta}_s$ is defined, a potential α_s (see equation [3.214]) can be introduced, such that:

$$\boldsymbol{\beta}_s = -\mathbf{grad}\alpha_s,\ \alpha_s\big|_{\Gamma_{h1}} = \alpha_{h1},\ \alpha_s\big|_{\Gamma_{h2}} = \alpha_{h2},\ \alpha_{h1} - \alpha_{h2} = 1$$
$$\text{i.e. } \alpha_s \in H(\mathbf{grad}, \Omega) \qquad [3.414]$$

For the choice of constants, a simple solution is to consider $\alpha_{h1} = 1$ on Γ_{e1} and $\alpha_{h2} = 0$ on Γ_{e2}.

Based on equation [3.302], replacing $\mathbf{J}$ by its expression given by equation [3.409] and $\mathbf{H}$ by equation [3.411], and given that $\mathbf{H}_s \in H_{\Gamma h1 \cup \Gamma h2}(\mathbf{curl}0, \Omega)$, we can write:

$$\mathbf{curl}(\mathbf{H}' - \mathbf{T}) = 0 \qquad [3.415]$$

This relation allows for the introduction of the magnetic scalar potential φ such that:

$$\mathbf{H}' - \mathbf{T} = -\mathbf{grad}\varphi \text{ with } \varphi \in H_{\Gamma_{h1} \cup \Gamma_{h2}}(\mathbf{grad}, \Omega) \qquad [3.416]$$

On the gates Γ_{h1} and Γ_{h2}, the tangential component of the field $\mathbf{H}$ is equal to zero and the boundary conditions are supported by the field $\mathbf{H}_s$. This is why the magnetic scalar potential is equal to zero on these two equipotential surfaces.

Gathering equations [3.411], [3.412], [3.414] and [3.416] and rearranging them, we obtain the expression of the magnetic field in the domain Ω_c as follows:

$$\mathbf{H} = \mathbf{T} - f_m \mathbf{grad}\alpha_s - \mathbf{grad}\varphi \qquad [3.417]$$

Then, the magnetic flux density can be written using the behavior law [1.26] as follows:

$$\mathbf{B} = \mu(\mathbf{T} - f_m \mathbf{grad}\alpha_s - \mathbf{grad}\varphi) \qquad [3.418]$$

The magnetic flux density $\mathbf{B}$ can also be written as a function of the support field $\boldsymbol{\beta}_s$ (see equation [3.414]):

$$\mathbf{B} = \mu(\mathbf{T} + f_m \boldsymbol{\beta}_s - \mathbf{grad}\varphi) \qquad [3.419]$$

As for the electric field, using equation [3.409] and the behavior law [1.20], its expression in the conducting domain is:

$$\mathbf{E} = \sigma^{-1}\mathbf{curlT} \qquad [3.420]$$

At this level of development, the electric vector potential **T** and the expression of the magnetic field given by equation [3.417] verify, respectively, the expressions [3.304] and [3.302]. If in equation [3.301] the electric field is replaced by its expression given in equation [3.420] and the magnetic flux density by equation [3.418], then the equation below is obtained.

$$\mathbf{curl}\sigma^{-1}\mathbf{curlT} = -\frac{\partial}{\partial t}(\mu(\mathbf{T} - f_m\mathbf{grad}\alpha_s - \mathbf{grad}\varphi)) \qquad [3.421]$$

This expression can also be written using the support field $\boldsymbol{\beta}_s$ of equation [3.414] as follows:

$$\mathbf{curl}\sigma^{-1}\mathbf{curlT} = -\frac{\partial}{\partial t}(\mu(\mathbf{T} + f_m\boldsymbol{\beta}_s - \mathbf{grad}\varphi)) \qquad [3.422]$$

To obtain a full equation system, equation [3.303] should be verified. To this end, the magnetic flux density is replaced in this equation by its expression given in equation [3.418] as follows:

$$\text{div}(\mu(\mathbf{T} - f_m\mathbf{grad}\alpha_s - \mathbf{grad}\varphi)) = 0 \qquad [3.423]$$

In this equation, $\boldsymbol{\beta}_s$ [3.414] can also be introduced as follows:

$$\text{div}(\mu(\mathbf{T} + f_m\boldsymbol{\beta}_s - \mathbf{grad}\varphi)) = 0 \qquad [3.424]$$

Equations [3.421] and [3.423] or still [3.422] and [3.424] represent, for the magnetic formulation, the system to be solved in the conducting domain. Completeness requires taking into account the equations of magnetostatics in Ω_0 and verifying the conditions of continuity at the interface between subdomains Ω_c and Ω_0.

In the domain Ω_0, we have to solve equations [3.207] and [3.303] with the function spaces defined in equation [3.377]. Given the absence of current density in Ω_0, the electric vector potential is zero. Under these conditions, gathering equations [3.303] and [3.418] with $\mathbf{T} = 0$, we obtain:

$$\text{div}(\mu(f_m\mathbf{grad}\alpha_s + \mathbf{grad}\varphi)) = 0 \qquad [3.425]$$

As above, the support field $\boldsymbol{\beta}_s$ can also be introduced as follows:

$$\mathrm{div}(\mu(f_m\boldsymbol{\beta}_s - \mathbf{grad}\varphi)) = 0 \qquad [3.426]$$

which corresponds to the magnetic scalar potential formulation in magnetostatics.

Let us now verify the conditions of continuity at the interface between the subdomains Ω_0 and Ω_c. For the current density, its normal component is naturally equal to zero, considering the boundary conditions imposed on Γ_c for the magnetic vector potential $\mathbf{T}$ (see equation [3.409]). The electric field $\mathbf{E}$ is uniquely defined in the conducting domain (see equation [3.376]); therefore, there is no particular constraint on the boundary Γ_c. As for the conservation of the tangential component of the magnetic field, it is ensured via potentials α_s and φ (it should be recalled that $\mathbf{T} \wedge \mathbf{n} = 0$ on Γ_c). Finally, the conservation of the normal component of $\mathbf{B}$ is ensured via equations [3.423] or [3.424] associated with [3.425].

The study of the problem represented in Figure 3.24 involves solving equations [3.422], [3.424] and [3.426]. Nevertheless, as the vector potential $\mathbf{T}$ is defined on the entire domain Ω (see equation [3.409]), it is equal to zero on Ω_0. It is therefore possible to gather equations [3.424] and [3.426]. Under these conditions, the system of equations to be solved has the following form:

$$\mathbf{curl}\sigma^{-1}\mathbf{curl}\mathbf{T} + \frac{\partial}{\partial t}(\mu(\mathbf{T} + f_m\boldsymbol{\beta}_s - \mathbf{grad}\varphi)) = 0 \text{ on } \Omega_c \qquad [3.427]$$

$$\mathrm{div}(\mu(\mathbf{T} + f_m\boldsymbol{\beta}_s - \mathbf{grad}\varphi)) = 0 \text{ on } \Omega \qquad [3.428]$$

3.6.2.2.2. Imposed magnetic flux

When the source term is the magnetic flux $\phi(t)$, the configuration of the problem is the same as that given in Figure 3.24. The unknowns are then $\mathbf{T}$ in the domain Ω_c, φ throughout the domain Ω and the magnetomotive force f_s. Similar to section 3.6.2.1.2, an additional equation is needed in this case to express the flux ϕ as a function of potentials $\mathbf{T}$-φ and of the magnetomotive force. To this end, a similar approach to that in section 3.6.2.1.2 is used, and a power balance is written. Expression [3.395] is recalled below, taking into account that the current density is zero in Ω_0, as follows:

$$p = f_m\frac{d\phi}{dt} = \iiint_{\Omega_c} \mathbf{E}.\mathbf{J}d\tau + \iiint_{\Omega} \frac{\partial \mathbf{B}}{\partial t}.\mathbf{H}d\tau \qquad [3.429]$$

Expressing the current density as a function of the vector potential **T** (see equation [3.409]) and replacing the field **H** by its expression [3.417], we obtain:

$$f_m \frac{d\phi}{dt} = \iiint_{\Omega_c} \mathbf{E}.\mathbf{curlT}d\tau + \iiint_{\Omega} \frac{\partial \mathbf{B}}{\partial t}.(\mathbf{T} - f_m \mathbf{grad}\alpha_s - \mathbf{grad}(\varphi)d\tau \qquad [3.430]$$

Applying formula [2.27] to the first term on the right-hand side, we have:

$$\iiint_{\Omega_c} \mathbf{E}.\mathbf{curlT}d\tau = \iiint_{\Omega_c} \mathbf{curlE}.\mathbf{T}d\tau - \oiint_{\Gamma_c} (\mathbf{n} \wedge \mathbf{E}).\mathbf{T}dS \qquad [3.431]$$

Taking into account the properties of the vector potential **T** on the boundary Γ_c (see equation [3.409]), the surface integral is zero. This can be readily proven using the mixed product. Let us consider now the second integral term, on the right-hand side of equation [3.430], and decompose it using the fact that the vector potential **T** is zero on Ω_0. Then, the following can be written as:

$$\iiint_{\Omega} \frac{\partial \mathbf{B}}{\partial t}.(\mathbf{T} - f_m \mathbf{grad}\alpha_s - \mathbf{grad}(\varphi)d\tau = \iiint_{\Omega_c} \frac{\partial \mathbf{B}}{\partial t}.\mathbf{T}d\tau - \iiint_{\Omega} \frac{\partial \mathbf{B}}{\partial t}.(f_m \mathbf{grad}\alpha_s + \mathbf{grad}(\varphi)d\tau \qquad [3.432]$$

Grouping equations [3.430], [3.431] and [3.432], we obtain:

$$f_m \frac{d\phi}{dt} = \iiint_{\Omega_c} \mathbf{curlE}.\mathbf{T}d\tau + \iiint_{\Omega_c} \frac{\partial \mathbf{B}}{\partial t}.\mathbf{T}d\tau - \iiint_{\Omega} \frac{\partial \mathbf{B}}{\partial t}.(f_m \mathbf{grad}\alpha_s + \mathbf{grad}(\varphi)d\tau \qquad [3.433]$$

This equation is simplified if **curlE** is replaced by its expression given by equation [3.301]. Then, we obtain:

$$f_m \frac{d\phi}{dt} = -\iiint_{\Omega} \frac{\partial \mathbf{B}}{\partial t}.(f_m \mathbf{grad}\alpha_s + \mathbf{grad}(\varphi)d\tau \qquad [3.434]$$

Let us now consider the last term of the integral on the right-hand side of equation [3.434], the formulas of vector operators (see equation [2.23]) allowing us to write:

$$\iiint_{\Omega} \frac{\partial \mathbf{B}}{\partial t}.\mathbf{grad}\,\varphi d\tau = -\iiint_{\Omega} \varphi \mathrm{div}(\frac{\partial \mathbf{B}}{\partial t})d\tau + \oiint_{\Gamma} \varphi \frac{\partial \mathbf{B}}{\partial t}.\mathbf{n}dS \qquad [3.435]$$

This equation is equal to zero. Indeed, as the magnetic flux density is divergence free, the first term on the right-hand side is zero. As for the second term, namely the surface integral, it is also zero, as the boundary Γ consists of the boundaries Γ_b, Γ_{h1} and Γ_{h2}. But on Γ_b, we have $\mathbf{B.n} = 0$ and on Γ_{h1} and Γ_{h2}, we have $\varphi = 0$ (see equation [3.416]). Under these conditions, equation [3.434] is written as:

$$f_m \frac{d\phi}{dt} = -f_m \iiint_\Omega \frac{\partial \mathbf{B}}{\partial t} . \mathbf{grad}\alpha_s d\tau \qquad [3.436]$$

As the studied domain Ω is not subject to deformation over time and the term α_s is time-independent, the above equation can be rewritten as follows:

$$f_m \frac{d\phi}{dt} = -f_m \frac{d}{dt} \iiint_\Omega \mathbf{B}.\mathbf{grad}\alpha_s d\tau \qquad [3.437]$$

By identification, the magnetic flux ϕ can be expressed using the potential α_s or the source field $\boldsymbol{\beta}_s$ (see equation [3.414]) up to a constant. This constant is considered to be equal to zero, as $\phi = 0$ when $\mathbf{B} = 0$ on the entire domain. Hence, we have:

$$\phi = -\iiint_\Omega \mathbf{B}.\mathbf{grad}\alpha_s d\tau = \iiint_\Omega \mathbf{B}.\boldsymbol{\beta}_s d\tau \qquad [3.438]$$

This expression can also be written by replacing the magnetic flux density $\mathbf{B}$ with its expression given in equation [3.419]:

$$\phi = -\iiint_\Omega \mu(\mathbf{T} + f_m \boldsymbol{\beta}_s - \mathbf{grad}\varphi).\mathbf{grad}\alpha_s d\tau \qquad [3.439]$$

In conclusion, for a magnetodynamics problem, with the ($\mathbf{T}$-φ) magnetic formulation and when the source term is the magnetic flux, we need to solve equations [3.427] and [3.428], to which expression [3.439] must be added.

3.6.3. *Summarizing tables*

This section offers a synthetic presentation of the main results when the source terms are global quantities imposed on the boundary of the domain.

For electric quantities (see the studied geometry in Figure 3.23), namely the electromotive force e(t) or the current density flux I(t), Table 3.7 summarizes the properties of the source terms and the equations to be solved for the electric formulation. Again for the imposed electric quantities, Table 3.8 summarizes the properties of source terms and the equations to be solved for the magnetic formulation.

Let us now consider the case when magnetic quantities ($\phi(t)$ and $f_m(t)$) are imposed on the boundary of the domain. The studied geometry is the one in Figure 3.24. Table 3.9 summarizes the properties of the source fields and the equations to be solved with the electric formulation. Table 3.10 presents the synthesis for the magnetic formulation.

Global electric quantities imposed on the boundary		
(A-V) electric formulation	Source term: electromotive force e(t)	
	Source field support two possibilities: $\boldsymbol{\beta}_e$ or $\boldsymbol{\alpha}_e$	Decomposition of the electric field: $\mathbf{E} = \mathbf{Es} + \mathbf{E'}$
		$\mathbf{E}_s = e\boldsymbol{\beta}_e,\ \boldsymbol{\beta}_e \in H_{\Gamma_{e1}\cup\Gamma_{e2}}(\mathbf{curl}0, \Omega_c)$
		$\mathbf{E}_s = -e\,\mathbf{grad}\alpha_e,\ \alpha_e \in H(\mathbf{grad}, \Omega_c)$
	Properties of the unknowns **B** and **E'**, introduction of potentials **A**, V	$\mathrm{div}\mathbf{B} = 0, \mathbf{B} \in H_0(\mathrm{div}0, \Omega)$, $\mathbf{B} = \mathbf{curl}\mathbf{A}, \mathbf{A} \in H_0(\mathbf{curl}, \Omega)$
		$\mathbf{curl}\mathbf{E}' = -\frac{\partial \mathbf{B}}{\partial t},\ \mathbf{E}' \in H_{\Gamma_{e1}\cup\Gamma_{e2}}(\mathbf{curl}, \Omega_c)$, $\mathbf{E}' = -(\frac{\partial A}{\partial t} + \mathbf{grad}V),\ V \in H_{\Gamma_{e1}\cup\Gamma_{e2}}(\mathbf{grad}, \Omega_c)$
	Equation to be solved: two forms are possible $\boldsymbol{\beta}_e$ and $\boldsymbol{\alpha}_e$ or for the function $\boldsymbol{\beta}_e$	$\mathbf{curl}\mu^{-1}(\mathbf{curl}\mathbf{A}) + \sigma(\frac{\partial \mathbf{A}}{\partial t} + \mathbf{grad}V) - \sigma e\boldsymbol{\beta}_e = 0 \text{ on } \Omega$
		$\mathrm{div}(\sigma(\frac{\partial \mathbf{A}}{\partial t} + \mathbf{grad}V - e\boldsymbol{\beta}_e)) = 0 \text{ on } \Omega_c$
	Source term: current intensity I(t)	
	e becomes an unknown; an additional equation is needed	
	$I = \iiint_{\Omega_c} \boldsymbol{\beta}_e . \mathbf{J} d\tau$	

Table 3.7. *Summary of the equations to be solved in magnetodynamics with the electric formulation for electric global quantities imposed on the boundary (see Figure 3.23)*

Global electric quantities imposed on the boundary		
(T-φ) magnetic formulation	Source term: current intensity I(t)	
	Source fields, two possibilities: $\boldsymbol{\lambda}_I$ or $\boldsymbol{\chi}_I$	Decomposition of the current density: $\mathbf{J} = \mathbf{J}_s + \mathbf{J}'$
		$\mathbf{J}_S = I\boldsymbol{\lambda}_I,\ \boldsymbol{\lambda}_I \in H(\text{div}0, \Omega)$
		$\mathbf{J}_s = I\,\mathbf{curl}\boldsymbol{\chi}_I,\ \boldsymbol{\chi}_I \in H(\mathbf{curl}, \Omega)$
	Properties of the unknowns $\mathbf{J}'$ (extended in Ω_0) and $\mathbf{H}$, introduction of potentials $\mathbf{T}$, φ	$\text{div}\mathbf{J}' = 0,\ \mathbf{J}' \in H(\text{div}0, \Omega),$ $\mathbf{J}' = \mathbf{curl}\mathbf{T},\ \mathbf{T} \in H_{\Gamma_j}(\mathbf{curl}, \Omega_c)$
		$\mathbf{curl}\mathbf{H} = \mathbf{J},\ \mathbf{H} \in H(\mathbf{curl}, \Omega),$ $\mathbf{H} = \mathbf{T} + I\boldsymbol{\chi}_I - \mathbf{grad}\varphi,\ \varphi \in H(\mathbf{grad}, \Omega)$
	Equation to be solved: two forms are possible with $\boldsymbol{\lambda}_I$ and $\boldsymbol{\chi}_I$ or for the function $\boldsymbol{\chi}_I$	$\mathbf{curl}(\sigma^{-1}(I\,\mathbf{curl}\boldsymbol{\chi}_I + \mathbf{curl}\mathbf{T})) + \frac{\partial}{\partial t}(\mu_c(I\boldsymbol{\chi}_I + \mathbf{T} - \mathbf{grad}\varphi)) = 0 \text{ on } \Omega_c$
		$\text{div}(\mu(I\boldsymbol{\chi}_I + \mathbf{T} - \mathbf{grad}\varphi)) = 0 \text{ on } \Omega$
	Source term: electromotive force e(t)	
	I becomes an unknown; an additional equation is needed	
	$e = \iiint_{\Omega_c} \mathbf{E}.\boldsymbol{\lambda}_I d\tau + \iiint_{\Omega_c} \frac{\partial \mathbf{B}}{\partial t} \cdot \boldsymbol{\chi}_I d\tau$	

Table 3.8. *Summary of equations to be solved in magnetodynamics with the magnetic formulation for global electric quantities imposed on the boundary (see Figure 3.23)*

<table>
<tr><th colspan="3">Global magnetic quantities imposed on the boundary</th></tr>
<tr><td rowspan="10">(A-V) electric formulation</td><td colspan="2">Source term: magnetic flux $\phi(t)$</td></tr>
<tr><td rowspan="3">Source fields, two possibilities: $\boldsymbol{\lambda}_\phi$ or $\boldsymbol{\chi}_\phi$ with a cut on Γ_b not simply connected</td><td>Decomposition of the magnetic flux density: $\mathbf{B} = \mathbf{B}_s + \mathbf{B}'$</td></tr>
<tr><td>$\mathbf{B}_S = \phi\lambda_\phi,\ \lambda_\phi \in H_{\Gamma_b}(\mathrm{div}0, \Omega)$</td></tr>
<tr><td>$\mathbf{B}_s = \phi\mathbf{curl}\chi_\phi,\ \chi_\phi \in H^{\Delta}_{\Gamma_b}(\mathbf{curl}, \Omega)$</td></tr>
<tr><td rowspan="2">Properties of the unknowns $\mathbf{B}'$ and $\mathbf{E}$, introduction of potentials $\mathbf{A}$, V</td><td>$\mathrm{div}\mathbf{B}' = 0,\ \mathbf{B}' \in H_{\Gamma_b}(\mathrm{div}0, \Omega),$
$\mathbf{B}' = \mathbf{curl}\mathbf{A},\ \mathbf{A} \in H_{\Gamma_b}(\mathbf{curl}, \Omega)$</td></tr>
<tr><td>$\mathbf{curl}\mathbf{E} = -\frac{\partial}{\partial t}(\mathbf{curl}\mathbf{A} + \phi\mathbf{curl}\chi_\phi), \mathrm{E} \in H(\mathbf{curl}, \Omega_c),$
$\mathbf{E} = -(\frac{\partial \mathbf{A}}{\partial t} + \frac{\partial \phi\chi_\phi}{\partial t} + \mathbf{grad}\mathrm{V}),\ \mathrm{V} \in H(\mathbf{grad}, \Omega_c)$</td></tr>
<tr><td rowspan="2">Equation to be solved: two forms are possible $\boldsymbol{\lambda}_\phi$ and $\boldsymbol{\chi}_\phi$, or for the function $\boldsymbol{\chi}_\phi$</td><td>$\mathbf{curl}\mu^{-1}(\mathbf{curl}(\mathbf{A} + \phi\chi_\phi)$
$+\sigma(\frac{\partial \mathbf{A}}{\partial t} + \mathbf{grad}\mathrm{V} + \frac{\partial \phi\chi_\phi}{\partial t}) = 0 \text{ on } \Omega$</td></tr>
<tr><td>$\mathrm{div}(\sigma(\frac{\partial \mathbf{A}}{\partial t} + \frac{\partial \phi\chi_\phi}{\partial t} + \mathbf{grad}\mathrm{V})) = 0 \text{ on } \Omega_c$</td></tr>
<tr><td colspan="2">Source term: magnetomotive force $f_m(t)$</td></tr>
<tr><td colspan="2">$\phi(t)$ becomes an unknown; an additional equation is needed</td></tr>
<tr><td></td><td colspan="2">$f_m = \iiint_\Omega \lambda_\phi.\mathbf{H}d\tau - \iiint_\Omega \chi_\phi.\mathbf{J}d\tau$</td></tr>
</table>

Table 3.9. *Summary of equations to be solved in magnetodynamics with the electric formulation for global electric quantities imposed on the boundary (see Figure 3.24)*

Global magnetic quantities imposed on the boundary		
(T-φ) magnetic formulation	Source term: magnetomotive force f_m (t)	
		Decomposition of the magnetic field: $\mathbf{H} = \mathbf{H}_s + \mathbf{H}'$
	Source field support two possibilities: $\boldsymbol{\beta}_s$ or α_s	$\mathbf{H}_s = f_m \boldsymbol{\beta}_s, \; \boldsymbol{\beta}_s \in H_{\Gamma_{h1} \cup \Gamma_{h2}}(\mathbf{curl}0, \Omega)$
		$\mathbf{H}_s = -f_m \mathbf{grad}\alpha_s, \; \alpha_s \in H(\mathbf{grad}, \Omega)$
	Properties of the unknowns **J** and **H'**, introduction of potentials **T**, φ	$\text{div}\mathbf{J} = 0 \;\; \mathbf{J} \in H_0(\text{div}0, \Omega_c)$, $\mathbf{J} = \mathbf{curlT}, \; \mathbf{T} \in H_0(\mathbf{curl}, \Omega)$
		$\mathbf{H}' = \mathbf{T} - \mathbf{grad}\varphi, \qquad \varphi \in H(\mathbf{grad}, \Omega)$
	Equation to be solved: two forms are possible $\boldsymbol{\beta}_s$ and α_s or for the function $\boldsymbol{\beta}_s$	$\mathbf{curl}\sigma^{-1}\mathbf{curlT} + \frac{\partial}{\partial t}(\mu(\mathbf{T} + f_m \boldsymbol{\beta}_s - \mathbf{grad}\varphi)) = 0$ on Ω_c
		$\text{div}(\mu(\mathbf{T} + f_m \boldsymbol{\beta}_s - \mathbf{grad}\varphi)) = 0 \;\; \text{sur} \;\; \Omega$
	Source term: magnetic flux $\phi(t)$	
	$f_m(t)$ becomes an unknown; an additional equation is needed	
	$\phi = \iiint_{\Omega_c} \mathbf{B}.\boldsymbol{\beta}_s d\tau$	

Table 3.10. *Summary of the equations to be solved in magnetodynamics with the magnetic formulation for the global electric quantities imposed on the boundary (see Figure 3.24)*

3.6.4. *Tonti diagram*

The structure of the Tonti diagram is equivalent to that proposed for static problems (electrostatics, electrokinetics and magnetostatics). However, the notion of time must be introduced, which leads to a split in the sequences of function spaces, as can be noted in Figure 3.25. A three-dimensional structure is then obtained (Bossavit 1997). The front plane supports the diagram of magnetostatics, while the back plane supports the case of electrostatics. For electrokinetics, the work involves the diagonal part of the diagram. It should be noted that the front and back planes are linked by a time derivative. The function spaces with the boundary conditions can then be positioned on the set of physical quantities **E**, **D**, **J**, **H** and **B**. Then, the set of source fields and vector and scalar potentials are placed.

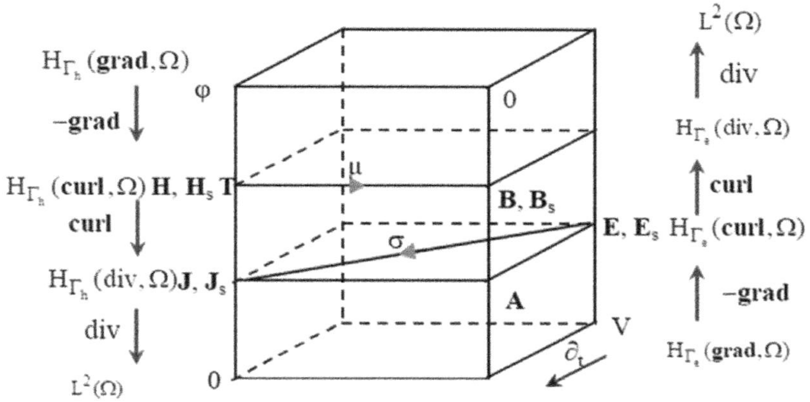

Figure 3.25. *Tonti diagram for magnetodynamics*

4

Formulations in the Discrete Domain

4.1. Introduction

The analytical solution to the formulations proposed in Chapter 3 is not accessible unless the geometry is extremely simple, but very often these are academic cases. Since the solution in the continuous domain, which is referred to as the "exact solution" cannot be obtained, numerical methods can be used. The application of these numerical methods leads to a discretized model. The resulting discretized model leads to an approximation to the exact solution, known as the "discrete solution". Among the many existing numerical methods, this chapter focuses on the method that is currently most widely used for solving low-frequency electromagnetism problems, namely the finite element method.

As a first step, this chapter applies the weighted residual method to the equations presented in Chapter 3. This method leads to weak formulations, used for the discretization of Maxwell's equations. Discrete function spaces are then introduced, to which the discretized electromagnetic fields belong. These spaces are generated by basis functions, namely Whitney elements. Space discretization using the finite element method will then be addressed.

Finally, for the sake of lighter notations, we decided to omit space dependency in the continuous domain. However, in this chapter, in order to avoid any confusion in the notations, the quantities in the discrete domain will be marked by an index "d", and space dependency will be indicated. For example, the discretized form of the electric field $\mathbf{E}$ will be denoted by $\mathbf{E}_d(x)$.

It should be recalled that this book focuses on the theoretical foundations leading to the construction of a system of equations. For the computer implementation, specific books can be referred to (Russenschuck 2010; Dhatt et al. 2012; Bastos and Sadowski 2014; Cardoso 2016). Electromagnetic quantities, such as the force or the torque, besides the Maxwell tensor, can often be determined using the virtual work method (Coulomb 1983). The same is valid for the use of software in electrical engineering (Hameyer et al. 1999).

4.2. Weighted residual method: weak form of Maxwell's equations

4.2.1. *Methodology*

Depending on the type of problem posed, either in electrostatics, electrokinetics, magnetostatics or magnetodynamics and according to the chosen potential formulation, the systems of partial differential equations to be solved, referred to as "strong formulations", were largely developed in Chapter 3. Nevertheless, as indicated in section 4.1, for many applications, it is impossible to solve these equations directly and obtain the exact solution. Moreover, based on the strong formulation, an approximation to the numerical solution cannot be readily obtained.

A more global approach, referred to as the "weighted residual method", can be considered. It allows us to pose the problem in an integral form, and not locally, as for strong formulations. The resulting formulations are referred to as the "weak formulation". This approach is expected to facilitate the work in a finite dimension space of approximation, which makes it easier to use numerical methods. The objective is to find in this space a solution approximating the exact solution that minimizes a residue, which leads to the solution of a system of equations. This section explains the general approach to building weak formulations.

Consider an operator "$\mathcal{L}$", applied to a vector function $\mathbf{U}$, with, for example, $\mathbf{U} \in \mathrm{H}(\mathcal{L},\Omega)$ and a source term $\mathbf{f}_s$ and associated boundary conditions. The strong formulation is written as:

$$\mathcal{L}(\mathbf{U}) - \mathbf{f}_S = 0 \qquad [4.1]$$

This reflects a general form of potential formulations already noted in Chapter 3, for example those of magnetostatics, with equation [3.220] or [3.281]. For these formulations, operator $\mathcal{L}$ is the divergence (for the first one) and the curl (for the second one).

Consider now the integral on the entire domain Ω of equation [4.1], weighted by a function $\mathbf{\Psi}$, known as the "weighting function". Then, it can be written as:

$$\iiint_{\Omega} (\mathcal{L}(\mathbf{U}) - \mathbf{f}_S).\mathbf{\Psi})d\tau = 0 \quad [4.2]$$

As will be seen in sections 4.2.2–4.2.5, $\mathbf{\Psi}$ will be chosen in the adjoint function space of $H(\mathcal{L},\Omega)$.

The solution $\mathbf{U}$ to equation [4.1], known as the "exact solution", implies, in fact, that the integral form [4.2] is zero. On the contrary, there is no reciprocity: if the integral form is zero for any function $\mathbf{\Psi}$, equation [4.1] is not automatically verified at any point of the space. This is referred to as a "weak solution".

As an illustration of the notion of solution in the weak sense, a specific case is presented. Consider a function $\mathbf{U}$, defined inside a contractible domain Ω of boundary Γ. The properties of function $\mathbf{U}$ are:

$$\text{div}\mathbf{U} = 0, \ \ \mathbf{U}.\mathbf{n}|_{\Gamma} = 0 \ \text{ i.e. } \ \mathbf{U} \in H_0(\text{div}0, \Omega) \quad [4.3]$$

The condition [4.3] can be expressed in an integral form using the weighted residual method. The choice of the weighting functions is generally made, as mentioned above, in the adjoint function space of the differential operator of the equation to be solved. In this case, the function space of the differential operator of equation [4.3] is $H_0(\text{div}0, \Omega)$. Consequently, as shown by equation [2.62], the weighting functions are $\psi \in H(\mathbf{grad}, \Omega)$. The weighted residual method leads to the integral form, which must be verified for each weighting function of $H(\mathbf{grad}, \Omega)$, such that:

$$\iiint_{\Omega} \psi \text{div}\mathbf{U} d\tau = 0 \quad [4.4]$$

Based on equation [4.4], formula [2.23] related to vector operators is used. Considering the boundary conditions on the boundary Γ of function $\mathbf{U}$, the weak form can be written as follows:

$$\iiint_{\Omega} \mathbf{U}.\mathbf{grad}\psi d\tau = 0 \quad [4.5]$$

In what follows, $\mathbf{U}'$ is the notation for a solution field to equation [4.5], for all the functions ψ of $H(\mathbf{grad}, \Omega)$ and we will see to what extent it verifies condition [4.3].

Since **U**' verifies [4.5], the reverse operation is performed and, using equation [2.23], the following can be written:

$$\iiint_{\Omega} \mathbf{U}'.\mathbf{grad}\psi d\tau = -\iiint_{\Omega} \psi \mathrm{div}\mathbf{U}' d\tau + \iint_{\Gamma} \psi \mathbf{U}'.\mathbf{n} dS = 0 \qquad [4.6]$$

Consider now, as shown in Figure 4.1, a subdomain Ω' of Ω and a scalar weighting function ψ' ∈ H(**grad**, Ω), such that ψ' is non-zero in Ω' ⊂ Ω and is zero in Ω – Ω'. On the small size domain Ω', let us pose the function ψ' is equal to 1 except for the vicinity of the boundary Γ', where it varies continuously from 1 to zero in a transition zone *Z* of thickness "z". It can be verified that the function ψ' is continuous. It belongs to H(**grad**, Ω). Since ψ' is zero on the boundary Ω, we have:

$$\iiint_{\Omega} \psi' \mathrm{div}\mathbf{U}' d\tau = \iiint_{\Omega'-\mathcal{Z}} \mathrm{div}\mathbf{U}' d\tau + \iiint_{\mathcal{Z}} \psi' \mathrm{div}\mathbf{U}'.d\tau = 0 \qquad [4.7]$$

If *Z* tends to zero, the integral on the transition zone tends to zero and we have:

$$\iiint_{\Omega'-\mathcal{Z}} \mathrm{div}\mathbf{U}' d\tau = 0 \qquad [4.8]$$

The left-hand side term of equation [4.8] yields, up to a factor (which is the inverse of the volume of Ω'), the average of div**U**' on Ω'. Since Ω' can be taken as small as needed, and the average of div**U**' is zero on Ω', then div**U**' is locally zero. Consider now a family of subdomains Ω_k' covering the entire domain Ω with which the functions ψ_k' are associated. The set of solutions **U**' locally verifies relation [4.8] on all the subdomains Ω_k' (the subdomains Ω_k' can be chosen as small as needed), but not at any point of the domain. The solution is then considered weakly verified on average on the domain.

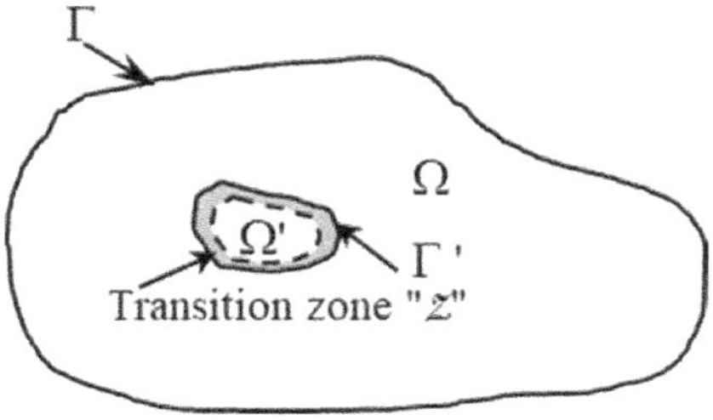

Figure 4.1. *Example of subdomain Ω'⊂ Ω with a transition zone*

Although leading to a weak solution, the weighted residual method has the following essential advantages:

– under specific conditions (that are met in the case of potential formulations), it makes it possible to alleviate the constraints on the function **U** by transferring them to the weighting function;

– as will be noted in sections 4.2.2–4.2.5, it makes it possible to naturally introduce the boundary conditions;

– it highly facilitates the construction of a numerical model leading to an approximation to the solution to equation [4.1].

For our applications, the scalar or vector weighting functions, denoted by ψ or $\boldsymbol{\Psi}$, respectively, will be chosen in the adjoint spaces of the operators **grad**, **curl** and div that were introduced in section 2.6.1. The integral formulation [4.2] will not be used as such in practice, but under a modified form, following an integration by parts that features an adjoint operator. This will lead to the weak formulation that imposes fewer differentiability constraints on the solution.

Sections 4.2.2–4.2.5 present the application of the weighted residual method to various strong potential formulations presented in Chapter 3. This will be referred to as the "weak form of the formulation" or the weak formulation. For the sake of lighter notations, the fact that the solution resulting from the weak formulation is an approximation to the solution will not be indicated.

4.2.2. *Weak form of the equations of electrostatics*

For the development of the weighted residual method in electrostatics, let us consider again the example presented in Figure 3.11. For this example, we are in the presence of a domain Ω composed of two electrodes in contact with the external environment, denoted by $\mathcal{E}_1$ and $\mathcal{E}_2$ and an internal electrode $\mathcal{E}_3$. It should be recalled that for the studied structure, the studied domain Ω' is defined by $\Omega' = \Omega - \Omega_{\mathcal{E}3}$. The source terms can be circulations "f_{ij}" of the electric field (see equation [3.108]) or the total charges Q_k on the electrodes (see equation [3.110]). Section 4.2.2.1 will focus on the scalar potential formulation, and section 4.2.2.2 will focus on the vector potential formulation.

It is important to recall that the equations to be solved are given in equations [3.104] and [3.105] with the dielectric behavior law [1.19]. The boundary conditions on fields **E** and **D** are defined in equation [3.106], and the function spaces to which they belong are given by relation [3.107].

4.2.2.1. *Scalar potential V formulation*

For the scalar potential formulation, in the presence of two source terms f_{13} and f_{23} (see section 3.3.2.1.1), the electric field can be written in the form [3.116]:

$$\mathbf{E} = f_{13}\boldsymbol{\beta}_{13} + f_{23}\boldsymbol{\beta}_{23} - \mathbf{grad}\mathrm{V}, \quad \mathrm{V} \in \mathrm{H}_{\Gamma_{e1} \cup \Gamma_{e2} \cup \Gamma_{e3}}(\mathbf{grad}, \Omega') \qquad [4.9]$$

with $\boldsymbol{\beta}_{ij}$ defined by relation [3.113]. The electric displacement field is given by equation [3.117] as follows:

$$\mathbf{D} = \varepsilon(f_{13}\boldsymbol{\beta}_{13} + f_{23}\boldsymbol{\beta}_{23} - \mathbf{grad}\mathrm{V}) \qquad [4.10]$$

To obtain the scalar potential formulation, in the presence of the two source terms, the electric displacement field is introduced as defined by equation [4.10] based on equation [3.105]. This yields:

$$\mathrm{div}(\varepsilon(f_{13}\boldsymbol{\beta}_{13} + f_{23}\boldsymbol{\beta}_{23} - \mathbf{grad}\mathrm{V})) = 0 \qquad [4.11]$$

To apply the weighted residual method, the function space of the weighting functions must be chosen. As indicated in section 4.2.1, it is chosen in the adjoint space of the differential operator. In equation [4.11], the divergence operator is applied to the expression of the electric displacement field with, as indicated by relation [3.107], $\mathbf{D} \in \mathrm{H}_{\Gamma d}(\mathrm{div}0, \Omega')$. According to Table 2.1, it can be shown that the adjoint operator is the gradient and the associated space $\mathrm{H}_{\Gamma e1 \cup \Gamma e2 \cup \Gamma e3}(\mathbf{grad}, \Omega')$. The weighting functions are therefore scalar such that $\psi \in \mathrm{H}_{\Gamma e1 \cup \Gamma e2 \cup \Gamma e3}(\mathbf{grad}, \Omega')$. These various points are summarized in Table 4.1.

Vector operator	Associated function space	Adjoint operator	Weighting function
div	$\mathrm{H}_{\Gamma d}(\mathrm{div}0, \Omega')$	$-\mathbf{grad}$	$\psi \in \mathrm{H}_{\Gamma e1 \cup \Gamma e2 \cup \Gamma e3}(\mathbf{grad}, \Omega')$

Table 4.1. *Electrostatics, scalar potential V formulation; vector operator and function space of potential and weighting functions*

Applying the weighted residual method to equation [4.11] yields:

$$\iiint_{\Omega'} \mathrm{div}(\varepsilon(f_{13}\boldsymbol{\beta}_{13} + f_{23}\boldsymbol{\beta}_{23} - \mathbf{grad}\mathrm{V}))\psi d\tau = 0 \qquad [4.12]$$

Having defined the gradient of the weighting function ψ, it is then possible to use formula [2.23] related to vector operators. Equation [4.12] can then be rewritten as follows:

$$-\iiint_{\Omega'} (\varepsilon(\mathbf{grad}V - f_{13}\boldsymbol{\beta}_{13} - f_{23}\boldsymbol{\beta}_{23})).\mathbf{grad}\psi d\tau + \oint\!\!\oint_{\Gamma} \psi\varepsilon(f_{13}\boldsymbol{\beta}_{13} + f_{23}\boldsymbol{\beta}_{23} - \mathbf{grad}V).\mathbf{n}dS = 0 \quad [4.13]$$

It can be noted that the transition from equation [4.12] to equation [4.13] alleviates the differentiability conditions on the term "div($\varepsilon(f_{13}\boldsymbol{\beta}_{13}+f_{23}\boldsymbol{\beta}_{23}-\mathbf{grad}V)$)". The unknown V should be differentiated twice in [4.12], whereas it should be done only once in [4.13] "by transferring" the differentiation order on the weighting function ψ.

As indicated in section 4.2.1, it can be readily verified that any solution to equation [4.11] is a solution to equation [4.13]. On the other hand, as mentioned previously in section 4.2.1, a solution to equation [4.13] does not automatically verify equation [4.11]. It can then be said that the solution to equation [4.13] is a weak solution.

The integral on the boundary Γ in equation [4.13] can be decomposed by introducing terms related to boundary conditions of a different nature:

$$\oint\!\!\oint_{\Gamma} \psi\varepsilon\,(f_{13}\boldsymbol{\beta}_{13} + f_{23}\boldsymbol{\beta}_{23} - \mathbf{grad}V).\mathbf{n}dS = \iint_{\Gamma_{e1}\cup\Gamma_{e2}\cup\Gamma_{e3}} \psi\,\varepsilon(f_{13}\boldsymbol{\beta}_{13} + f_{23}\boldsymbol{\beta}_{23} - \mathbf{grad}V).\mathbf{n}dS + \iint_{\Gamma_d} \psi\,\varepsilon(f_{13}\boldsymbol{\beta}_{13} + f_{23}\boldsymbol{\beta}_{23} - \mathbf{grad}V).\mathbf{n}dS \quad [4.14]$$

To analyze the contributions of each term, in relation to the boundary conditions on the boundaries $\Gamma_{e1}\cup\Gamma_{e2}\cup\Gamma_{e3}$ and Γ_d, we consider the relations given in equation [3.106] and the domain of definition of the weighting function ψ. As for the contribution of the terms related to the boundaries $\Gamma_{e1}\cup\Gamma_{e2}\cup\Gamma_{e3}$, it is zero, since $\psi \in H_{\Gamma e1\cup\Gamma e2\cup\Gamma e3}(\mathbf{grad}, \Omega)$. For the boundary Γ_d, we have $\mathbf{D}.\mathbf{n}\,|_{\Gamma d} = 0$ (see equation [4.10]), which in our case is $\varepsilon(f_{13}\boldsymbol{\beta}_{13}+f_{23}\boldsymbol{\beta}_{23}-\mathbf{grad}V).\mathbf{n} = 0$. Under these conditions, the surface term related to this boundary is set to zero.

The weak form of the scalar potential formulation is written as:

$$\iiint_{\Omega'} \varepsilon\mathbf{grad}V.\mathbf{grad}\psi d\tau = \iiint_{\Omega'} \varepsilon(f_{13}\boldsymbol{\beta}_{13} + f_{23}\boldsymbol{\beta}_{23}).\mathbf{grad}\psi d\tau \quad [4.15]$$

Let us analyze this expression in terms of the equations to be solved. The choice of the electric scalar potential V, which is the unknown of the problem, and that of the fields $\boldsymbol{\beta}_{ij}$, supporting the source terms with their properties, implies that equation [3.104] is automatically strongly verified. The same is true for the boundary conditions of the electric field on the boundaries $\Gamma_{e1} \cup \Gamma_{e2} \cup \Gamma_{e3}$ via the function spaces to which V and the source fields defined by $\boldsymbol{\beta}_{ij}$ belong. On the other hand, it should be noted that equation [3.60] is weakly verified via the weighted residual method. For the electric displacement field, the boundary condition on Γ_d is also weakly verified. Indeed, it is imposed by the surface integral form of equation [4.14] considered zero in equation [4.15].

Considering now as a source term the total charges Q_k on the electrodes, the circulations f_{ij} become the unknowns of the problem. To obtain a full equation system, relations [3.125] and [3.126], which are recalled below, are added to equation [4.15]:

$$Q_1 = \iiint_{\Omega'} \varepsilon \boldsymbol{\beta}_{13} \cdot (f_{13} \boldsymbol{\beta}_{13} + f_{23} \boldsymbol{\beta}_{23} - \mathbf{grad} V) d\tau \qquad [4.16]$$

$$Q_2 = \iiint_{\Omega'} \varepsilon \boldsymbol{\beta}_{23} \cdot (f_{13} \boldsymbol{\beta}_{13} + f_{23} \boldsymbol{\beta}_{23} - \mathbf{grad} V) d\tau \qquad [4.17]$$

This leads to a system of equations to be solved, in which the unknowns are the scalar potential V and the circulations f_{ij}.

NOTE.– The functions $\boldsymbol{\beta}_{13}$ and $\boldsymbol{\beta}_{23}$ play the role of weighting functions in expressions [4.16] and [4.17], respectively, in the same way as the field $\mathbf{grad}\psi$ in the weak formulation [4.15]. These functions are in both cases curl free. On the other hand, it can be noted that the circulations of $\boldsymbol{\beta}_{13}$ and $\boldsymbol{\beta}_{23}$ differ from zero between two electrodes, which is not the case for the function $\mathbf{grad}\psi$. This shows that adding the two equations [4.16] and [4.17] enriches the space of weighting functions, which becomes $H_{\Gamma e1 \cup \Gamma e2 \cup \Gamma e3}(\mathbf{grad}, \Omega') \cup \{ \boldsymbol{\beta}_{13}, \boldsymbol{\beta}_{23} \}$. In general, and as will be seen in what follows for other cases (electrokinetics, magnetostatics, magnetodynamics), taking into account source terms that do not appear naturally in the potential formulations leads to an enrichment of the space of the weighting functions.

Hybrid source terms can also be considered, namely the total charges Q_k on an electrode and a circulation f_{ij}. Assume that the imposed source terms are the circulation f_{13} and the total charges Q_2. Based on equation [4.11], it can be noted that the unknowns are the scalar potential V and the circulation f_{23}. The system of equations to be solved is then composed of the integral form [4.15] and equation [4.17]. Conversely, if the source terms are Q_1 and f_{23}, to have a full equation system, equation [4.15] is added to equation [4.16].

4.2.2.2. *Vector potential P formulation*

For the vector potential formulation, when the source terms are the total charges Q_1 and Q_2, the superposition theorem is applied to express the electric displacement field. Based on equations [3.138] and [3.139], the electric displacement field **D** is then written as:

$$\mathbf{D} = Q_1 \boldsymbol{\lambda}_{13} + Q_2 \boldsymbol{\lambda}_{23} + \mathbf{curl P} \quad \text{with} \quad \mathbf{P} \in H_{\Gamma_d}(\mathbf{curl}, \Omega') \qquad [4.18]$$

where the fields $\boldsymbol{\lambda}_{13}$ and $\boldsymbol{\lambda}_{23}$ are defined, respectively, by relations [3.131] and [3.132]. Based on the electric displacement field **D** defined by equation [4.18], the electric field can be expressed using the behavior law [1.19] as follows:

$$\mathbf{E} = \varepsilon^{-1}(\mathbf{curl P} + Q_1 \boldsymbol{\lambda}_{13} + Q_2 \boldsymbol{\lambda}_{23}) \quad \text{with} \quad \mathbf{E} \in H_{\Gamma_{e1} \cup \Gamma_{e2} \cup \Gamma_{e3}}(\mathbf{curl}0, \Omega') \qquad [4.19]$$

Introducing the expression of **E** thus obtained in equation [3.104], the equation to be solved has the following form:

$$\mathbf{curl}(\varepsilon^{-1}(Q_1 \boldsymbol{\lambda}_{13} + Q_2 \boldsymbol{\lambda}_{23} + \mathbf{curl P})) = 0 \qquad [4.20]$$

To apply the weighted residual method, the weighting functions are chosen in the adjoint space of the differential operator. In the case of equation [4.20], the differential operator is the curl, applied to the expression of the electric field, which is $\mathbf{E} \in H_{\Gamma e1 \cup \Gamma e2 \cup \Gamma e3}(\mathbf{curl}, \Omega')$. In this case (see Table 2.1), the adjoint operator is the curl and the chosen vector weighting function is $\boldsymbol{\Psi} \in H_{\Gamma d}(\mathbf{curl}, \Omega')$. These various points are summarized in Table 4.2.

Vector operator	Associated function space	Adjoint operator	Weighting function
curl	$H_{\Gamma e1 \cup \Gamma e2 \cup \Gamma e3}(\mathbf{curl}0, \Omega')$	**curl**	$\boldsymbol{\Psi} \in H_{\Gamma d}(\mathbf{curl}, \Omega')$

Table 4.2. *Electrostatics; vector potential P formulation; vector operator and function space of the potential and weighting functions*

Under these conditions, the application of the weighted residual method to equation [4.20] yields the following expression:

$$\iiint_{\Omega} \mathbf{curl}(\varepsilon^{-1}(Q_1 \boldsymbol{\lambda}_{13} + Q_2 \boldsymbol{\lambda}_{23} + \mathbf{curl P})).\boldsymbol{\Psi} d\tau = 0 \qquad [4.21]$$

In order to lower the level of differentiability of the potential **P**, formula [2.27] related to vector operators is applied. The Ostrogradski theorem can be used to write equation [4.21] by introducing the Γ boundary integral as follows:

$$\iiint_{\Omega'} \varepsilon^{-1}(Q_1\boldsymbol{\lambda}_{13} + Q_2\boldsymbol{\lambda}_{23} + \mathbf{curlP}).\mathbf{curl}\boldsymbol{\Psi}d\tau$$
$$-\oiint_{\Gamma}(\varepsilon^{-1}(Q_1\boldsymbol{\lambda}_{13} + Q_2\boldsymbol{\lambda}_{23} + \mathbf{curlP}) \wedge \mathbf{n}).\boldsymbol{\Psi}dS = 0 \qquad [4.22]$$

Based on this equation, the boundary integral term can be decomposed by introducing two terms related to the boundaries Γ_d and $\Gamma_{e1} \cup \Gamma_{e2} \cup \Gamma_{e3}$ in the following form:

$$\oiint_{\Gamma}(\varepsilon^{-1}(Q_1\boldsymbol{\lambda}_{13} + Q_2\boldsymbol{\lambda}_{23} + \mathbf{curlP}) \wedge \mathbf{n}).\boldsymbol{\Psi}dS =$$
$$\iint_{\Gamma_d}(\varepsilon^{-1}(Q_1\boldsymbol{\lambda}_{13} + Q_2\boldsymbol{\lambda}_{23} + \mathbf{curlP}) \wedge \mathbf{n}).\boldsymbol{\Psi}dS \qquad [4.23]$$
$$+\iint_{\Gamma_{e1} \cup \Gamma_{e2} \cup \Gamma_{e3}}(\varepsilon^{-1}(Q_1\boldsymbol{\lambda}_{13} + Q_2\boldsymbol{\lambda}_{23} + \mathbf{curlP}) \wedge \mathbf{n}).\boldsymbol{\Psi}dS$$

In this expression, as the function $\boldsymbol{\Psi}$ belongs to $H_{\Gamma d}(\mathbf{curl}, \Omega')$, the term related to the boundary Γ_d is zero. Moreover, the contributions on the boundaries $\Gamma_{e1} \cup \Gamma_{e2} \cup \Gamma_{e3}$ are also zero. Indeed, the tangential component of the electric field, therefore the term $\varepsilon^{-1}(\mathbf{curlP} + Q_1\boldsymbol{\lambda}_{13} + Q_2\boldsymbol{\lambda}_{23})$, is equal to zero on these boundaries. Under these conditions, equation [4.22], which represents the weak form of the vector potential formulation, can be expressed as follows:

$$\iiint_{\Omega'} \varepsilon^{-1}(Q_1\boldsymbol{\lambda}_{13} + Q_2\boldsymbol{\lambda}_{23} + \mathbf{curlP}).\mathbf{curl}\boldsymbol{\Psi}d\tau = 0 \qquad [4.24]$$

Let us consider again the initial problem defined by equations [3.104], [3.105] and the boundary conditions [3.106]. The definition of support fields $\boldsymbol{\lambda}_{ij}$, the equations [3.131] and [3.132] and the introduction of the vector potential **P** allow for the strong verification of equation [3.105]. The boundary conditions on Γ_d are also strongly verified via the properties of source fields $\boldsymbol{\lambda}_{ij}$ and the function space to which the vector potential **P** of equation [3.138] belongs. As for equation [3.104], it is weakly verified on the domain Ω', via the weighted residual method. The same is true for the boundary condition of the electric field on the boundary $\Gamma_{e1} \cup \Gamma_{e2} \cup \Gamma_{e3}$.

If the source terms are the circulations f_{13} and f_{23} of the electric field, the application of the weighted residual method is unchanged. On the other hand, as seen in section 3.3.2.2.2, equations [3.146] and [3.147] should be added. The unknowns of the problem are then the total charges Q_1, Q_2 and the vector potential **P**.

As already indicated (see note in section 4.2.2.1), this leads to an enrichment of the weighting function space $H_{\Gamma d}(\mathbf{curl}, \Omega')$ by functions $\boldsymbol{\lambda}_{ij}$.

Hybrid source terms can also be considered, namely total charges and a circulation (see section 3.3.2.2.3). As an example, let us consider as source terms Q_1 and f_{23}. Besides the vector potential, we have as unknown (see equation [4.20]) the total charges Q_2. In this case, the equation to be solved is equation [4.24] to which equation [3.147] must be added. Conversely, if the source terms are Q_2 and f_{13}, equation [3.146] must be added to the integral form [4.24].

4.2.3. *Weak form of the equations of electrokinetics*

The case to be studied for electrokinetics is the multisource case presented in Figure 3.14. The equations to be solved are defined in equations [3.149] and [3.150], with the electrical behavior law [1.20]. As for the boundary conditions, defined in equation [3.191], they are given by the expressions of equation [3.151]. As source terms, it is possible to impose on the boundaries Γ_{ek} with $k \in \{1,2,3\}$ two electromotive forces, two currents or a combination of both.

The equation can be written using the scalar potential or the vector potential formulation. It should be recalled that the scalar potential formulation is suitable when electromotive forces are imposed on the boundaries between the surfaces Γ_{e1}, Γ_{e2} and Γ_{e3}. On the other hand, if the flux "I" of the current density is imposed on the boundaries Γ_{e1}, Γ_{e2} and Γ_{e3}, in this case, the vector potential formulation is the most suitable.

The weighted residual method will be applied to the scalar potential formulation in section 4.2.3.1 and to the vector potential formulation in section 4.2.3.2.

4.2.3.1. *Scalar potential V formulation*

With the scalar potential V formulation, when the electromotive forces e_{13} and e_{23} are imposed, the equation to be solved can be written in the form of equation [3.196] with $V \in H_{\Gamma e1 \cup \Gamma e2 \cup \Gamma e3}(\mathbf{grad}, \Omega)$. This equation is as follows:

$$\mathrm{div}\sigma(e_{13}\boldsymbol{\beta}_{13} + e_{23}\boldsymbol{\beta}_{23} - \mathbf{grad}V) = 0 \qquad [4.25]$$

with, as shown by equation [3.192], $\boldsymbol{\beta}_{i,j} \in H_{\Gamma e1 \cup \Gamma e2 \cup \Gamma e3}(\mathbf{curl}, \Omega)$.

To apply the weighted residual method, we must define the weighting functions, which are chosen in the adjoint space of the differential operator of the unknown field. For equation [4.25], the vector operator is the divergence and the associated function space, which corresponds to that of the current density, is therefore $H_{\Gamma j}(\text{div}0, \Omega)$ (see equation [3.195]). Based on Table 2.1, it can be deduced that the adjoint operator is the gradient, and the weighting function, which is a scalar, is defined by $\psi \in H_{\Gamma e1 \cup \Gamma e2 \cup \Gamma e3}(\mathbf{grad}, \Omega)$. These various points are summarized in Table 4.3.

Vector operator	Associated function space	Adjoint operator	Weighting function
div	$H_{\Gamma j}(\text{div}0, \Omega)$	$-\mathbf{grad}$	$\psi \in H_{\Gamma e1 \cup \Gamma e2 \cup \Gamma e3}(\mathbf{grad}, \Omega)$

Table 4.3. *Electrokinetics, scalar potential V formulation; vector operator and function space of potential and weighting functions*

Applying the weighted residual method to equation [4.25], we obtain:

$$\iiint_{\Omega} \text{div}(\sigma(e_{13}\boldsymbol{\beta}_{13} + e_{23}\boldsymbol{\beta}_{23} - \mathbf{grad}V))\psi d\tau = 0 \qquad [4.26]$$

Equation [4.26] can be integrated by parts using formula [2.23] related to the vector operators. This allows for the introduction of boundary conditions as follows:

$$-\iiint_{\Omega} \sigma(\mathbf{grad}V - (e_{13}\boldsymbol{\beta}_{13} + e_{23}\boldsymbol{\beta}_{23})).\mathbf{grad}\psi d\tau + \oiint_{\Gamma} \psi\sigma(e_{13}\boldsymbol{\beta}_{13} + e_{23}\boldsymbol{\beta}_{23} - \mathbf{grad}V).\mathbf{n}dS = 0 \qquad [4.27]$$

It can again be noted that the weighted residual method allows for the lowering of the level of differentiability of the unknown V of the problem by transferring it to the weighting function ψ.

The surface integral can be decomposed into several terms in order to highlight the boundary conditions. Then, we have:

$$\begin{aligned}\oiint_{\Gamma} \psi\sigma(e_{13}\boldsymbol{\beta}_{13} + e_{23}\boldsymbol{\beta}_{23} - \mathbf{grad}V).\mathbf{n}dS = \\ \iint_{\Gamma_{e1}\cup\Gamma_{e2}\cup\Gamma_{e3}} \psi\sigma(e_{13}\boldsymbol{\beta}_{13} + e_{23}\boldsymbol{\beta}_{23} - \mathbf{grad}V).\mathbf{n}dS \\ + \iint_{\Gamma j} \psi\sigma(e_{13}\boldsymbol{\beta}_{13} + e_{23}\boldsymbol{\beta}_{23} - \mathbf{grad}V).\mathbf{n}dS\end{aligned} \qquad [4.28]$$

This surface integral is equal to zero. Indeed, the term related to boundary $\Gamma_{e1}\cup\Gamma_{e2}\cup\Gamma_{e3}$ is naturally zero, taking into account the function space to which the weighting function ψ belongs. As for the second integral, it is a function of the normal component of the current density on the boundary Γ_j (see equation [3.161]).

As the latter is equal to zero, the integral is also zero. Under these conditions, equation [4.27] has the following form:

$$\iiint_{\Omega} \sigma(\mathbf{grad}V.\mathbf{grad}\psi)d\tau = \iiint_{\Omega} \sigma(e_{13}\boldsymbol{\beta}_{13} + e_{23}\boldsymbol{\beta}_{23}).\mathbf{grad}\psi d\tau \qquad [4.29]$$

Let us consider again the initial problem defined by equations [3.149], [3.150], [3.191] and [3.151]. The properties of the scalar potential V and of the support fields $\boldsymbol{\beta}_{ij}$ allow for the strong verification of equation [3.149] and of the boundary conditions on $\Gamma_{e1} \cup \Gamma_{e2} \cup \Gamma_{e3}$. On the other hand, the weighted residual method makes it possible to impose equation [3.150] in the weak sense. The boundary conditions of the current density on Γ_j are also imposed in the weak sense via the integral term of equation [4.28], considered to be zero in equation [4.29].

If the source terms are the current density fluxes I_1 and I_2, as already seen in section 3.4.2.1, the electromotive forces e_{13} and e_{23} become additional unknowns. Following the same approach as that detailed for the electrostatics at the end of section 4.2.2.1, a system of equations can be built from relation [4.29] to which equations [3.198] and [3.199] are added. If an electromotive force and a current are now imposed, to obtain a full equation system, the equation expressing the imposed current is added, namely equation [3.198] or [3.199].

4.2.3.2. *Vector potential T formulation*

For the vector potential formulation, with the imposed current density fluxes I_1 and I_2, equation [3.204] must be solved with $\mathbf{T} \in H_{\Gamma j}(\mathbf{curl}, \Omega)$. This yields:

$$\mathbf{curl}(\sigma^{-1}(I_1\boldsymbol{\lambda}_{13} + I_2\boldsymbol{\lambda}_{23} + \mathbf{curl}\mathbf{T})) = 0 \qquad [4.30]$$

To apply the weighted residual method, the function space to which the weighting function belongs must be determined. As already seen in section 4.2.1, it is defined in the adjoint space of the differential operator of the unknown field. Equation [4.30] shows that the differential operator is the curl and it applies to the electric field (see equation [3.203]) with $\mathbf{E} \in H_{\Gamma e1 \cup \Gamma e2 \cup \Gamma e3}(\mathbf{curl}0, \Omega)$. Consequently, using Table 2.1, it can be shown that the adjoint operator is also the curl and the weighting function is a vector field such that $\boldsymbol{\Psi} \in H_{\Gamma j}(\mathbf{curl}, \Omega)$. These various data are gathered in Table 4.4.

Applying the weighted residual method to equation [4.30], we obtain:

$$\iiint_{\Omega} \mathbf{curl}(\sigma^{-1}(I_1\boldsymbol{\lambda}_{13} + I_2\boldsymbol{\lambda}_{23} + \mathbf{curl}\mathbf{T})).\boldsymbol{\Psi} d\tau = 0 \qquad [4.31]$$

Vector operator	Associated function space	Adjoint operator	Weighting function
curl	$H_{\Gamma e1 \cup \Gamma e2 \cup \Gamma e3}$ (**curl**0, Ω)	**curl**	$\boldsymbol{\Psi} \in H_{\Gamma j}$(**curl**, Ω)

Table 4.4. *Electrokinetics; vector potential T formulation; vector operator and function space of potential and weighting functions*

Using formula [2.27], related to the vector operators, the differentiability conditions on the vector potential **T** are lowered and the boundary conditions are naturally introduced:

$$\iiint_{\Omega} (\sigma^{-1}(\mathbf{curl}\mathbf{T} + I_1\boldsymbol{\lambda}_{13} + I_2\boldsymbol{\lambda}_{23})).\mathbf{curl}\boldsymbol{\Psi} d\tau - \oiint_{\Gamma} (\sigma^{-1}(\mathbf{curl}\mathbf{T} + I_1\boldsymbol{\lambda}_{13} + I_2\boldsymbol{\lambda}_{23}) \wedge \mathbf{n}).\boldsymbol{\Psi} dS = 0 \qquad [4.32]$$

The surface integral of this equation can be decomposed into two terms following the boundaries $\Gamma_{e1} \cup \Gamma_{e2} \cup \Gamma_{e3}$ and Γ_j as follows:

$$\oiint_{\Gamma} (\sigma^{-1}(\mathbf{curl}\mathbf{T} + I_1\boldsymbol{\lambda}_{13} + I_2\boldsymbol{\lambda}_{23}) \wedge \mathbf{n}).\boldsymbol{\Psi} dS = \iint_{\Gamma_j} (\sigma^{-1}(\mathbf{curl}\mathbf{T} + I_1\boldsymbol{\lambda}_{13} + I_2\boldsymbol{\lambda}_{23}) \wedge \mathbf{n}).\boldsymbol{\Psi} dS + \iint_{\Gamma e_1 \cup \Gamma_{e2} \cup \Gamma_{e3}} (\sigma^{-1}(\mathbf{curl}\mathbf{T} + I_1\boldsymbol{\lambda}_{13} + I_2\boldsymbol{\lambda}_{23}) \wedge \mathbf{n}).\boldsymbol{\Psi} dS \qquad [4.33]$$

The first term on the right-hand side is equal to zero, as $\boldsymbol{\Psi} \in H_{\Gamma j}$(**curl**, Ω). As for the second term, it corresponds to the surface integral of the tangential component of the electric field that is equal to zero on $\Gamma_{e1} \cup \Gamma_{e2} \cup \Gamma_{e3}$ (see equation [3.203]). Under these conditions, equation [4.32], which represents the weak form of the equation to be solved, can be written as follows:

$$\iiint_{\Omega} (\sigma^{-1}\mathbf{curl}\mathbf{T}.\mathbf{curl}\boldsymbol{\Psi} d\tau = -\iiint_{\Omega} \sigma^{-1}(I_1\boldsymbol{\lambda}_{13} + I_2\boldsymbol{\lambda}_{23}).\mathbf{curl}\boldsymbol{\Psi} dS \qquad [4.34]$$

Let us compare the integral form [4.34] with equations [3.149], [3.150], [3.191] and [3.151] of the initial problem. The choice of the vector potential **T** and of the support fields $\boldsymbol{\lambda}_{13}$ and $\boldsymbol{\lambda}_{23}$ allows the properties of the current density (see equation [3.150] and its boundary conditions on Γ_j [3.151]) to be strongly imposed. As for the electric field, equation [3.149] is weakly imposed with the weighted residual method. The boundary condition of the electric field on the boundary $\Gamma_{e1} \cup \Gamma_{e2} \cup \Gamma_{e3}$ (see equation [3.151]) is also weakly imposed. Indeed, the boundary integral on $\Gamma_{e1} \cup \Gamma_{e2} \cup \Gamma_{e3}$ of equation [4.33] is considered zero in equation [4.34].

Let us now consider that the source terms are the electromotive forces e_{13} and e_{23} imposed on the boundaries Γ_{e1}, Γ_{e2} and Γ_{e3}. In this case, currents I_1 and I_2 become unknowns of the problem. To obtain a full equation system, relations [3.205] and [3.206] introducing the electromotive forces are added. If a current and an electromotive force are imposed as source terms, to have a well-posed problem the equation expressing the imposed electromotive force is added, namely equation [3.205] or [3.206].

4.2.4. *Weak form of the equations of magnetostatics*

In the case of magnetostatics, the two (scalar potential φ and vector potential $\mathbf{A}$) formulations will be studied, for the example presented in Figure 3.20. It should be recalled that in this case three different kinds of source terms appear. A magnetic quantity imposed on the boundary (magnetomotive force or magnetic flux) and two source terms inside the domain (current intensity in a stranded conductor and a permanent magnet). As far as the inductor is concerned (see section 3.5.2.3), knowing the intensity of current I, the current density $\mathbf{J}_0$ is given by expression [1.62].

For the two formulations (see sections 4.2.4.1 and 4.2.4.2) and depending on various source terms, the equation of equilibrium and the weighted residual method will be reviewed. Finally, the equations and the boundary conditions imposed in the strong sense and in the weak sense will be summarized.

4.2.4.1. *Scalar potential φ formulation*

The scalar potential formulation relies on the equations developed in section 3.5.2.5. The first studied case is the one where the source terms are the magnetomotive force f_m, the current intensity I (which represents the current density flux $\mathbf{J}_0$) and the coercive field $\mathbf{H}_c$ of a permanent magnet (see equation [1.64]). Then, as the source term on the boundaries, the magnetomotive force will be replaced by the magnetic flux ϕ.

4.2.4.1.1. Imposed source terms f_m, I, H_c

For the scalar potential formulation, having as source terms f_m and I, the magnetic field is expressed as an equation [3.263]. The scalar potential, thus defined, belongs to the function space $H_{\Gamma h1 \cup \Gamma h2}(\mathbf{grad}, \Omega)$. The magnetic flux density is obtained via the magnetic behavior law that allows for the introduction of the source term related to the permanent magnet, the coercive field $\mathbf{H}_c$ (see equation [3.264]). The equation to be solved [3.266], written in the form of an equation of equilibrium, can then be written as follows:

$$\begin{aligned}&\text{div}(\mu(\mathbf{grad}\varphi - f_m\boldsymbol{\beta}_s - I\chi_I + \mathbf{H}_c)) = 0\\&\text{with } \varphi \in H_{\Gamma_{h1}\cup\Gamma_{h2}}(\mathbf{grad},\Omega)\end{aligned} \quad [4.35]$$

The function space associated with this expression is defined by the magnetic flux density, i.e. $\mathbf{B} \in H_{\Gamma b}(\text{div}0, \Omega)$, as shown by equation [3.261]. According to Table 2.1, the adjoint operator is the gradient and the weighting function is a scalar function such that $\psi \in H_{\Gamma h1 \cup \Gamma h2}(\mathbf{grad}, \Omega)$. These various results are summarized in Table 4.5.

Vector operator	**Associated function space**	**Adjoint operator**	**Weighting function**
div	$H_{\Gamma b}(\text{div}0, \Omega)$	$-$ **grad**	$\psi \in H_{\Gamma h1 \cup \Gamma h2}(\mathbf{grad}, \Omega)$

Table 4.5. *Magnetostatics, scalar potential φ formulation; vector operator and function space of the potential and of the weighting functions*

Under these conditions, the weighted residual method, applied to equation [4.35], has the following form:

$$\begin{aligned}&\iiint_\Omega \text{div}(\mu(\mathbf{grad}\varphi - f_m\boldsymbol{\beta}_s - I\chi_I + \mathbf{H}_c))\psi d\tau = 0\\&\text{with } \psi \in H_{\Gamma_{h1}\cup\Gamma_{h2}}(\mathbf{grad},\Omega)\end{aligned} \quad [4.36]$$

Using formula [2.23], related to the vector operators, the "boundary" term is introduced:

$$\begin{aligned}&-\iiint_\Omega (\mu(\mathbf{grad}\varphi - f_m\boldsymbol{\beta}_s - I\chi_I + \mathbf{H}_c)).\mathbf{grad}\psi d\tau\\&+\oiint_\Gamma \psi(\mu(\mathbf{grad}\varphi - f_m\boldsymbol{\beta}_s - I\chi_I + \mathbf{H}_c)).\mathbf{n}dS = 0\end{aligned} \quad [4.37]$$

As noted in section 4.2.1, the weighted residual method makes it possible to decrease the constraints of differentiability on the scalar potential φ. This constraint is transferred to the weighting function. Similarly, the problems of discontinuity of the field $\mathbf{H}_c$, on some surfaces of the permanent magnet, mentioned in section 3.5.2.4, are also lifted.

Let us now decompose the surface term by introducing the various boundary conditions as follows:

$$\oiint_{\Gamma} \psi(\mu(\mathbf{grad}\varphi - f_m \boldsymbol{\beta}_s - I\boldsymbol{\chi}_I + \mathbf{H}_c)).\mathbf{n}dS =$$
$$\iint_{\Gamma_{h1}\cup\Gamma_{h2}} \psi(\mu(\mathbf{grad}\varphi - f_m \boldsymbol{\beta}_s - I\boldsymbol{\chi}_I + \mathbf{H}_c)).\mathbf{n}dS \qquad [4.38]$$
$$+\iint_{\Gamma_b} \psi(\mu(\mathbf{grad}\varphi - f_m \boldsymbol{\beta}_s - I\boldsymbol{\chi}_I + \mathbf{H}_c)).\mathbf{n}dS$$

For this equation, the first term on the right-hand side is equal to zero, as $\psi \in H_{\Gamma h1 \cup \Gamma h2}(\mathbf{grad}, \Omega)$. The second integral on Γ_b is also equal to zero. Indeed, this expression contains the normal component of the magnetic flux density (see equation [3.260]) which is equal to zero on Γ_b. Consequently, the surface integral [4.38] is equal to zero and equation [4.37], which represents the weak formulation of the problem, is written as:

$$\iiint_{\Omega} \mu(\mathbf{grad}\varphi.\mathbf{grad}\psi)d\tau = \iiint_{\Omega} (\mu(f_m \boldsymbol{\beta}_s + I\boldsymbol{\chi}_I - \mathbf{H}_c)).\mathbf{grad}\psi d\tau \qquad [4.39]$$

Let us analyze this formulation with respect to the initial problem, defined by equations [3.258]–[3.261]. The choice of the scalar potential φ and of the support fields $\boldsymbol{\beta}_s$ and $\boldsymbol{\chi}_I$ allows, considering their respective properties, for the strong verification of equation [3.258] and the imposition of boundary conditions of the magnetic field on $\Gamma_{h1}\cup\Gamma_{h2}$. On the other hand, equation [3.259] is weakly imposed, via the weighted residual method. Concerning the boundary conditions on the magnetic flux density (see equation [3.260]), they are also weakly verified via the integral term on Γ_b of equation [4.38] considered zero in the expression [4.39].

4.2.4.1.2. Imposed source terms ϕ, I, H_c

In the case of the scalar potential formulation, if the source term is the flux ϕ of the magnetic flux density imposed through the boundaries Γ_{h1} and Γ_{h2}, the magnetomotive force f_m becomes an unknown of the problem. The approach is then similar to the one in section 4.2.2.1. To obtain a full problem, equation [3.267] is added to the integral form [4.39].

4.2.4.2. *Vector potential A formulation*

For the vector potential formulation, in the presence of several source terms, the developments have been introduced in section 3.5.3.5. As a first step, the source terms are considered the flux ϕ, the current intensity I in the inductor and the remanent magnetic flux density $\mathbf{B}_r$ to represent the permanent magnet. It should be noted (see equation [1.63]) that $\mathbf{B}_r$ and $\mathbf{H}_c$ are linked by relation $\mathbf{B}_r = -\mu_A \mathbf{H}_c$. In section 4.2.4.2.2, the magnetic flux is replaced by the magnetomotive force f_m. It should be recalled that the studied structure is "represented" in Figure 3.20 and that equations [3.258]–[3.261] should be solved.

4.2.4.2.1. Imposed source terms ϕ, I, B_r

With the vector potential **A** formulation and in the presence of the source terms ϕ, I and $\mathbf{B}_r$, the expressions of the magnetic flux density and magnetic field are given, respectively, by equations [3.295] and [3.296]. The function spaces of **H** and **B** are defined in equation [3.261], and the equation to be solved (see equation [3.298]) is rewritten in the form of an equilibrium equation

$$\mathbf{curl}\mu^{-1}(\mathbf{curlA} + \phi\boldsymbol{\lambda}_\phi + \mathbf{B}_r) - I\boldsymbol{\lambda}_I = 0 \quad [4.40]$$

In this expression, $\boldsymbol{\lambda}_\phi$ and $\boldsymbol{\lambda}_I$ represent, respectively, the support fields of the flux ϕ and of the current density $\mathbf{J}_0$. These fields, as shown in relations [3.273] and [3.289], are defined in the function spaces, such that $\boldsymbol{\lambda}_\phi \in H_{\Gamma b}(\mathrm{div}0, \Omega)$ and $\boldsymbol{\lambda}_I \in H_0(\mathrm{div}0, \Omega)$.

In expression [4.40], the vector operator is the curl and the associated function space corresponds to the magnetic field, namely $\mathbf{H} \in H_{\Gamma h1 \cup \Gamma h2}(\mathbf{curl}, \Omega)$ (see equation [3.261]). Under these conditions (see Table 2.1), the adjoint operator is also the curl and the space of weighting functions is a field of vectors such that $\boldsymbol{\Psi} \in H_{\Gamma h1 \cup \Gamma h2}(\mathbf{curl}, \Omega)$. These various results are summarized in Table 4.6.

Vector operator	Associated function space	Adjoint operator	Weighting function
curl	$H_{\Gamma h1 \cup \Gamma h2}(\mathbf{curl}, \Omega)$	**curl**	$\Psi \in H_{\Gamma b}(\mathbf{curl}, \Omega)$

Table 4.6. *Magnetostatics; vector potential A formulation; vector operator and function space of potential and of weighting functions*

Applying the weighted residual method to equation [4.40] leads to the following expression:

$$\iiint_\Omega (\mathbf{curl}\mu^{-1}(\mathbf{curlA} + \phi\boldsymbol{\lambda}_\phi + \mathbf{B}_r) - I\boldsymbol{\lambda}_I).\boldsymbol{\Psi}d\tau = 0 \quad [4.41]$$

Applying formula [2.27], the surface integral appears naturally and the following can be written:

$$\begin{aligned}&\iiint_{\Omega}\mu^{-1}\mathbf{curlA}.\mathbf{curl\Psi}d\tau+\iiint_{\Omega}\mu^{-1}(\phi\boldsymbol{\lambda}_{\phi}+\mathbf{B}_{r}).\mathbf{curl\Psi}dS\\&-\oiint_{\Gamma}\mu^{-1}(\mathbf{curlA}+\phi\boldsymbol{\lambda}_{\phi}+\mathbf{B}_{r})\wedge\mathbf{n}).\boldsymbol{\Psi}dS-\iiint_{\Omega}I\boldsymbol{\lambda}_{I}.\boldsymbol{\Psi}d\tau=0\end{aligned} \quad [4.42]$$

NOTE.– After equation [3.293], we noted that, since the normal component of $\mathbf{B}_r$ was discontinuous on the domain, this led to constraints on the strong form. In the case of the weak formulation, it can be noted (see equation [4.42]) that these constraints are lifted.

Comparing equations [4.41] and [4.42], it can be noted that the differentiability conditions on the vector potential **A** are alleviated. Moreover, the surface integral can be decomposed into two terms related to the boundaries $\Gamma_{h1}\cup\Gamma_{h2}$ and Γ_b, which yields:

$$\begin{aligned}&\oiint_{\Gamma}(\mu^{-1}(\mathbf{curlA}+\phi\boldsymbol{\lambda}_{\phi}+\mathbf{B}_{r})\wedge\mathbf{n}).\boldsymbol{\Psi}dS=\\&\iint_{\Gamma_{h1}\cup\Gamma_{h2}}(\mu^{-1}(\mathbf{curlA}+\phi\boldsymbol{\lambda}_{\phi}+\mathbf{B}_{r})\wedge\mathbf{n}).\boldsymbol{\Psi}dS\\&+\iint_{\Gamma_{b}}(\mu^{-1}(\mathbf{curlA}+\phi\boldsymbol{\lambda}_{\phi}+\mathbf{B}_{r})\wedge\mathbf{n}).\boldsymbol{\Psi}dS\end{aligned} \quad [4.43]$$

As $\boldsymbol{\Psi} \in H_{\Gamma b}(\mathbf{curl}, \Omega)$, the surface integral on Γ_b is equal to zero. The same is true for the surface integral on $\Gamma_{h1}\cup\Gamma_{h2}$ where the tangential component of the magnetic field is zero (see its domain of definition, equation [3.261]). Consequently, equation [4.42], which represents the weak form of the equations to solve, takes the following form:

$$\iiint_{\Omega}\mu^{-1}\mathbf{curlA}.\mathbf{curl\Psi}d\tau=-\iiint_{\Omega}\mu^{-1}(\phi\boldsymbol{\lambda}_{\phi}.-\mathbf{B}_{r}).\mathbf{curl\Psi}+\iiint_{\Omega}I\boldsymbol{\lambda}_{I}.\boldsymbol{\Psi}d\tau \quad [4.44]$$

Let us consider again the initial problem with equations [3.258]–[3.261]. The choice of vector potential **A**, associated with the properties of support fields $\boldsymbol{\lambda}_\phi$ and $\boldsymbol{\lambda}_I$, allows for the strong verification of equation [3.259]. The boundary conditions on Γ_b of the magnetic flux density (see equation [3.260]) are also strongly verified thanks to the function spaces of potential **A** and of fields $\boldsymbol{\lambda}_\phi$ and $\boldsymbol{\lambda}_I$. On the other hand, relation [3.258] is weakly imposed via the weighted residual method. As for the boundary conditions, on the magnetic field (see equation [3.260]), they are also weakly verified via the surface integral on $\Gamma_{h1}\cup\Gamma_{h2}$ of equation [4.42], which is considered zero in equation [4.44].

4.2.4.2.2. Imposed source terms f_m, J_0, B_r

If the source term imposed on the boundary $\Gamma_{h1} \cup \Gamma_{h2}$ is the magnetomotive force f_m, the flux ϕ proves to be an unknown. The approach is similar to that in section 4.2.2.1. To obtain a full equation system, the integral form [4.44] is added to equation [3.300], which allows for the expression of the magnetomotive force as a function of the vector potential **A**, the flux ϕ and the remanent magnetic flux density.

4.2.5. *Weak form of the equations of magnetodynamics*

This section uses the same approach as that followed in the introduction of potential formulations in section 3.6. The studied structure, presented in the general case of Figure 3.22, is a domain Ω of boundary Γ, composed of two subdomains denoted by Ω_c and Ω_0. The subdomain Ω_c, of boundary Γ_c, is a conductor and holds two gates Γ_{n1} and Γ_{n2} in contact with the external environment. Electric quantities e(t) or I(t) or magnetic quantities $f_m(t)$ or $\phi(t)$ can be imposed on these two gates. The remaining boundary of the conducting subdomain Ω_c, denoted by Γ_j, is in contact with the subdomain Ω_0 and represents a wall for the current density. The not simply connected subdomain Ω_0 is not conducting, meaning that conductivity σ is equal to zero (see equation [3.307]). The part of its boundary, in contact with the external environment, is a wall for the magnetic flux density and is denoted by Γ_b.

In what follows, as introduced in section 3.6, we reconsider the various supply modes with the **A**-V and **T**-φ formulations and apply the weighted residual method.

4.2.5.1. *Imposed electrical quantities*

In this case, the studied domain is represented in Figure 3.23, and we have to solve equations [3.301]–[3.304]. For the given example, when the electrical quantities e(t) or I(t) are imposed, the boundary conditions and the function spaces of fields **E**, **J**, **H** and **B** are defined by expressions [3.310]–[3.312].

4.2.5.1.1. Electric formulation A-V

In the case of the electric formulation, the developments rely on the magnetic vector potential **A** and the electrical scalar potential V. For the given example, we have (see equations [3.316] and [3.319]), respectively, $\mathbf{A} \in H_0(\mathbf{curl}, \Omega)$ and $V \in H_{\Gamma e1 \cup \Gamma e2}(\mathbf{grad}, \Omega_c)$. The system to be solved, when the source term is the electromotive force "e", is composed of equations [3.327] and [3.328] for, respectively, the domain Ω and the subdomain Ω_c. These relations presented in the form of equilibrium equations are:

$$\mathbf{curl}\mu^{-1}(\mathbf{curlA})+\sigma(\frac{\partial\mathbf{A}}{\partial\mathbf{t}}+\mathbf{grad}V)-\sigma e\boldsymbol{\beta}_e=0 \text{ on } \Omega \quad [4.45]$$

$$\text{div}(\sigma(\frac{\partial\mathbf{A}}{\partial t}+\mathbf{grad}V-e\boldsymbol{\beta}_e))=0 \text{ on } \Omega_c \quad [4.46]$$

It should be recalled that $\boldsymbol{\beta}_e$ is the support field of the electromotive force, which is defined by relation [3.313]. In order to develop the weighted residual method for these equations, weighting functions should be defined. For equation [4.45], the vector operator is the curl and the associated space is defined by the magnetic field, i.e. (see relation [3.312]) $\mathbf{H}\in H(\mathbf{curl}, \Omega)$. The adjoint operator (see Table 2.1) is the curl and the space of weighting functions $\boldsymbol{\Psi}\in H_0(\mathbf{curl}, \Omega)$. As for equation [4.46], the vector operator is the divergence and the associated space is defined by the current density (relation [3.310]), with $\mathbf{J}\in H_{\Gamma j}(\mathbf{curl}, \Omega_c)$. In this case (see Table 2.1), the adjoint operator is the gradient and the space of weighting functions $\psi\in H_{\Gamma e1\cup\Gamma e2}(\mathbf{grad},\Omega_c)$. Table 4.7 summarizes these various results.

Equation	Vector operator	Associated function space	Adjoint operator	Weighting function
[4.45]	**curl**	$H(\mathbf{curl}, \Omega)$	**curl**	$\boldsymbol{\Psi} \in H_0(\mathbf{curl}, \Omega)$
[4.46]	div	$H_{\Gamma j}(\text{div}0, \Omega_c)$	$-$ **grad**	$\psi \in H_{\Gamma e1\cup\Gamma e2}(\mathbf{grad}, \Omega_c)$

Table 4.7. *Magnetodynamics, imposed electrical quantities, A-V electrical formulation; vector operator and function space of potentials and weighting functions*

As a first step, the weighted residual method is applied to equation [4.45], which is integrated over the domain Ω. Then, we obtain:

$$\iiint_\Omega(\mathbf{curl}\mu^{-1}(\mathbf{curlA}).\boldsymbol{\Psi}+\sigma(\frac{\partial\mathbf{A}}{\partial t}+\mathbf{grad}V-e\boldsymbol{\beta}_e).\boldsymbol{\Psi})d\tau=0 \quad [4.47]$$

It should be noted that the scalar potential V and the source field $\boldsymbol{\beta}_e$ are only defined on the domain Ω_c. Nevertheless, since the conductivity σ is zero on Ω_0 (see equation [3.307]), the integral of $\sigma(\partial\mathbf{A}/\partial t + \mathbf{grad}V - e\boldsymbol{\beta}_e)$ can be extended to the entire domain Ω. The first term of the volume integral can be rewritten using expression [2.27] that introduces the surface integral on Γ as follows:

$$\begin{aligned}&\iiint_{\Omega}\mathbf{curl}(\mu^{-1}\mathbf{curlA}).\mathbf{\Psi}d\tau\\&=\iiint_{\Omega}\mu^{-1}\mathbf{curlA}.\mathbf{curl\Psi}d\tau-\oiint_{\Gamma}(\mu^{-1}\mathbf{curlA}\wedge\mathbf{n}).\mathbf{\Psi}dS\end{aligned}\qquad[4.48]$$

Given the function space to which the weighting function $\mathbf{\Psi}$ belongs, the integral on the boundary Γ is equal to zero. Under these conditions, gathering equations [4.47] and [4.48], we obtain:

$$\iiint_{\Omega}(\mu^{-1}(\mathbf{curlA}.\mathbf{curl\Psi})+\sigma(\frac{\partial\mathbf{A}}{\partial t}+\mathbf{grad}V-e\boldsymbol{\beta}_e).\mathbf{\Psi})d\tau=0\qquad[4.49]$$

Let us again apply the weighted residual method for equation [4.46] with the scalar weighting function ψ defined in Table 4.7. The following can be written as:

$$\iiint_{\Omega_c}\mathrm{div}(\sigma(\frac{\partial\mathbf{A}}{\partial t}+\mathbf{grad}V-e\boldsymbol{\beta}_e))\psi d\tau=0\qquad[4.50]$$

Formula [2.23], related to the vector operators, leads to the following:

$$\begin{aligned}&\iiint_{\Omega_c}\mathrm{div}(\sigma(\frac{\partial\mathbf{A}}{\partial t}+\mathbf{grad}V-e\boldsymbol{\beta}_e))\psi d\tau=\\&\quad-\iiint_{\Omega_c}\sigma(\frac{\partial\mathbf{A}}{\partial t}+\mathbf{grad}V-e\boldsymbol{\beta}_e).\mathbf{grad}\psi d\tau\\&\qquad+\oiint_{\Gamma_c}\psi(\sigma(\frac{\partial\mathbf{A}}{\partial t}+\mathbf{grad}V-e\boldsymbol{\beta}_e)).\mathbf{n}dS=0\end{aligned}\qquad[4.51]$$

The surface integral on Γ_c can be decomposed by introducing the terms related to the boundaries Γ_j and $\Gamma_{e1}\cup\Gamma_{e2}$. Then, we have:

$$\begin{aligned}&\oiint_{\Gamma_c}\psi(\sigma(\frac{\partial\mathbf{A}}{\partial t}+\mathbf{grad}V-e\boldsymbol{\beta}_e).\mathbf{n})dS=\\&\quad\iint_{\Gamma_{e1}\cup\Gamma_{e2}}\psi(\sigma(\frac{\partial\mathbf{A}}{\partial t}+\mathbf{grad}V-e\boldsymbol{\beta}_e).\mathbf{n})dS\\&\qquad+\iint_{\Gamma_j}\psi(\sigma(\frac{\partial\mathbf{A}}{\partial t}+\mathbf{grad}V-e\boldsymbol{\beta}_e).\mathbf{n}dS\end{aligned}\qquad[4.52]$$

The first term on the right-hand side of the equality is equal to zero, as (see Table 4.7) the weighting function ψ is zero on the boundary $\Gamma_{e1}\cup\Gamma_{e2}$. As for the second term, it is also zero, since $\mathbf{J.n}=0$ on Γ_j. Equation [4.51] can then be rewritten as follows:

$$\iiint_{\Omega_c}(\sigma\mathbf{grad}\psi\mathbf{grad}V+\sigma(\frac{\partial\mathbf{A}}{\partial t}-e\boldsymbol{\beta}_e).\mathbf{grad}\psi)d\tau=0 \qquad [4.53]$$

The weak form of the magnetodynamics problem represented in Figure 3.23, with the electric formulation **A**-V, is expressed using equations [4.49] and [4.53]. It should be noted that the basis equations [3.301] and [3.303] are strongly verified via the properties of the potentials **A**, V and of the support field $\boldsymbol{\beta}_e$. Considering the function spaces to which these potentials and also $\boldsymbol{\beta}_e$ belong, the boundary conditions on $\Gamma_{e1}\cup\Gamma_{e2}$ and Γ_b on the fields **E** and **B** are also strongly verified. On the other hand, equations [3.302] and [3.304] are weakly verified, via the weighted residual method. The boundary conditions of the current density on the boundary Γ_j are also weakly imposed via the boundary integral of equation [4.52]. Indeed, the latter is considered zero in expression [4.53].

If the flux of the current density is imposed through the boundaries Γ_{e1} and Γ_{e2}, the electromotive force becomes an unknown of the problem. The process used is the same as in section 4.2.2.1. For a full equation system, relation [3.338] or [3.339] (see section 3.6.1.1.2) is added to equations [4.49] and [4.53].

4.2.5.1.2. Magnetic formulation T-φ

The magnetic formulation uses the electric vector potential **T** and the magnetic scalar potential φ. For the example in Figure 3.23, with the current density flux I(t) as a source term, the potentials **T** and φ are defined, as shown by relations [3.348] and [3.354], in the function spaces $H_{\Gamma j}(\mathbf{curl}, \Omega_c)$ and $H(\mathbf{grad}, \Omega)$, respectively. In this case, the equations to be solved are given by relations [3.362] and [3.363], which are recalled below:

$$\begin{aligned}&\mathbf{curl}(\sigma^{-1}(\mathbf{Icurl}\boldsymbol{\chi}_I+\mathbf{curlT}))\\&+\frac{\partial}{\partial t}(\mu(I\boldsymbol{\chi}_I+\mathbf{T}-\mathbf{grad}\varphi))=0 \text{ on } \Omega_c\end{aligned} \qquad [4.54]$$

$$\operatorname{div}(\mu(I\boldsymbol{\chi}_I+\mathbf{T}-\mathbf{grad}\varphi))=0 \text{ on } \Omega \qquad [4.55]$$

It should be recalled (see equation [3.346]) that the vector potential $\boldsymbol{\chi}_I$ is defined throughout the domain Ω, and it belongs to the function space H(**curl**, Ω).

According to the weighted residual method, the function spaces to which the weighting functions belong should be determined. For equation [4.54], the vector operator is the curl and the associated space is defined by the electric field (see relation [3.310]), i.e. $\mathbf{E} \in H_{\Gamma e1 \cup \Gamma e2}(\mathbf{curl}, \Omega_c)$. The adjoint operator is then the curl (see Table 2.1) and the weighting functions $\boldsymbol{\Psi}$ belong to $H_{\Gamma j}(\mathbf{curl}, \Omega_c)$.

NOTE.– The space of weighting functions generally corresponds to the space to which unknown potentials belong. But the function space of the vector potential **T** is H(**curl**, Ω) (see equation [3.347]) which a priori does not correspond to that of weighting functions $\boldsymbol{\Psi}$. It should nevertheless be recalled that **T** is equal to zero on Ω-Ω_c. This potential should therefore be determined only on Ω_c, and the restriction of **T** to Ω_c belongs to the same space $H_{\Gamma j}(\mathbf{curl}, \Omega_c)$ as the weighting functions $\boldsymbol{\Psi}$ (see equation [3.348]).

For equation [4.55], the vector operator is the divergence and the associated function space is defined by the magnetic flux density with (see equation [3.312]) $\mathbf{B} \in H_0(\mathrm{div}0, \Omega)$. As shown in Table 2.1, the adjoint operator of the divergence is the gradient and the weighting functions ψ belong to H(**grad**, Ω). Table 4.8 summarizes these various results.

Equation	Vector operator	Associated function space	Adjoint operator	Weighting function
[4.54]	**curl**	$H_{\Gamma e1 \cup \Gamma e2}(\mathbf{curl}, \Omega)$	**curl**	$\Psi \in H_{\Gamma j}(\mathbf{curl}, \Omega_c)$
[4.55]	div	$H_0(\mathrm{div}0, \Omega)$	– **grad**	$\psi \in H(\mathbf{grad}, \Omega)$

Table 4.8. *Magnetodynamics, imposed electric quantities, electric formulation T-φ; vector operator and function space of potentials and weighting functions*

Let us apply the weighted residual method to equation [4.54]. This yields:

$$\iiint_{\Omega_c} (\mathbf{curl}\sigma^{-1}(\mathbf{curl}\mathbf{T} + \mathrm{I}\mathbf{curl}\boldsymbol{\chi}_I).\boldsymbol{\Psi} + \frac{\partial}{\partial t}\mu(\mathbf{T} + \mathrm{I}\boldsymbol{\chi}_I - \mathbf{grad}(\varphi).\boldsymbol{\Psi})d\tau = 0 \quad [4.56]$$

Formula [2.27] related to the vector operators, applied to the first term of the volume integral of equation [4.56], allows us to introduce the surface term. Using the properties of the mixed product, this yields:

$$\iiint_{\Omega_c} \mathbf{curl}(\sigma^{-1}\mathbf{curlT} + \mathbf{I}\mathbf{curl}\chi_I).\mathbf{curl}\mathbf{\Psi}d\tau =$$
$$\iiint_{\Omega_c} \sigma^{-1}(\mathbf{curlT} + \mathbf{I}\mathbf{curl}\chi_I).\mathbf{curl}\mathbf{\Psi}d\tau \qquad [4.57]$$
$$-\oiint_{\Gamma_c} (\mathbf{n} \wedge \mathbf{\Psi}).\sigma^{-1}(\mathbf{curlT} + \mathbf{I}\mathbf{curl}\chi_I)dS$$

The surface integral on Γ_c can be decomposed on the boundaries Γ_j and $\Gamma_{e1} \cup \Gamma_{e2}$ of the conducting domain. Then, we have:

$$\oiint_{\Gamma_c} (\mathbf{n} \wedge \mathbf{\Psi}).\sigma^{-1}(\mathbf{curlT} + \mathbf{I}\mathbf{curl}\chi_I)dS$$
$$= \iint_{\Gamma_j} (\mathbf{n} \wedge \mathbf{\Psi}).\sigma^{-1}(\mathbf{curlT} + \mathbf{I}\mathbf{curl}\chi_I)dS \qquad [4.58]$$
$$+\iint_{\Gamma_{e1} \cup \Gamma_{e2}} (\mathbf{n} \wedge \mathbf{\Psi}).\sigma^{-1}(\mathbf{curlT} + \mathbf{I}\mathbf{curl}\chi_I)dS$$

It can be readily shown that the surface integral on Γ_c is equal to zero. Indeed, the contribution on Γ_j is zero as shown by the function space to which the weighting function $\mathbf{\Psi}$ belongs (see Table 4.8). Using the mixed product, it can be shown that the second integral on the right-hand side represents the tangential component of the electric field (see equation [3.351]) on the boundary $\Gamma_{e1} \cup \Gamma_{e2}$. As this component is zero, the integral is also zero. Under these conditions, equation [4.56] has the following form:

$$\iiint_{\Omega_c} (\sigma^{-1}(\mathbf{curlT} + \mathbf{I}\mathbf{curl}\chi_I).\mathbf{curl}\mathbf{\Psi}$$
$$+\frac{\partial}{\partial t}\mu(\mathbf{T} + \mathrm{I}\chi_I - \mathbf{grad}\varphi).\mathbf{\Psi})d\tau = 0 \qquad [4.59]$$

This expression is the weak form of equation [4.54]. Applying the weighted residual method to equation [4.55] yields:

$$\iiint_{\Omega} \mathrm{div}(\mu(\mathbf{T} + \mathrm{I}\chi_I - \mathbf{grad}\varphi))\psi d\tau = 0 \qquad [4.60]$$

In order to introduce the boundary terms, formula [2.23] is applied to the volume integral. Then, we obtain:

$$\begin{aligned}&\iiint_{\Omega}(\operatorname{div}\mu(\mathbf{T}+\mathrm{I}\chi_{\mathrm{I}}-\mathbf{grad}\varphi))\psi d\tau=\\&-\iiint_{\Omega}\mu(\mathbf{T}+\mathrm{I}\chi_{\mathrm{I}}-\mathbf{grad}\varphi).\mathbf{grad}\psi d\tau\\&+\oiint_{\Gamma}\psi\mu(\mathbf{T}+\mathrm{I}\chi_{\mathrm{I}}-\mathbf{grad}\varphi).\mathbf{n}dS=0\end{aligned}\qquad[4.61]$$

The surface integral on the boundary Γ of the domain introduces the magnetic flux density (see equation [3.355]) and more particularly its normal component. But as shown by relation [3.312], this normal component is zero on the boundary. Considering this property, we can write:

$$\iiint_{\Omega}\mu(\mathbf{T}+\mathrm{I}\chi_{\mathrm{I}}-\mathbf{grad}\varphi).\mathbf{grad}\psi d\tau=0\qquad[4.62]$$

It should be recalled (see equations [3.347] and [3.348]) that the vector potential **T** is defined throughout the domain Ω, but it is equal to zero on Ω_0.

When applying the weighted residual method, for the **T**-φ formulation, the system to be solved is composed of equations [4.59] and [4.62]. In this case, equations [3.302] and [3.304] are strongly verified by the use of potentials **T** and φ and of the associated potential $\boldsymbol{\chi}_{\mathrm{I}}$. The same is true for the boundary condition of the current density on the boundary Γ_j, taking into account the function space to which the potential **T** belongs (see equation [3.348]). On the other hand, the use of the weighted residual method verifies equations [3.301] and [3.303] in the weak sense. This is also valid for the boundary conditions of the electric field on $\Gamma_{e1} \cup \Gamma_{e2}$ and the magnetic flux density on Γ_b. Indeed, the corresponding boundary integrals are considered zero in equations [4.58] and [4.61].

If the electromotive force "e" between the boundaries Γ_{e1} and Γ_{e2} is now imposed as a source term, then the current density flux becomes an unknown. To obtain a full equation system, relation [3.373] is added to the system composed of equations [4.59] and [4.62] (see section 3.6.1.2.2).

4.2.5.2. *Imposed magnetic quantities*

In this section, we again consider the example in Figure 3.24, having the flux φ(t) of the magnetic flux density or the magnetomotive force $f_m(t)$ as the imposed global quantities. Equations [3.301]–[3.304] should be solved. The boundary conditions on the boundaries are given in equations [3.374] and [3.375] and the function spaces to which the fields **E**, **J**, **H** and **B** belong are defined by relations [3.376] and [3.377]. For this study, the two (electric and magnetic) formulations have been introduced in section 3.6.2. In what follows, the weighted residual method is applied to these formulations.

4.2.5.2.1. Electric formulation (A-V)

As a first step, the magnetic flux is imposed on the boundaries Γ_{h1} and Γ_{h2}. For this configuration, the vector potential **A** belongs, as shown in relation [3.382], to the function space $H_{\Gamma b}(\mathbf{curl}, \Omega)$ and the scalar potential V (see equation [3.387]) to the function space $H(\mathbf{grad}, \Omega_c)$. It should be recalled (see section 3.6.2) that the boundary Γ_b is not simply connected. When the source term is the flux ϕ, the system to be solved is composed of equation [3.393] for the domain Ω and equation [3.394] for the subdomain Ω_c. These two equations are recalled as follows:

$$\mathbf{curl}\mu^{-1}(\mathbf{curl}(\mathbf{A}+\phi\chi_\phi)+\sigma(\frac{\partial \mathbf{A}}{\partial t}+\mathbf{grad}V+\frac{\partial \phi \chi_\phi}{\partial t})=0 \text{ on } \Omega \quad [4.63]$$

$$\text{div}(\sigma(\frac{\partial \mathbf{A}}{\partial t}+\frac{\partial \phi \chi_\phi}{\partial t}+\mathbf{grad}V))=0 \text{ on } \Omega_c \quad [4.64]$$

In these expressions, the associated potential $\boldsymbol{\chi}_\phi$ is defined on the entire domain Ω. Given that the boundary Γ_b is not simply connected, $\boldsymbol{\chi}_\phi$ belongs (see equation [3.381]) to the function space $H^{\Delta}{}_{\Gamma b}(\mathbf{curl}, \Omega)$.

The function space to which the weighting functions belong must be determined to apply the weighted residual method to equations [4.63] and [4.64]. For equation [4.63], the vector operator is the curl and the associated space is defined by the magnetic field (see equation [3.377]), i.e. $\mathbf{H} \in H_{\Gamma h1 \cup \Gamma h2}(\mathbf{curl}, \Omega)$. In this case (see Table 2.1), the adjoint operator is also the curl and the weighting functions $\boldsymbol{\Psi}$ belong to $H_{\Gamma b}(\mathbf{curl}, \Omega)$. For equation [4.64], the vector operator is the divergence and the associated space is defined by the current density **J** (see expression [3.376]), with $\mathbf{J} \in H_0(\text{div}0, \Omega_c)$. In this case, as shown in Table 2.1, the adjoint operator is the gradient and the weighting functions ψ belong to $H(\mathbf{grad}, \Omega_c)$. Table 4.9 summarizes these various results.

Equation	**Vector operator**	**Associated function space**	**Adjoint operator**	**Weighting function**
[4.63]	**curl**	$H_{\Gamma h1 \cup \Gamma h2}(\mathbf{curl}, \Omega)$	**curl**	$\Psi \in H_{\Gamma b}(\mathbf{curl}, \Omega)$
[4.64]	div	$H_0(\text{div}0, \Omega_c)$	– **grad**	$\psi \in H(\mathbf{grad}, \Omega_c)$

Table 4.9. *Magnetodynamics, imposed magnetic quantities, electric formulation A-φ; vector operator and function space of potentials and weighting functions*

Having defined the weighting functions and their associated function spaces, the weighted residual method is now applied. Concerning equation [4.63], the following can be written:

$$\iiint_{\Omega} (\mathbf{curl}\,(\mu^{-1}(\mathbf{curl}\mathbf{A} + \phi\chi_{\phi})) + \sigma(\frac{\partial \mathbf{A}}{\partial t} + \mathbf{grad}\mathrm{V} + \frac{\partial \phi\chi_{\phi}}{\partial t}).\boldsymbol{\Psi})\mathrm{d}\tau = 0 \qquad [4.65]$$

As already mentioned, after equation [4.47], relying on conductivity, which is zero in Ω_0, it can be shown that the integral of term $\sigma(\partial\mathbf{A}/\partial t + \mathbf{grad}\mathrm{V} + \phi\partial\chi_{\phi}/\partial t)$ can be extended to the entire domain Ω. The application of formula [2.27] to the first term of integral [4.65] allows us to write:

$$\begin{aligned} &\iiint_{\Omega} \mathbf{curl}(\mu^{-1}(\mathbf{curl}\mathbf{A} + \phi\chi_{\phi})).\boldsymbol{\Psi}\mathrm{d}\tau = \\ &\quad \iiint_{\Omega} \mu^{-1}\mathbf{curl}(\mathbf{A} + \phi\chi_{\phi}).\mathbf{curl}\boldsymbol{\Psi}\mathrm{d}\tau \\ &\qquad - \oint\!\!\oint_{\Gamma} \mu^{-1}(\mathbf{curl}(\mathbf{A} + \phi\chi_{\phi}) \wedge \mathbf{n}).\boldsymbol{\Psi}\mathrm{dS} = 0 \end{aligned} \qquad [4.66]$$

The surface integral on Γ can be decomposed (see equation [3.374]) following the boundaries $\Gamma_{h1} \cup \Gamma_{h2}$ and Γ_b. This yields the following formula:

$$\begin{aligned} &\oint\!\!\oint_{\Gamma} \mu^{-1}(\mathbf{curl}(\mathbf{A} + \phi\chi_{\phi}) \wedge \mathbf{n}).\boldsymbol{\Psi}\mathrm{dS} \\ &\quad = \iint_{\Gamma_{h1} \cup \Gamma_{h2}} \mu^{-1}(\mathbf{curl}(\mathbf{A} + \phi\chi_{\phi}) \wedge \mathbf{n}.\boldsymbol{\Psi})\mathrm{dS} \\ &\qquad + \iint_{\Gamma_b} \mu^{-1}(\mathbf{curl}(\mathbf{A} + \phi\chi_{\phi}) \wedge \mathbf{n}\ .\boldsymbol{\Psi})\mathrm{dS} \end{aligned} \qquad [4.67]$$

In this expression, for the first integral on the right-hand side, the tangential component of the magnetic field is integrated on the boundary $\Gamma_{h1} \cup \Gamma_{h2}$. Considering the function space to which the field **H** belongs, this integral is equal to zero. The second integral on Γ_b is also equal to zero. Indeed, using the properties of the mixed product, the tangential component of the weighting function $\boldsymbol{\Psi}$, which is zero on Γ_b (see Table 4.9), is introduced. Under these conditions, equation [4.66] is written as:

$$\iiint_{\Omega} (\mu^{-1}(\mathbf{curl}(\mathbf{A} + \phi\chi_{\phi}).\mathbf{curl}\boldsymbol{\Psi} + \sigma(\frac{\partial \mathbf{A}}{\partial t} + \mathbf{grad}\mathrm{V} + \frac{\partial \phi\chi_{\phi}}{\partial t}).\boldsymbol{\Psi})\mathrm{d}\tau = 0 \qquad [4.68]$$

Let us now consider equation [4.64] to which we apply the weighted residual method considering as the weighting function the scalar function ψ defined in Table 4.9. Then, we have:

$$\iiint_{\Omega_c} \mathrm{div}(\sigma(\frac{\partial \mathbf{A}}{\partial t} + \frac{\partial \phi \chi_\phi}{\partial t} + \mathbf{grad}V))\psi d\tau = 0 \qquad [4.69]$$

We then apply formula [2.23] to this equation, and the boundary conditions can be introduced. Then, the following can be written:

$$\begin{aligned}\iiint_{\Omega_c} \mathrm{div}(\sigma(\frac{\partial \mathbf{A}}{\partial t} + \frac{\partial \phi \chi_\phi}{\partial t} + \mathbf{grad}V))\psi d\tau = \\ - \iiint_{\Omega_c} \sigma(\frac{\partial \mathbf{A}}{\partial t} + \frac{\partial \phi \chi_\phi}{\partial t} + \mathbf{grad}V).\mathbf{grad}\psi d\tau \\ + \oiint_{\Gamma_c} \psi\sigma(\frac{\partial \mathbf{A}}{\partial t} + \frac{\partial \phi \chi_\phi}{\partial t} + \mathbf{grad}V).\mathbf{n}dS = 0\end{aligned} \qquad [4.70]$$

The surface integral term on Γ_c makes the normal component of the current density appear. But this component is zero on the boundary Γ_c (see equation [3.376]). Under these conditions, equation [4.70] has the form:

$$\iiint_{\Omega_c} \sigma(\frac{\partial \mathbf{A}}{\partial t} + \frac{\partial \phi \chi_\phi}{\partial t} + \mathbf{grad}V).\mathbf{grad}\psi d\tau = 0 \qquad [4.71]$$

The weak form of magnetodynamic equations, with the electric formulation and where the source term is the magnetic flux, is given by equations [4.68] and [4.71]. The choice of **A**-V formulation, associated with potential χ_ϕ, allows for the strong verification of the basis equations [3.301] and [3.303]. The same is true for the boundary conditions of the magnetic flux density **B** on the boundary Γ_b, as shown by the properties of **A** and χ_ϕ (see relations [3.381] and [3.382]). On the other hand, the weighted residual method leads to verifying equations [3.302] and [3.304] in the weak sense, respectively, on domains Ω and Ω_c. Similarly, the boundary conditions of fields **H** and **J** on, respectively, the boundaries $\Gamma_{h1} \cup \Gamma_{h2}$and Γ_j, are also weakly verified. This is due to the boundary integrals of equations [4.67] and [4.70] considered zero in expressions [4.68] and [4.71], respectively.

Considering now the case in which the source term is a magnetomotive force f_m imposed on the boundaries Γ_{hk}, the approach is the same as that used in section 4.2.2.1. As shown in section 3.6.2.1.2, the flux ϕ becomes an unknown of the problem. To obtain a full equation system, relation [3.407] is added to equations [4.68] and [4.71].

4.2.5.2.2. Magnetic formulation T-φ

With the formulation **T**-φ, involving magnetic quantities, imposing the magnetomotive force f_m is quite natural (see section 3.6.2.2). For the example presented in Figure 3.24, the electric vector potential **T** belongs to the function space $H_0(\mathbf{curl}, \Omega)$, as shown by equation [3.410]. It should, however, be noted that $\mathbf{T} = 0$ in $\Omega\text{-}\Omega_c$. As for the magnetic scalar potential φ, it belongs (see equation [3.416]) to $H_{\Gamma h1 \cup \Gamma h2}(\mathbf{grad}, \Omega)$. The equations to be solved are given by relations [3.427] and [3.428]. They are recalled as follows:

$$\mathbf{curl}\sigma^{-1}\mathbf{curl}(\mathbf{T}) + \frac{\partial}{\partial t}(\mu(\mathbf{T} + f_m\boldsymbol{\beta}_s - \mathbf{grad}\varphi)) = 0 \text{ on } \Omega_c \quad [4.72]$$

$$\mathrm{div}(\mu(\mathbf{T} + f_m\boldsymbol{\beta}_s - \mathbf{grad}\varphi)) = 0 \text{ on } \Omega \quad [4.73]$$

In these expressions, $\boldsymbol{\beta}_s$ represents the support field of the magnetomotive force. It is defined by relation [3.412].

In order to apply the weighted residual method, the weighting functions must be defined. In the case of equation [4.72], the vector operator is a curl and the associated function space is defined by the electric field (see equation [3.376]), namely $H(\mathbf{curl}, \Omega_c)$. In this case, the adjoint operator (see Table 2.1) is the curl and the weighting functions $\boldsymbol{\Psi}$ belong to the function space $H_{\Gamma c}(\mathbf{curl}, \Omega_c)$. As for equation [4.73], the vector operator is the divergence and the function space is defined by the magnetic flux density with $\mathbf{B} \in H_{\Gamma b}(\mathrm{div}0, \Omega)$, as shown by equation [3.377]. The adjoint operator is then (see Table 2.1) the gradient, and the weighting functions ψ belong to $H_{\Gamma h1 \cup \Gamma h2}(\mathbf{grad}, \Omega)$. These various results are summarized in Table 4.10.

Equation	**Vector operator**	**Associated function space**	**Adjoint operator**	**Weighting function**
[4.72]	**curl**	$H(\mathbf{curl}, \Omega_c)$	**curl**	$\boldsymbol{\Psi} \in H_{\Gamma c}(\mathbf{curl}, \Omega_c)$
[4.73]	div	$H_{\Gamma b}(\mathrm{div}0, \Omega_c)$	– **grad**	$\psi \in H_{\Gamma h1 \cup \Gamma h2}(\mathbf{grad}, \Omega_c)$

Table 4.10. *Magnetodynamics, imposed magnetic quantities, electric formulation T-φ; vector operator and function space of potentials and weighting functions*

Let us now apply the weighted residual method to these two equations. For equation [4.72], this yields:

$$\iiint_{\Omega_c} (\mathbf{curl}(\sigma^{-1}\mathbf{curlT}) + \frac{\partial}{\partial t}(\mu(\mathbf{T} + f_m\boldsymbol{\beta}_s - \mathbf{grad}\varphi))).\boldsymbol{\Psi}d\tau = 0 \quad [4.74]$$

Let us introduce the boundary conditions by means of relation [2.27] applied to the first term of the integral as follows:

$$\iiint_{\Omega_c} \mathbf{curl}(\sigma^{-1}\mathbf{curlT}).\boldsymbol{\Psi}d\tau = \iiint_{\Omega_c} \sigma^{-1}\mathbf{curlT}.\mathbf{curl}\boldsymbol{\Psi}d\tau - \oiint_{\Gamma_c} (\mathbf{n} \wedge \sigma^{-1}\mathbf{curlT}).\boldsymbol{\Psi}d\tau \quad [4.75]$$

Considering the vector space to which the functions $\boldsymbol{\Psi}$ belong, the surface integral is equal to zero. Under these conditions, the weighted residual method applied to equation [4.72] has the form:

$$\iiint_{\Omega_c} (\sigma^{-1}\mathbf{curlT}.\mathbf{curl}\boldsymbol{\Psi} + \frac{\partial}{\partial t}\mu(\mathbf{T} + f_m\boldsymbol{\beta}_s - \mathbf{grad}\varphi).\boldsymbol{\Psi})d\tau = 0 \quad [4.76]$$

Let us now apply the weighted residual method to equation [4.73]. The following equation is then obtained:

$$\iiint_{\Omega} \psi \mathrm{div}(\mu(\mathbf{T} + f_m\boldsymbol{\beta}_s - \mathbf{grad}\varphi))d\tau = 0 \quad [4.77]$$

Using relation [2.23], the boundary conditions are introduced, differentiating between the boundaries $\Gamma_{h1} \cup \Gamma_{h2}$ and Γ_b, as follows:

$$\begin{aligned} &\iiint_{\Omega} \psi \mathrm{div}(\mu(\mathbf{T} + f_m\boldsymbol{\beta}_s - \mathbf{grad}\varphi))d\tau = \\ &\quad -\iiint_{\Omega} \mu(\mathbf{T} + f_m\boldsymbol{\beta}_s - \mathbf{grad}\varphi).\mathbf{grad}\psi d\tau \\ &\quad\quad + \iint_{\Gamma_{h1} \cup \Gamma_{h2}} \psi\mu(\mathbf{T} + f_m\boldsymbol{\beta}_s - \mathbf{grad}\varphi).\mathbf{n}dS \\ &\quad\quad\quad + \iint_{\Gamma_b} \psi\mu(\mathbf{T} + f_m\boldsymbol{\beta}_s - \mathbf{grad}\varphi).\mathbf{n}dS = 0 \end{aligned} \quad [4.78]$$

First of all, consider the surface integral on $\Gamma_{h1} \cup \Gamma_{h2}$; the properties of the weighting function ψ (see Table 4.10) show that it is equal to zero. On the other hand, the second integral on Γ_b introduces the normal component of the field **B** (see equation [3.418]). But as shown by equation [3.377], this component is equal to zero on Γ_b.

Under these conditions, equation [4.78] is written as:

$$\iiint_\Omega \mu(\mathbf{T} + f_m \boldsymbol{\beta}_s - \mathbf{grad}\varphi).\mathbf{grad}\psi d\tau = 0 \quad [4.79]$$

When the source term is the magnetomotive force, with the magnetic formulation, the weak form of our magnetodynamics problem is described by equations [4.72] and [4.73]. The properties of potentials **T** and φ (see equations [3.409] and [3.416]), associated with field $\boldsymbol{\beta}_s$, support of the source field (see equation [3.412]) allow for the strong verification of equations [3.302] and [3.304]. The same is true for the boundary conditions on fields **H** and **J** (boundaries Γ_c and $\Gamma_{h1} \cup \Gamma_{h2}$) via the function spaces of definition of **T**, φ and $\boldsymbol{\beta}_s$. On the other hand, equations [3.301] and [3.303] are weakly verified with the weighted residual method on Ω_c and Ω, respectively. The boundary condition of the magnetic flux density on Γ_b is also weakly verified. This is due to the boundary integral of equation [4.78] considered zero in expression [4.79].

Let us now consider the case in which the source term is the magnetic flux ϕ imposed on the boundaries $\Gamma_{h1} \cup \Gamma_{h2}$: the approach is the same as in section 4.2.2.1. As shown in section 3.6.2.2.2, the magnetomotive force f_m becomes an unknown of the problem. To obtain a full equation system, relation [3.439] is added to equations [4.76] and [4.79].

4.2.6. *Synthesis of results*

In this section, the weighted residual method was applied to the examples presented in Chapter 3. With respect to electrostatics, electrokinetics and magnetostatics, we considered the multisource systems. For magnetodynamics, we applied the weighted residual method to the configurations introduced in section 3.6.

For each of the applications, the use of the weighted residual method, associated with the potential formulation, leads to the strong verification of an equilibrium equation and a weak verification of the other.

In all the cases, the strongly and weakly imposed properties are independent of the source terms; they depend only on the potential formulation used and the boundary conditions imposed on the boundary.

In this context, Table 4.11 summarizes for various formulations the properties that were strongly and weakly verified with the weighted residual method.

	Formulation	Strong properties	Weak properties
Electrostatics	Scalar potential V	$\mathbf{curlE} = 0,\ \mathbf{E} \wedge \mathbf{n}\|_{\Gamma_e} = 0$	$\mathrm{div}\,\mathbf{D} = 0,\ \mathbf{D}\,.\,\mathbf{n}\|_{\Gamma d} = 0$
	Vector potential **P**	$\mathrm{div}\mathbf{D} = 0,\ \mathbf{D}.\mathbf{n}\|_{\Gamma_d} = 0$	$\mathbf{curlE} = 0,\ \mathbf{E} \wedge \mathbf{n}\|_{\Gamma_e} = 0$
Electrokinetics	Scalar potential V	$\mathbf{curlE} = 0,\ \mathbf{E} \wedge \mathbf{n}\|_{\Gamma_e} = 0$	$\mathrm{div}\mathbf{J} = 0,\ \mathbf{J}\,.\mathbf{n}\|_{\Gamma_j} = 0$
	Vector potential **T**	$\mathrm{div}\mathbf{J} = 0,\ \mathbf{J}\,.\mathbf{n}\|_{\Gamma_j} = 0$	$\mathbf{curlE} = 0,\ \mathbf{E} \wedge \mathbf{n}\|_{\Gamma_e} = 0$
Magnetostatics	Scalar potential φ	$\mathbf{curlH} = \mathbf{J}_0,\ \mathbf{H} \wedge \mathbf{n}\|_{\Gamma_h} = 0$	$\mathrm{div}\mathbf{B} = 0,\ \mathbf{B}.\mathbf{n}\|_{\Gamma_b} = 0$
	Vector potential **A**	$\mathrm{div}\mathbf{B} = 0,\ \mathbf{B}.\mathbf{n}\|_{\Gamma_b} = 0$	$\mathbf{curlH} = \mathbf{J}_0,\ \mathbf{H} \wedge \mathbf{n}\|_{\Gamma_h} = 0$
Magnetodynamics	Electric (**A-V**)	$\mathbf{curlE} = -\frac{\partial \mathbf{B}}{\partial t}, \mathbf{E} \wedge \mathbf{n}\|_{\Gamma_e} = 0$ $\mathrm{div}\mathbf{B} = 0,\ \mathbf{B}.\mathbf{n}\|_{\Gamma_b} = 0$	$\mathbf{curlH} = \mathbf{J}, \mathbf{H} \wedge \mathbf{n}\|_{\Gamma_h} = 0$ $\mathrm{div}\mathbf{J} = 0,\ \mathbf{J}\,.\mathbf{n}\|_{\Gamma_j} = 0$
	Magnetic (**T**-φ)	$\mathbf{curlH} = \mathbf{J}, \mathbf{H} \wedge \mathbf{n}\|_{\Gamma_h} = 0$ $\mathrm{div}\mathbf{J} = 0,\ \mathbf{J}\,.\mathbf{n}\|_{\Gamma_j} = 0$	$\mathbf{curlE} = -\frac{\partial \mathbf{B}}{\partial t}, \mathbf{E} \wedge \mathbf{n}\|_{\Gamma_e} = 0$ $\mathrm{div}\mathbf{B} = 0,\ \mathbf{B}.\mathbf{n}\|_{\Gamma_b} = 0$

Table 4.11. *Potential formulations; properties of strongly and weakly verified solutions*

4.3. Finite element discretization

4.3.1. *The need for discretization*

Let us consider, as an example, the scalar potential formulation in the case of electrokinetics. The objective is to find a scalar potential V belonging to H(**grad**, Ω) and verifying equation [4.29] for any ψ belonging to H(**grad**, Ω). Assuming there is a basis υ_k ($k \in \mathbb{N}$) of H(**grad**, Ω), then we can write:

$$V = \sum_{k=1}^{\infty} \upsilon_k V_k \qquad [4.80]$$

where the scalar potentials V_k are the unknowns to be determined.

The number of terms to be calculated to weakly define the solution is infinite. Such a process is naturally impossible in practice, given that a computer can only process a finite amount of operations. There are cases where it is possible to determine an analytical or semi-analytical solution, but they concern structures with an extremely simple geometry. Consequently, we have to work with a basis of finite dimension υ'_k ($1 \leq k \leq N$) of a subspace of dimension N (denoted by H'(**grad**, Ω)) of H(**grad**, Ω). Then, we obtain an approximation V' to the "exact" solution in the weak sense, as the space H'(**grad**, Ω) is less "rich" than H(**grad**, Ω). This solution is written as:

$$V' = \sum_{k=1}^{N} \upsilon'_k V'_k \qquad [4.81]$$

Since there are N unknowns V'_k to be determined, we must find N independent equations. For this purpose, we can use the weighted residual method and apply the weak formulation [4.29]. As already noted, the weighting functions ψ belong to the same space H(**grad**, Ω) as the scalar potential V. The approach can be similar in the discrete case and V' and ψ' can be taken in the same space H'(**grad**, Ω). This is referred to as the Ritz–Galerkin method. Applying to equation [4.29] the N basis functions $\psi' = \upsilon'_k$, we build a system of N equations with N unknowns V'_k. The solution to this system of equations, coupled with enforcing the boundary conditions, makes it possible to determine the N coefficients V'_k. The quality of the approximation V' is expected to strongly depend on the choice of the basis υ'_k, referred to as "discretization", which will be used.

The finite element method is often used, in the case of low-frequency electromagnetism, to build these subspaces of discretization. The domain Ω is then decomposed into elementary geometric units. The domain thus discretized with a mesh $\mathcal{M}$ is denoted by Ω_d.

The geometric elements generally used in 3D are tetrahedra, hexahedra, prisms, etc. Based on this spatial division, a finite number of interpolation functions is defined, forming the basis of subspaces where the approximation to the solution is to be found.

Section 4.3.2 provides the expressions of the interpolation functions of the first order for tetrahedral elements that are widely used in practice. Moreover, most of the properties presented can subsequently be generalized to other types of elements. Section 4.3.3 will show how the vector operators **grad**, **curl** and div are discretized by introducing the notion of incidence matrix. Then, section 4.3.4 will present the discretization of physical fields and associated fields like potential and source fields as well as the introduction of gauges and boundary conditions. Finally, section 3.4.4 presents the Ritz–Galerkin method, which can be used to build a system of equations whose solution leads to the approximation to the solution of a given problem.

4.3.2. *Approximation functions*

A finite element is built from a geometric form, denoted by $\mathcal{k}$, which in 2D can be a triangle, a quadrangle, etc., or in 3D it can be a tetrahedron, a hexahedron, etc. As noted above, this book considers the three-dimensional case, namely tetrahedrons. This type of element remains the one that is actually most commonly used in practice, as it allows for the meshing of complex geometries that are met in low-frequency electromagnetism applications. It should moreover be noted that most of the properties that will be presented in detail in what follows can be generalized to other types of elements.

This element is associated with a space of $n_{\mathcal{k}}$ scalar or vector interpolation functions that correspond to degrees of freedom. Under these conditions, if a function f(x) is considered, defined on the element $\mathcal{k}$, it can be approximated by a discretized function $f_d(x)$ as follows:

$$f_d(x) = \sum_{i=1}^{n_{\mathcal{k}}} \omega_i(x) \mathcal{f}_i \cong f(x) \qquad [4.82]$$

where $\omega_i(x)$ represents the interpolation functions and $\mathcal{f}_i$ represents the coefficients that allow for the "best" approximation of the function f(x). In electromagnetism, Whitney elements (Bossavit 1997) are used, as will be seen, to create function spaces that have properties similar to those of the continuous domain, which were introduced in Chapter 2. The interpolation functions are then associated with geometric entities of the element, such as nodes, edges, facets and volumes.

For various types of interpolation functions (nodes, edges, facets and volumes), we present, in the case of tetrahedrons, the shape functions, their properties and the associated subspaces.

4.3.2.1. *Node elements*

In the case of node elements, for each node "i" of a mesh $\mathcal{M}$, a scalar function $\omega_{ni}(x)$ is defined. As an illustration, Figure 4.2(a) presents a tetrahedron and the four nodes associated with its vertices.

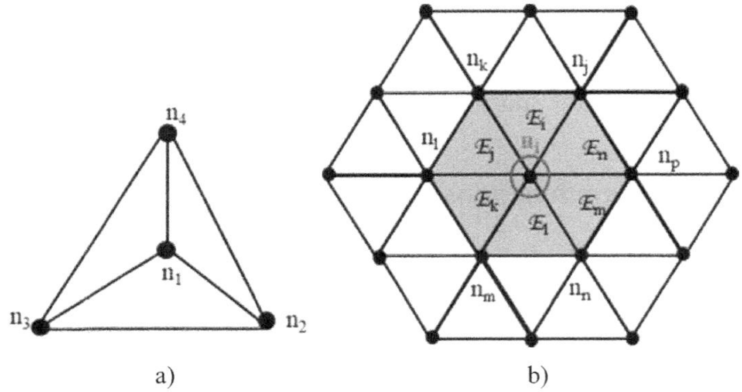

Figure 4.2. *First order node elements: a) case of a tetrahedron; b) 2D case, triangular elements related to node "n_0"*

The node interpolation functions $\omega_{ni}(x)$ are also known as "nodal elements". They are continuous on the domain Ω_d and vary linearly from one node to its nearest neighbors, in the case of first-order elements. Let us consider the example of an interpolation function $\omega_{ni}(x)$. It is equal to the unit at node "i" and it is equal to zero at other nodes of the mesh as follows:

$$\omega_{n_i}(x_j) = \delta(i, j) \qquad [4.83]$$

where x_j represents the coordinates of node "j" and $\delta(i,j)$ is the Kronecker symbol.

For a better illustration of this definition, let us consider, for the sake of simplicity, the 2D example consisting of triangular elements in Figure 4.2(b). The function associated with node n_i satisfies the property $\omega_{ni}(x_{ni}) = 1$. On the other hand, at node n_j, for example, we have $\omega_{ni}(x_{nj}) = 0$. Moreover, as mentioned above, the interpolation function $\omega_{ni}(x)$ varies linearly on the elements containing the node n_i. Hence, $\omega_{ni}(x)$ is equal to 1 at the node n_i and is equal to zero at the other nodes. In the case of Figure 4.2(b), the function $\omega_{ni}(x)$ is not equal to zero on the gray elements ($\mathcal{E}_i$, $\mathcal{E}_j$, $\mathcal{E}_k$, $\mathcal{E}_l$, $\mathcal{E}_m$ and $\mathcal{E}_n$). On the other elements of mesh $\mathcal{M}$, the function $\omega_{ni}(x)$ is equal to zero. On a given element k, only the functions associated with the nodes of $\mathcal{k}$ are not zero. Considering these properties, if $\mathcal{N}$ is the number of nodes of mesh $\mathcal{M}$, it can be shown that:

$$\sum_{i=1}^{\mathcal{N}} \omega_{n_i}(x) = 1 \quad \forall x \in \Omega_d \qquad [4.84]$$

Moreover, it can be noted that the function $\omega_{ni}(x)$ is continuous throughout the mesh.

If we now consider a function $u_d(x)$, belonging to the space defined by the set of functions $\omega_{ni}(x)$ on a mesh $\mathcal{M}$, this function is written as:

$$u_d(x) = \sum_{i=1}^{\mathcal{N}} \omega_{n_i}(x) u_i \qquad [4.85]$$

It can then be noted that, given the property [4.83], we have:

$$u_d(x_i) = u_i \qquad [4.86]$$

In other terms, the scalar values u_i, which will subsequently be the unknowns of the problem, have a physical meaning as they correspond to the values of field $u_d(x)$ at the various nodes of mesh $\mathcal{M}$.

Equation [4.85] can also be written using a vector of dimension $\mathcal{N}$ and denoted by $[\omega_{\mathcal{N}}(x)]$ (each term represents a nodal interpolation function) and a vector $[u_{\mathcal{N}}]$ that represents the $\mathcal{N}$ discrete values u_i of the function $u_d(x)$ at the nodes of the mesh $\mathcal{M}$. Then, we have:

$$u_d(x) = [\omega_{\mathcal{N}}(x)]^t [u_{\mathcal{N}}] \qquad [4.87]$$

Considering the continuity properties of the node interpolation functions, at the interface between two elements, the function $u_d(x)$ is naturally continuous. The set of nodal interpolation functions generates a discrete subspace, of finite dimension, that will be denoted by $W^0(\Omega_d)$. It can be verified that $W^0(\Omega_d)$ is a subspace of $H(\mathbf{grad}, \Omega_d)$, as the nodal functions are continuous on the domain and their gradient exists. We can then write:

$$W^0(\Omega_d) = \{u_d(x) \in H(\mathbf{grad}, \Omega_d);\ u_d(x) = \sum_{i=1}^{\mathcal{N}} \omega_{n_i}(x) u_i,\ u_i \in \Re\} \qquad [4.88]$$

4.3.2.2. *Edge elements*

Each of the $\mathcal{A}$ edges of the mesh $\mathcal{M}$ is associated with a vector interpolation function, denoted by $\boldsymbol{\omega}_{ai}(x)$. In the case of a tetrahedron, these functions are built based on interpolation functions of nodes located at the ends of the edge. Figure 4.3(a) shows the six edges of the tetrahedron with arbitrarily chosen orientation.

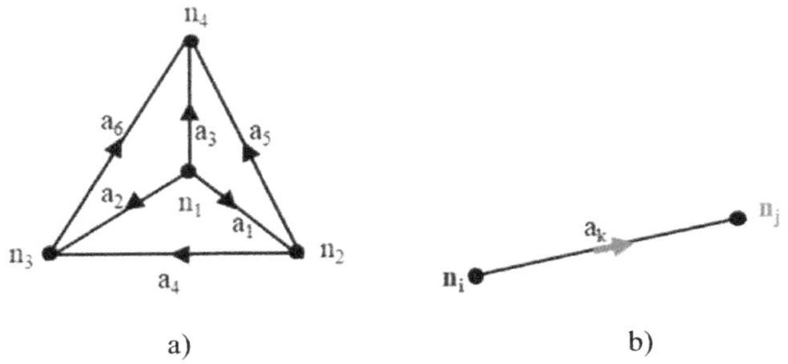

Figure 4.3. *a) Edge elements of a tetrahedron; b) orientation of an edge with i < j*

Considering now the edge a_k in Figure 4.3(b), the interpolation function $\boldsymbol{\omega}_{ak}(x)$ is built from the nodal functions associated with the nodes $\omega_{ni}(x)$ and $\omega_{nj}(x)$ as follows (Dular and Piriou 2008):

$$\boldsymbol{\omega}_{a_k}(x) = \omega_{n_i}(x)\,\mathbf{grad}\omega_{n_j}(x) - \omega_{n_j}(x)\,\mathbf{grad}\omega_{n_i}(x) \qquad [4.89]$$

Let us focus on the properties of this function and, for this, let us consider the edge a_1 of the tetrahedron in Figure 4.3(a). The nodes at the edge extremities are n_1 and n_2. Let us now consider the facet defined by nodes $\{n_2, n_3, n_4\}$. On this facet, the function $\omega_{n1}(x)$ is zero ($\omega_{n1}(x)$ is zero at nodes n_2, n_3 and n_4 (see equation [4.83]), and varies linearly on the facet). Consequently, the interpolation function $\boldsymbol{\omega}_{a1}(x)$ has the following form:

$$\boldsymbol{\omega}_{a_1}(x) = \omega_{n_2}(x)\,\mathbf{grad}\omega_{n_1}(x) \qquad [4.90]$$

Moreover, as $\omega_{n1}(x)$ is zero on the facet defined by the nodes $\{n_2, n_3, n_4\}$, its gradient is normal to this facet. The circulation of $\boldsymbol{\omega}_{a1}(x)$ on any path belonging to this facet is therefore zero and therefore on its three edges, i.e. a_4, a_5 and a_6. Applying the same reasoning to the facet defined by the nodes $\{n_1, n_3, n_4\}$, it can be shown that the circulation of $\boldsymbol{\omega}_{a1}(x)$ is also zero on the edges a_2 and a_3. Moreover, we have the circulation of $\boldsymbol{\omega}_{a1}(x)$ that is equal to 1 on its own edge. It can be noted that, except for the edge a_1 with which it is associated, the circulation of $\boldsymbol{\omega}_{a1}(x)$ is zero on all the edges of the elements to which it belongs and therefore on the entire mesh. This result can be generalized to the set of edge interpolation functions and we have the following property:

$$\int_{a_j} \boldsymbol{\omega}_{a_k}(x).\mathbf{dl} = \delta(j,k) \qquad [4.91]$$

Let us now consider a function $\mathbf{v}_d(x)$ belonging to the space defined by the edge elements, which can be written as:

$$\mathbf{v}_d(x) = \sum_{i=1}^{\mathcal{A}} \boldsymbol{\omega}_{a_i}(x)v_i \qquad [4.92]$$

In this expression, due to the property [4.91], the scalar v_i represents the value of the circulation of $\mathbf{v}_d(x)$ on the edge "I" of the mesh. Expression [4.92] of the vector function $\mathbf{v}_d(x)$ can also be written using a vector $[\boldsymbol{\omega}_{\mathcal{A}}(x)]$, composed of $\mathcal{A}$ vector interpolation functions of the mesh $\mathcal{M}$. The vector $[v_{\mathcal{A}}]$ is then defined, corresponding to circulations v_i of function $\mathbf{v}_d(x)$, as follows:

$$\mathbf{v}_d(x) = [\boldsymbol{\omega}_{\mathcal{A}}(x)]^t [v_{\mathcal{A}}] \qquad [4.93]$$

Finally, according to equation [4.89], since nodal functions and also the tangential component of their gradient are continuous at the interface between two elements, the tangential component of interpolation functions $\omega_{ai}(x)$ and therefore $\mathbf{v}_d(x)$ are conserved.

The set of edge interpolation functions generates the discrete subspace, of finite dimension, denoted by $W^1(\Omega_d)$. It can be shown that $W^1(\Omega_d)$ belongs to the space H(**curl**, Ω_d):

$$W^1(\Omega_d) = \{v_d(x) \in H(\mathbf{curl}, \Omega_d);\ \ v_d(x) = \sum_{i=1}^{\mathcal{A}} \omega_{a_i}(x)v_i,\ v_i \in \Re\} \qquad [4.94]$$

4.3.2.3. *Facet elements*

Each of the $\mathcal{F}$ facets of the mesh is associated with a vector interpolation function denoted by $\boldsymbol{\omega}_{fk}(x)$. Similar to the edge elements, these interpolation functions are built from interpolation functions of nodes belonging to the facet. Figure 4.4(a) shows the four facets of a tetrahedron.

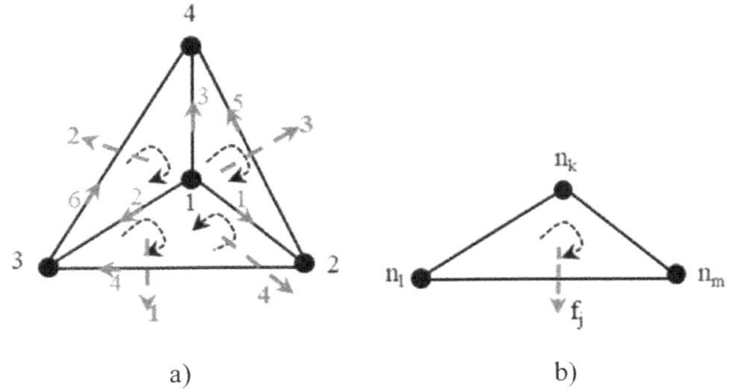

Figure 4.4. *a) Facet elements of a tetrahedron; b) facet orientation*

The orientation of facets is defined by their normal (using the right-hand rule). It should be noted that these orientations are completely arbitrary but their definition remains essential. As an illustration, Table 4.12 presents for each facet of the tetrahedron in Figure 4.4(a), the succession of nodes corresponding to their orientation.

Facet elements; succession of nodes			
$f_1 = \{1, 2, 3\}$	$f_2 = \{1, 3, 4\}$	$f_3 = \{1, 4, 2\}$	$f_4 = \{2, 4, 3\}$

Table 4.12. *Facet elements; succession of nodes of the tetrahedron in Figure 4.4(a)*

Let us consider the facet f_j, represented in Figure 4.4(b). Following its orientation, we have the succession of nodes $\{n_k, n_m, n_l\}$. The corresponding interpolation function can then be written as (Dular and Piriou 2008):

$$\begin{aligned}\omega_{f_j}(x) = 2(&\omega_{n_k}(x)\,\mathbf{grad}\,\omega_{n_m}(x) \wedge \mathbf{grad}\,\omega_{n_l}(x) \\ &+ \omega_{n_l}(x)\,\mathbf{grad}\,\omega_{n_k}(x) \wedge \mathbf{grad}\,\omega_{n_m}(x) \\ &+ \omega_{n_m}(x)\,\mathbf{grad}\,\omega_{n_l}(x) \wedge \mathbf{grad}\,\omega_{n_k}(x)).\end{aligned} \quad [4.95]$$

As can be noted in the above expression, the three terms are obtained by the circular permutation of the indices of node functions.

Let us now consider the facet f_1, of Figure 4.4(a), defined by the nodes {1, 2, 3}, its interpolation function is written based on equation [4.95]:

$$\boldsymbol{\omega}_{f_1}(x) = 2(\omega_{n_1}(x)\,\mathbf{grad}\omega_{n_2}(x) \wedge \mathbf{grad}\omega_{n_3}(x) + \omega_{n_3}(x)\,\mathbf{grad}\omega_{n_1}(x) \wedge \mathbf{grad}\omega_{n_2}(x) + \omega_{n_2}(x)\,\mathbf{grad}\omega_{n_3}(x) \wedge \mathbf{grad}\omega_{n_1}(x)) \quad [4.96]$$

Consider the flux of the function $\boldsymbol{\omega}_{f1}(x)$ through the facet f_2 defined by the nodes{1, 3, 4}. On the one hand, as the function $\omega_{n2}(x)$ is zero on the facet f_2, the flux of term $\omega_{n2}(x)\mathbf{grad}\omega_{n3}(x) \wedge \mathbf{grad}\omega_{n1}(x)$ is zero. On the other hand, as noted in the case of edge interpolation functions, given that $\omega_{n2}(x)$ is constant on the facet f_2, vector $\mathbf{grad}\omega_{n2}(x)$ is normal to it. Under these conditions, since the two remaining terms $\omega_{n1}(x)\mathbf{grad}\omega_{n2}(x) \wedge \mathbf{grad}\omega_{n3}(x)$ and $\omega_{n3}(x)\mathbf{grad}\omega_{n1}(x) \wedge \mathbf{grad}\omega_{n2}(x)$ are in a plane perpendicular to $\mathbf{grad}\omega_{n2}(x)$, they are tangential to the facet f_2. Consequently, the flux of these two terms through f_2 is also equal to zero. A similar type of reasoning can be applied to facets f_3 and f_4 leading to the fact that the flux of $\boldsymbol{\omega}_{f1}(x)$ is also zero through these facets. The same approach is taken on the facets of the adjacent element (not represented in Figure 4.4(a)) containing the facet f_1. Then, it can be noted that the flux of $\boldsymbol{\omega}_{f1}(x)$ through f_1 is equal to 1 and is zero on the other facets of this element.

Let us now consider all the other facets of the mesh that do not belong to the two adjacent elements containing the facet f_1. It proves that these nodal functions associated with the nodes of f_1 (namely n_1, n_2 and n_3) are zero on these facets and therefore the flux of $\boldsymbol{\omega}_{f1}(x)$ through them. In conclusion, the flux of $\boldsymbol{\omega}_{f1}(x)$ is zero through all the facets of the mesh, except for f_1 where it is equal to 1. This result can be generalized to the set of facets of the mesh and the following property is deduced:

$$\iint_{f_i} \boldsymbol{\omega}_{f_j}(x).\mathbf{n}dS = \delta(i, j) \quad [4.97]$$

Moreover, at the interface between two elements, the normal component of the function $\mathbf{w}_d(x)$ is conserved.

Consider now a vector function $\mathbf{w}_d(x)$ belonging to the space defined by the elements of the facet, which is written as:

$$\mathbf{w}_d(x) = \sum_{i=1}^{\mathcal{F}} \boldsymbol{\omega}_{f_i}(x) w_i \quad [4.98]$$

In this expression, given the property [4.97], w_i represents the flux of the function $\mathbf{w}_d(x)$ through the facet "i". This expression can also be written using a vector $[\boldsymbol{\omega}_{\mathcal{F}}(x)]$, composed of $\mathcal{F}$ vector interpolation functions of mesh $\mathcal{M}$, and a vector $[w_{\mathcal{F}}]$ which entries are the functions w_i, as follows:

$$\mathbf{w}_d(x) = [\boldsymbol{\omega}_{\mathcal{F}}(x)]^t \, [w_{\mathcal{F}}] \tag{4.99}$$

The set of facet interpolation functions generates a discrete space, of finite dimension, that is denoted by $W^2(\Omega_d)$. This space is a subspace of H(div, Ω_d), such that:

$$W^2(\Omega_d) = \{\mathbf{w}_d(x) \in H(\mathrm{div}, \Omega_d);\ \ \mathbf{w}_d(x) = \sum_{i=1}^{\mathcal{F}} \boldsymbol{\omega}_{f_i}(x) w_i,\ \ w_i \in \Re\} \tag{4.100}$$

4.3.2.4. *Volume elements*

The volume elements associate with each element of the mesh $\mathcal{V}$, a scalar interpolation function $\omega_{vi}(x)$. These volume elements are such that for a given element "v_i", its value is constant and is equal to (Dular and Piriou 2008):

$$\omega_{v_i}(x) = \frac{1}{\mathrm{vol}(v_i)} \text{ on element i and } \omega_{v_i} = 0 \text{ elsewhere} \tag{4.101}$$

where vol(v_i) represents the volume of the element "i". Under these conditions, integrating on the volume, the following property is found:

$$\iiint_{v_i} \omega_{v_i}(x) d\tau = 1 \tag{4.102}$$

Likewise, the integral of the function $\omega_{vi}(x)$ on an element v_j that differs from v_i is equal to zero. The following property is found:

$$\iiint_{v_i} \omega_{v_j}(x) d\tau = \delta(i, j) \tag{4.103}$$

Considering a scalar function $p_d(x)$, belonging to the set of volume elements, it can be written as:

$$p_d(x) = \sum_{i=1}^{\mathcal{V}} \omega_{v_i}(x) p_i \tag{4.104}$$

where p_i represents, for the element "i", the value of the volume integral of $p_d(x)$. Introducing the vector $[\omega_{\mathcal{V}}(x)]$ of dimension "$\mathcal{V}$" and the vector of $\mathcal{V}$ values of p_k (denoted $[p_{\mathcal{V}}]$), we have:

$$p_d(x) = [\omega_{\mathcal{V}}(x)]^t \, [p_{\mathcal{V}}] \qquad [4.105]$$

Taking into account the properties of functions ω_{vi}, at the interface between two elements, the discretized scalar quantity $p_d(x)$ is discontinuous. The set of volume interpolation functions generates the discrete subspace, of finite dimension, denoted by $W^3(\Omega_d)$. It is a subspace of $L^2(\Omega_d)$:

$$W^3(\Omega_d) = \left\{ p_d(x) \in L^2(\Omega_d);\ p_d(x) = \sum_{i=1}^{\mathcal{V}} \omega_{v_i}(x) p_i,\ p_i \in \Re \right\} \qquad [4.106]$$

4.3.2.5. *Synthesis of properties*

The main properties of the nodal, edge, facet and volume elements are summarized in Table 4.13. This presents the notations that will be used to define the interpolation functions, the continuity properties of discretized quantities at the interface between two elements, as well as the discrete subspaces generated.

Functions	**Properties**	**Continuity at the interface of elements**	**Generated space**
$\omega_n(x)$	$\omega_{n_i}(x_j) = \delta_{ij}$	Continuous	$W^0(\Omega_d)$
$\omega_a(x)$	$\int_{a_j} \boldsymbol{\omega}_{a_k}(x).\mathbf{dl} = \delta(j,k)$	$\omega_a(x) \wedge \mathbf{n}$, continuous	$W^1(\Omega_d)$
$\omega_f(x)$	$\int_{f_j} \boldsymbol{\omega}_{f_k}(x).\mathbf{n}dS = \delta(j,k)$	$\omega_f(x).\mathbf{n}$, continuous	$W^2(\Omega_d)$
$\omega_v(x)$	$\int_{v_j} \omega_{v_k}(x).dv = \delta(j,k)$	Discontinuous	$W^3(\Omega_d)$

Table 4.13. *Summary of the properties of the node, edge, facet and volume elements*

4.3.3. *Discretization of vector operators*

4.3.3.1. *Incidence matrices*

This section focuses on the edge–node, facet–edge and volume–facet incidences for a given mesh $\mathcal{M}$. Matrices of incidence between the various geometric entities of the mesh will be built. These matrices will allow us to define the discrete forms of gradient, curl and divergence vector operators. Then, these discrete operators will be applied to functions belonging to subspaces $W^0(\Omega_d)$, $W^1(\Omega_d)$ and $W^2(\Omega_d)$. Similar to section 4.3.2, this section will present the expressions of matrices in the case of tetrahedral elements. Nevertheless, for other types of elements (hexahedra, pyramids, etc.), the expressions can be deduced in a similar manner (Dular 1996; Geuzaine 2001).

4.3.3.2. *Node–edge incidence*

For a given mesh, the incidence of a node "n" on an edge "a" is denoted by i(a,n). This incidence is equal to 1 if the node "n" corresponds to the end of the edge, and −1 if it corresponds to the origin (it should be recalled that an edge is oriented, allowing for the definition of an origin and an end). If the node does not belong to the edge, this incidence is equal to zero. For a mesh composed of $\mathcal{A}$ edges and $\mathcal{N}$ nodes, the $\mathcal{A}\times\mathcal{N}$ incidence matrix will be denoted by $[\mathbf{G}_{\mathcal{A},\mathcal{N}}]$. As an illustration, consider the extremely simple case of the tetrahedral element in Figure 4.4(a). The matrix $[\mathbf{G}_{\mathcal{A},\mathcal{N}}]$ composed of six rows each associated with an edge and four columns each associated with one node is represented (see Table 4.14).

I(a,n)	**1**	**2**	**3**	**4**
1	−1	1	0	0
2	−1	0	1	0
3	−1	0	0	1
4	0	−1	1	0
5	0	−1	0	1
6	0	0	−1	1

Table 4.14. *$[G_{\mathcal{A}\mathcal{N}}]$ edge–node incidence matrix of a tetrahedron*

Consider now a function $u_d(x)$ discretized in $W^0(\Omega_d)$, namely the space of the nodal elements. Then, we have (see equations [4.85] and [4.87]):

$$u_d(x) = \sum_{i=1}^{\mathcal{N}} \omega_{n_i}(x) u_i = [\omega_{\mathcal{N}}(x)]^t [u_{\mathcal{N}}] \qquad [4.107]$$

If we now apply the gradient operator to the function $u_d(x)$, we have:

$$\mathbf{v}_d(x) = \mathbf{grad}\, u_d(x) = \sum_{i=1}^{\mathcal{N}} (\mathbf{grad}\, \omega_{n_i}(x)) u_i = (\mathbf{grad}[\omega_{\mathcal{N}}(x)]^t)[u_{\mathcal{N}}] \qquad [4.108]$$

Based on this expression, following the developments given in the note at the end of this section, we can write:

$$\mathbf{grad}\,\omega_{n_i}(x) = \sum_{k=1}^{\mathcal{A}} i(k,i)\boldsymbol{\omega}_{a_k}(x) = \lfloor \boldsymbol{\omega}_{\mathcal{A}}(x) \rfloor^t \lfloor i_{\mathcal{A},i} \rfloor \quad [4.109]$$

where the vector $[i_{\mathcal{A},i}]$ corresponds to the incidence of the node "i" on the $\mathcal{A}$ edges of the mesh.

Using expression [4.109], the following can be deduced for the set of $\mathcal{N}$ nodes of the mesh (Bossavit 1991):

$$\mathbf{grad}\,\lfloor \omega_{\mathcal{N}}(x) \rfloor^t = \lfloor \boldsymbol{\omega}_{\mathcal{A}}(x) \rfloor^t \lfloor \mathbf{G}_{\mathcal{A},\mathcal{N}} \rfloor \quad [4.110]$$

From expression [4.108], we can write, using relations [4.110] and [4.93], the succession of equations:

$$\mathbf{v}_d(x) = (\mathbf{grad}\lfloor \omega_{\mathcal{N}}(x) \rfloor^t)\lfloor u_{\mathcal{N}} \rfloor = \lfloor \boldsymbol{\omega}_{\mathcal{A}}(x) \rfloor^t \lfloor \mathbf{G}_{\mathcal{A},\mathcal{N}} \rfloor \lfloor u_{\mathcal{N}} \rfloor = \lfloor \boldsymbol{\omega}_{\mathcal{A}}(x) \rfloor^t \lfloor v_{\mathcal{A}} \rfloor \quad [4.111]$$

First of all, it can be noted that the gradient $\mathbf{v}_d(x)$ of a function $u_d(x)$ of $W^0(\Omega_d)$ belongs to $W^1(\Omega_d)$. Then, we have:

$$\mathbf{grad}(W^0(\Omega_d)) \subset W^1(\Omega_d) \quad [4.112]$$

By identification, from equation [4.111], the following relation can be deduced:

$$\lfloor v_{\mathcal{A}} \rfloor = \lfloor \mathbf{G}_{\mathcal{A},\mathcal{N}} \rfloor \lfloor u_{\mathcal{N}} \rfloor \quad [4.113]$$

Therefore, the $[\mathbf{G}_{\mathcal{A},\mathcal{N}}]$ incidence matrix allows for the direct determination of the vector$[v_{\mathcal{A}}]$ of the gradient of a function $u_d(x)$ based on components $[u_{\mathcal{N}}]$ in $W^0(\Omega_d)$. It can be noted that this expression does not depend on the shape functions. Hence, the $[\mathbf{G}_{\mathcal{A},\mathcal{N}}]$ matrix is the discrete equivalent of the gradient operator. If the domain is connected, the rank of this matrix, denoted by $\mathcal{R}_{[G]}$, is given by the following relation:

$$\mathcal{R}_{[\mathbf{G}]} = \mathcal{N} - 1 \quad [4.114]$$

NOTE.– In what follows, our objective is to find the space to which function $\mathbf{grad}\omega_{ni}(x)$ belongs. Equation [4.84] is written as:

$$\sum_{j=1}^{\mathcal{N}} \omega_{n_j}(x) = 1 \qquad [4.115]$$

Applying the gradient operator to this equation, we have:

$$\sum_{j=1}^{\mathcal{N}} \mathbf{grad}\,\omega_{n_j}(x) = 0 \qquad [4.116]$$

Consider the term $\mathbf{grad}\omega_{ni}(x)$ of equation [4.108]. Taking into account the properties [4.115] and [4.116], the following equality can be written:

$$\begin{aligned}\mathbf{grad}\omega_{n_i}(x) = &\left(\sum_{j=1}^{\mathcal{N}} \omega_{n_j}(x)\right)\mathbf{grad}\omega_{n_i}(x)\\ &- \omega_{n_i}(x)\left(\sum_{j=1}^{\mathcal{N}} \mathbf{grad}\omega_{n_j}(x)\right)\end{aligned} \qquad [4.117]$$

The linearity properties of the sum operator allow this expression to be rewritten in the following form:

$$\mathbf{grad}\omega_{n_i}(x) = \sum_{j=1}^{\mathcal{N}} (\omega_{n_j}(x)\mathbf{grad}\omega_{n_i}(x) - \omega_{n_i}(x)\mathbf{grad}\omega_{n_j}(x)) \qquad [4.118]$$

It can be noted that if a node "j" is not connected, via an edge, to a node "i", then the function $\omega_{nj}(x)\mathbf{grad}\omega_{ni}(x) - \omega_{ni}(x)\mathbf{grad}\omega_{nj}(x)$ is zero on the domain Ω_d. Indeed, in this case, we have $\omega_{ni}(x)$ or $\omega_{nj}(x)$ that is equal to zero. Let us denote by $\mathcal{S}_{e\{i\}}$ the set of nodes "j" connected to node "i". Under these conditions, equation [4.118] can be rewritten as:

$$\mathbf{grad}\omega_{n_i}(x) = \sum_{j\in \mathcal{S}_{e\{i\}}} (\omega_{n_j}(x)\mathbf{grad}\omega_{n_i}(x) - \omega_{n_i}(x)\mathbf{grad}\omega_{n_j}(x)) \qquad [4.119]$$

If we consider a node "j" belonging to $\mathcal{S}_{e\{i\}}$, then there is an edge "k" with "i" and "j" as vertices. This is shown by the example in Figure 4.5 (extracted from Figure 4.2(b)), which reproduces the elements of $\mathcal{S}_{e\{i\}}$ with six edges having as one vertex the node "i".

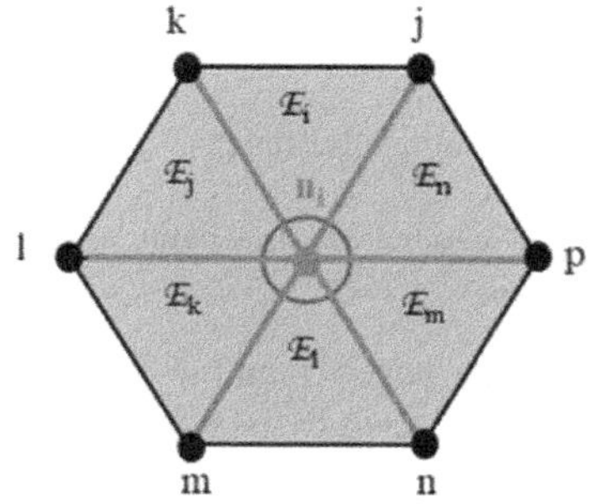

Figure 4.5. *Set Se{i} of nodes of Figure 4.2, connected to the node "i" via an edge of the mesh*

Moreover, relation [4.119] shows, up to a sign, the expression of edge functions $\boldsymbol{\omega}_{ak}(x)$ having the node "i" as one of their vertices (see equation [4.89]). The sign depends on the orientation of the considered edge "k" and, therefore, on the fact that the node "i" is either the origin, or the end. It is given by the edge–node incidence i(k,i). Moreover, if an edge "k" does not have the node "i" as a vertex, then the edge–node incidence i(k,i) is zero. Under these conditions, it can be verified that relation [4.118] can be rewritten in the following form using the incidence i(k,i):

$$\mathbf{grad}\,\omega_{n_i}(x) = \sum_{k=1}^{\mathcal{A}} i(k,i)\,\boldsymbol{\omega}_{a_k}(x) \qquad [4.120]$$

The vector of edge interpolation functions $[\boldsymbol{\omega}_{\mathcal{A}}(x)]^t$ can also be introduced as follows:

$$\mathbf{grad}\,\omega_{n_i}(x) = \sum_{k=1}^{\mathcal{A}} i(k,i)\boldsymbol{\omega}_{a_k}(x) = \left[\boldsymbol{\omega}_{\mathcal{A}}(x)\right]^t\left[i_{\mathcal{A},i}\right] \qquad [4.121]$$

This is similar to relation [4.109].

4.3.3.3. *Facet–edge incidence*

The facet–edge incidence is denoted by i(f, a). It should be recalled that the facets and edges have their own orientation. Hence, the incidence i(f, a) is equal to:

– "1" if the direction of circulation of edge "a" is the same as the one defining the orientation of the facet "f";

– "– 1" if the direction of circulation of edge "a" is opposite to the one defining the orientation of the facet "f";

– "0" if the edge does not belong to the facet.

Consider, as an example, Figure 4.4(a) and the sequence of nodes of the facet "f_1" of Table 4.12. Then, we have:

– i(1, 1) = 1, i(1, 4) = 1, as the edges 1 and 4 have the same orientation as the direction of circulation of facet 1 defined by its normal (right-hand rule);

– i(1, 2) = – 1, as the orientation of edge 2 is opposite to the direction of circulation of facet 1;

– i(1, 3) = 0, i(1, 5) = 0 and i(1, 6) = 0, as the edges 3, 5 and 6 do not belong to facet 1.

More generally, for a mesh composed of $\mathcal{F}$ facets and $\mathcal{A}$ edges, the dimension of the incidence matrix is $\mathcal{F}\times\mathcal{A}$. This matrix will be denoted by $[\mathbf{R}_{\mathcal{F}\mathcal{A}}]$. Considering again the case of the tetrahedron in Figure 4.4(a), the matrix $[\mathbf{R}_{\mathcal{F}\mathcal{A}}]$ is presented in Table 4.15.

i(f,a)	**1**	**2**	**3**	**4**	**5**	**6**
1	1	– 1	0	1	0	0
2	0	1	– 1	0	0	1
3	– 1	0	1	0	– 1	0
4	0	0	0	– 1	1	– 1

Table 4.15. *Facet–edge incidence matrix $[R_{\mathcal{F}\mathcal{A}}]$ of the tetrahedron in Figure 4.4(a)*

Let us now consider a field $\mathbf{v}_d(x)$, discretized in $W^1(\Omega_d)$. Equations [4.92] and [4.93] are written as:

$$\mathbf{v}_d(x) = \sum_{i=1}^{\mathcal{A}} \boldsymbol{\omega}_{a_i}(x) v_i = [\boldsymbol{\omega}_{\mathcal{A}}(x)]^t [v_{\mathcal{A}}] \qquad [4.122]$$

Applying to this equation the curl operator, we have:

$$\mathbf{w}_d(x) = \mathbf{curl}(\mathbf{v}_d(x)) = \sum_{k=1}^{\mathcal{A}} \mathbf{curl}(\boldsymbol{\omega}_{a_k}(x)) v_k \qquad [4.123]$$

Based on this expression, the next note will contain the developments that allow us to write the following equation:

$$\mathbf{curl}\left[\omega_{\mathcal{A}}(x)\right]^{t} = \left[\omega_{\mathcal{F}}(x)\right]^{t}\left[R_{\mathcal{F},\mathcal{A}}\right] \quad [4.124]$$

The function $\mathbf{w}_d(x)$, introduced in equation [4.123], is written as follows:

$$\mathbf{w}_d(x) = \mathbf{curl}\left[\omega_{\mathcal{A}}(x)\right]^{t}\left[v_{\mathcal{A}}\right] = \left[\omega_{\mathcal{F}}(x)\right]^{t}\left[\mathbf{R}_{\mathcal{F},\mathcal{A}}\right]\left[v_{\mathcal{A}}\right] \quad [4.125]$$

It can be noted that the curl $\mathbf{w}_d(x)$ of a function $\mathbf{v}_d(x)$ of $W^1(\Omega_d)$ belongs to $W^2(\Omega_d)$, which yields:

$$\mathbf{curl}(W^1(\Omega_d)) \subset W^2(\Omega_d) \quad [4.126]$$

By identification, the following property can be deduced from equations [4.99] and [4.125]:

$$\left[\mathbf{w}_{\mathcal{F}}\right] = \left[\mathbf{R}_{\mathcal{F},\mathcal{A}}\right]\left[v_{\mathcal{A}}\right] \quad [4.127]$$

Therefore, the matrix $[\mathbf{R}_{\mathcal{F},\mathcal{A}}]$ is the equivalent of the curl operator in the discrete domain. For a connected domain, the rank $\mathcal{R}_{[\mathbf{R}]}$ of matrix $[\mathbf{R}_{\mathcal{F},\mathcal{A}}]$ depends on the number of edges and nodes of the mesh and we have (Bossavit 1997):

$$\mathcal{R}[\mathbf{R}] = \mathcal{A} - (\mathcal{N} - 1) \quad [4.128]$$

Finally, the properties of the curl and the gradient in the discrete domain are equivalent to those in the continuous domain. In fact, for any field $u_d(x)$ belonging to $W^0(\Omega_d)$, we have $\mathbf{curl}(\mathbf{grad}u_d(x)) = 0$. Using expressions [4.87], [4.110] and [4.124], this property allows us to write:

$$\left[\omega_{\mathcal{F}}(x)\right]^{t}\left[\mathbf{R}_{\mathcal{F},\mathcal{A}}\right]\left[\mathbf{G}_{\mathcal{A},\mathcal{N}}\right]\left[u_{\mathcal{N}}\right] = 0 \quad [4.129]$$

And this for any function $u_d(x)$ hence any vector $[u_{\mathcal{N}}]$. Relation [4.129] allows us to write the property:

$$\left[\mathbf{R}_{\mathcal{F},\mathcal{A}}\right]\left[\mathbf{G}_{\mathcal{A},\mathcal{N}}\right] = 0 \quad [4.130]$$

We find here in a discrete form, with incidence matrices, the fact that the curl of a gradient is zero.

NOTE.– In order to determine the space to which functions $\mathbf{curl}\boldsymbol{\omega}_{ai}(x)$ belong, let us consider the simple example of the tetrahedron $\mathcal{V}$, represented in Figure 4.6. $\boldsymbol{\omega}_{ai}(x)$ is the notation for the interpolation function of edge i, whose vertices are the nodes j, k, and which is common to the referenced facets "n" and "o".

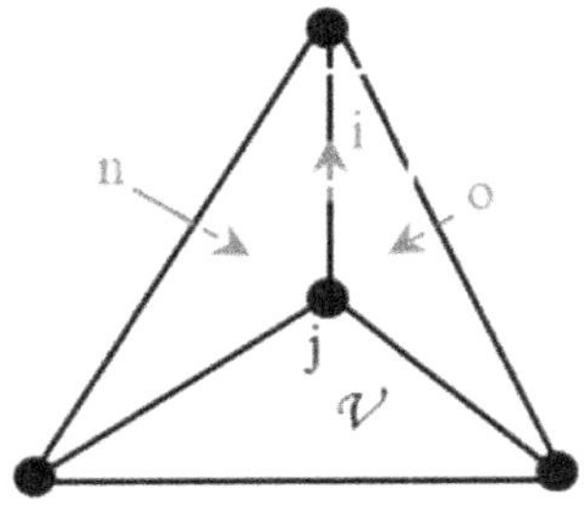

Figure 4.6. *Denomination of nodes j, k, l, m of a tetrahedron and of the edge i common to facets n and o*

Let us first recall a property of the curl operator:

$$\mathbf{curl}p\mathbf{v} = p\mathbf{curl}\mathbf{v} + \mathbf{grad}p \wedge \mathbf{v} \qquad [4.131]$$

where p is a scalar function and $\mathbf{v}$ is a vector function.

Let us now express the term $\mathbf{curl}\boldsymbol{\omega}_{ai}(x)$ in which the edge interpolation function is replaced by its expression given by equation [4.89]. Using formula [4.131] and after development, the following can be written:

$$\mathbf{curl}\omega_{a_i}(x) = 2\mathbf{grad}\omega_{n_j}(x) \wedge \mathbf{grad}\omega_{n_k}(x) \qquad [4.132]$$

where $\omega_{nj}(x)$ and $\omega_{nk}(x)$ represent the nodal interpolation functions of the two nodes of edge i.

Let us now consider (see Figure 4.6) the element $\mathcal{V}$ containing the edge i. The two other nodes of $\mathcal{V}$, differing from j and k, are the nodes l and m. Using the property [4.84], for the four nodes of element $\mathcal{V}$, equation [4.132] can be written in the following form:

$$\begin{aligned}\mathbf{curl}\omega_{a_i}(x) &= \sum_{h\in\{j,k,l,m\}} 2(\omega_{n_h}(x)\mathbf{grad}\omega_{n_j}(x)\wedge\mathbf{grad}\omega_{n_k}(x)) \\ &= 2\omega_{n_j}(x)\mathbf{grad}\omega_{n_j}(x)\wedge\mathbf{grad}\omega_{n_k}(x) \\ &+ 2\omega_{n_k}(x)\mathbf{grad}\omega_{n_j}(x)\wedge\mathbf{grad}\omega_{n_k}(x) \\ &+ 2\omega_{n_l}(x)\mathbf{grad}\omega_{n_j}(x)\wedge\mathbf{grad}\omega_{n_k}(x) \\ &+ 2\omega_{n_m}(x)2\mathbf{grad}\omega_{n_j}(x)\wedge\mathbf{grad}\omega_{n_k}(x)\end{aligned} \qquad [4.133]$$

Let us now apply the gradient operator to expression [4.84], as shown in equation [4.116], then we have:

$$\begin{aligned}\mathbf{grad}(\sum_{h\in\{j,k,l,m\}} \omega_{n_h}(x)) &= \mathbf{grad}\omega_{n_j}(x) + \mathbf{grad}\omega_{n_k}(x) \\ &+ \mathbf{grad}\omega_{n_l}(x) + \mathbf{grad}\omega_{n_m}(x) = 0\end{aligned} \qquad [4.134]$$

Based on equation [4.134], $\mathbf{grad}\omega_{nj}(x)$ can be expressed in relation [4.133] as a function of the gradient of three node functions $\omega_{nk}(x)$, $\omega_{nl}(x)$ and $\omega_{nm}(x)$. The same can be done for $\mathbf{grad}\omega_{nk}(x)$. These two functions are then replaced by their expression in relation [4.133]. It should be recalled that the vector product of a vector by itself is equal to zero. Then, after development, relation [4.133] has the form:

$$\begin{aligned}\mathbf{curl}\omega_{a_i}(x) &= -2\omega_{n_j}(x)\mathbf{grad}\omega_{n_l}(x)\wedge\mathbf{grad}\omega_{n_k}(x) \\ &- 2\omega_{n_j}(x)\mathbf{grad}\omega_{n_m}(x)\wedge\mathbf{grad}\omega_{n_k}(x) \\ &- 2\omega_{n_k}(x)\mathbf{grad}\omega_{n_j}(x)\wedge\mathbf{grad}\omega_{n_l}(x) \\ &- 2\omega_{n_k}(x)\mathbf{grad}\omega_{n_j}(x)\wedge\mathbf{grad}\omega_{n_m}(x) \\ &+ 2\omega_{n_l}(x)\mathbf{grad}\omega_{n_j}(x)\wedge\mathbf{grad}\omega_{n_k}(x) \\ &+ 2\omega_{n_m}(x)\mathbf{grad}\omega_{n_j}(x)\wedge\mathbf{grad}\omega_{n_k}(x)\end{aligned} \qquad [4.135]$$

To introduce the facet functions related to the facet n defined by the nodes j, l and k and to the facet o defined by the nodes j, m and k, the first three terms depending on the node functions associated with the nodes j, k and l are gathered with the three terms below depending on the nodes associated with the nodes j, m and k:

$$\begin{aligned}\mathbf{curl}\boldsymbol{\omega}_{a_i}(x) = & -2\omega_{n_j}(x)\mathbf{grad}\omega_{n_l}(x)\wedge\mathbf{grad}\omega_{n_k}(x) \\ & -2\omega_{n_k}(x)\mathbf{grad}\omega_{n_j}(x)\wedge\mathbf{grad}\omega_{n_l}(x) \\ & +2\omega_{n_l}(x)\mathbf{grad}\omega_{n_j}(x)\wedge\mathbf{grad}\omega_{n_k}(x) \\ & -2\omega_{n_j}(x)\mathbf{grad}\omega_{n_m}(x)\wedge\mathbf{grad}\omega_{n_k}(x) \\ & -2\omega_{n_k}(x)\mathbf{grad}\omega_{n_j}(x)\wedge\mathbf{grad}\omega_{n_m}(x) \\ & +2\omega_{n_m}(x)\mathbf{grad}\omega_{n_j}(x)\wedge\mathbf{grad}\omega_{n_k}(x)\end{aligned} \quad [4.136]$$

This equation can be written by inverting the terms of the vector product as follows:

$$\begin{aligned}\mathbf{curl}\boldsymbol{\omega}_{a_i}(x) = & -2\omega_{n_j}(x)\mathbf{grad}\omega_{n_l}(x)\wedge\mathbf{grad}\omega_{n_k}(x) \\ & -2\omega_{n_k}(x)\mathbf{grad}\omega_{n_j}(x)\wedge\mathbf{grad}\omega_{n_l}(x) \\ & -2\omega_{n_l}(x)\mathbf{grad}\omega_{n_k}(x)\wedge\mathbf{grad}\omega_{n_j}(x) \\ & -2\omega_{n_j}(x)\mathbf{grad}\omega_{n_m}(x)\wedge\mathbf{grad}\omega_{n_k}(x) \\ & -2\omega_{n_k}(x)\mathbf{grad}\omega_{n_j}(x)\wedge\mathbf{grad}\omega_{n_m}(x) \\ & -2\omega_{n_m}(x)\mathbf{grad}\omega_{n_k}(x)\wedge\mathbf{grad}\omega_{n_j}(x)\end{aligned} \quad [4.137]$$

This expression reveals (see equation [4.95]) the interpolation functions $\boldsymbol{\omega}_{fn}(x)$ and $\boldsymbol{\omega}_{fo}(x)$ of the facets "n" and "o" containing the nodes j, k, l and j, k, m (see Figure 4.6). These are counted as positive if the orientation of the facet corresponds to that of the edge and as negative otherwise. The defined edge–facet incidence can be naturally introduced here. Finally, on element $\mathcal{V}$, we have:

$$\mathbf{curl}\boldsymbol{\omega}_{a_i}(x) = i(n,i)\omega_{f_n}(x) + i(o,i)\boldsymbol{\omega}_{f_o}(x) \quad [4.138]$$

As this was defined, we have:

– the incidences i(n,i) of facets "n" that do not contain the edge "i" are zero;

– the function $\boldsymbol{\omega}_{fn}(x)$ is also zero on the element $\mathcal{V}$ if the facet n does not belong to $\mathcal{V}$.

Then, we can write, on the element $\mathcal{V}$:

$$\mathbf{curl}\boldsymbol{\omega}_{a_i}(x) = \sum_{m=1}^{\mathcal{F}} i(m,i)\,\boldsymbol{\omega}_{f_m}(x) \tag{4.139}$$

A similar reasoning can be applied to the set of elements $\mathcal{V}$ connected to the edge "i". On the elements that do not contain the edge "i", the function $\boldsymbol{\omega}_{ai}(x)$ is zero, and it can be verified that relation [4.139] is still true. It can therefore be concluded that relation [4.139] is verified on the entire domain Ω_d. It is then possible to complete equation [4.138] by introducing the vector of the facet interpolation functions $[\boldsymbol{\omega}_{\mathcal{F}}(x)]^t$ as follows:

$$\mathbf{curl}\boldsymbol{\omega}_{a_i}(x) = \sum_{m=1}^{\mathcal{F}} i(m,i)\,\boldsymbol{\omega}_{f_m}(x) = [\boldsymbol{\omega}_{\mathcal{F}}(x)]^t\left[i_{\mathcal{F},i}\right] \tag{4.140}$$

In this equation, the vector $[i_{\mathcal{F},i}]$ represents the incidence of the edge i on the $\mathcal{F}$ facets of the mesh. Introducing the matrix $[\mathbf{R}_{\mathcal{F},\mathcal{A}}]$, the property introduced in equation [4.124] can be deduced as follows:

$$\mathbf{curl}\left[\boldsymbol{\omega}_{\mathcal{A}}(x)\right]^t = [\boldsymbol{\omega}_{\mathcal{F}}(x)]^t\left[R_{\mathcal{F},\mathcal{A}}\right] \tag{4.141}$$

4.3.3.4. *Volume–facet incidence*

The volume–facet incidence is denoted by i(v,f). If the facet belongs to the element $\mathcal{V}$, this incidence is equal to 1 or –1 depending on the orientation of the normal to the facet (inward or outward with respect to the element). If the facet does not belong to the considered element, this incidence is equal to zero. For the tetrahedron in Figure 4.4(a), the incidence matrix is represented in Table 4.16. In this table, all the incidences are equal to 1, as all four facets have an outward orientation. In the general case of a mesh composed of $\mathcal{V}$ elements and $\mathcal{F}$ facets, the $\mathcal{V}\times\mathcal{F}$ incidence matrix will be denoted by $[D_{\mathcal{V},\mathcal{F}}]$.

i(v,f)	**1**	**2**	**3**	**4**
1	1	1	1	1

Table 4.16. *Volume–facet incidence matrix $[D_{\mathcal{V}\mathcal{F}}]$ of a tetrahedron*

Let us now consider a vector function $\mathbf{w}_d(x)$ of the space of facet elements. Using the notations of equations [4.98] and [4.99], we can write:

$$\mathbf{w}_d(x) = \sum_{i=1}^{\mathcal{F}} \boldsymbol{\omega}_{f_i}(x) w_i = [\boldsymbol{\omega}_{\mathcal{F}}(x)]^t [w_{\mathcal{F}}] \tag{4.142}$$

Applying the divergence operator to this equation, we have:

$$p_d(x) = \mathrm{div}(\mathbf{w}_d(x)) = \sum_{i=1}^{\mathcal{F}} \mathrm{div}(\boldsymbol{\omega}_{f_i}(x)) w_i = \mathrm{div}([\boldsymbol{\omega}_{\mathcal{F}}(x)]^t)[w_{\mathcal{F}}] \tag{4.143}$$

Based on this expression, and relying on the developments given in the note below, we can write:

$$\mathrm{div}(\boldsymbol{\omega}_{f_i}(x)) = \sum_{j=1}^{\mathcal{V}} i(j,i)\omega_{v_i}(x) = [\omega_{\mathcal{V}}(x)]^t [i_{\mathcal{V},f_i}] \tag{4.144}$$

where $[i_{\mathcal{V},fi}]$ represents the vector of incidences i(j,i) of the facet i on the set of $\mathcal{V}$ volumes j of the mesh. Based on this equation, and considering the set of $\mathcal{F}$ facets, if we introduce the matrix $[D_{\mathcal{V},\mathcal{F}}]$, we obtain the following property:

$$\mathrm{div}[\boldsymbol{\omega}_{\mathcal{F}}(x)]^t = [\omega_{\mathcal{V}}(x)]^t [D_{\mathcal{V},\mathcal{F}}] \tag{4.145}$$

Gathering equations [4.143] and [4.145], we can write:

$$p_d(x) = \mathrm{div}([\boldsymbol{\omega}_{\mathcal{F}}(x)]^t)[w_{\mathcal{F}}] = [\omega_{\mathcal{V}}(x)]^t [D_{\mathcal{V},\mathcal{F}}][w_{\mathcal{F}}] \tag{4.146}$$

This expression shows that the divergence $p_d(x)$ of a function of $W^2(\Omega_d)$ belongs to $W^3(\Omega_d)$. Therefore, we have:

$$\mathrm{div}(W^2(\Omega_d)) \subset W^3(\Omega_d) \tag{4.147}$$

Using equation [4.105], based on relation [4.146], the following property can be deduced:

$$[p_{\mathcal{V}}] = [D_{\mathcal{V},\mathcal{F}}][w_{\mathcal{F}}] \tag{4.148}$$

This relation allows us to state that the matrix $[D_{\mathcal{V},\mathcal{F}}]$ represents the equivalent of the divergence operator in the discrete domain. The rank $\mathcal{R}_{[D]}$ of matrix $[D_{\mathcal{V},\mathcal{F}}]$ depends on the number of facets, edges and nodes of the mesh. It is expressed by the relation (Dular 1996; Bossavit 1997):

$$\mathcal{R}_{[D]} = \mathcal{F} - (\mathcal{A} - (\mathcal{N} - 1)) \tag{4.149}$$

Finally, in the discrete domain, the properties of the divergence and curl are equivalent to those of the continuous domain. Indeed, considering a field $\mathbf{v}_d(x)$ of $W^1(\Omega_d)$, we then have $\text{div}(\mathbf{curlv}_d(x)) = 0$. The following can be readily deduced from equations [4.122], [4.124] and [4.145]:

$$[\boldsymbol{\omega}_{\mathcal{V}}(x)]^t [D_{\mathcal{V},\mathcal{F}}][\mathbf{R}_{\mathcal{F},\mathcal{A}}][v_{\mathcal{A}}] = 0 \tag{4.150}$$

which is valid for any vector $[v_{\mathcal{A}}]$ representing the components of the field $\mathbf{v}_d(x)$ in $W^1(\Omega_d)$, which is reflected by the property:

$$[D_{\mathcal{V},\mathcal{F}}]\ [\mathbf{R}_{\mathcal{F},\mathcal{A}}] = 0 \tag{4.151}$$

NOTE.– Consider now the term $\text{div}(\boldsymbol{\omega}_{fi}(x))$ associated with the facets whose vertices are the nodes {k, m, l} and let us find the space to which it belongs.

As a first step, using the properties of the divergence and curl vector operators, we show that the following can be written after development:

$$\begin{aligned} \text{div}(p\mathbf{u} \wedge \mathbf{v}) &= p\,\text{div}(\mathbf{u} \wedge \mathbf{v}) + (\mathbf{u} \wedge \mathbf{v}).\mathbf{grad}p \\ &= p(\mathbf{v}.\mathbf{curlu} - \mathbf{u}.\mathbf{curlv}) + (\mathbf{u} \wedge \mathbf{v}).\mathbf{grad}p \end{aligned} \tag{4.152}$$

In $\text{div}(\boldsymbol{\omega}_{fi}(x))$, $\boldsymbol{\omega}_{fi}(x)$ is replaced by equation [4.95] and formula [4.152] is applied. After development, we obtain:

$$\begin{aligned} \text{div}(\boldsymbol{\omega}_{f_i}(x) = {} & (\mathbf{grad}\omega_{n_m}(x) \wedge \mathbf{grad}\omega_{n_l}(x)).\mathbf{grad}\omega_{n_k}(x) + \\ & (\mathbf{grad}\omega_{n_k}(x) \wedge \mathbf{grad}\omega_{n_m}(x)).\mathbf{grad}\omega_{n_l}(x) + \\ & (\mathbf{grad}\omega_{n_l}(x) \wedge \mathbf{grad}\omega_{n_k}(x)).\mathbf{grad}\omega_{n_m}(x) \end{aligned} \tag{4.153}$$

Since the mixed product is unchanged by circular permutation, this equation can be rewritten as follows:

$$\text{div}(\boldsymbol{\omega}_{f_i}(x)) = 3(\mathbf{grad}\,\omega_{n_m}(x) \wedge \mathbf{grad}\,\omega_{n_l}(x)).\mathbf{grad}\,\omega_{n_k}(x)) \tag{4.154}$$

Consider now two adjacent elements o and p containing the facet i. The function $\mathrm{div}(\boldsymbol{\omega}_{fi}(x))$ is constant on the element o and equal, up to a sign, to the inverse of its volume. The same is true on element p. The sign depends on the orientation of the facet defined by the permutation of nodes k, m and l. It is therefore related to the notion of the facet–element incidence. On the other hand, the function $\mathrm{div}(\boldsymbol{\omega}_{fi}(x))$ is zero on any other element v of the mesh for which the facet–element incidence i(v,i) is zero. Using the definition of volume elements [4.101], we obtain the previously introduced relation [4.144] as follows:

$$\mathrm{div}(\boldsymbol{\omega}_{f_i}(x)) = \sum_{j=1}^{\mathcal{V}} i(v_j, f_i)\omega_{v_i}(x) = \left[\omega_{\mathcal{V}}(x)\right]^t \left[i_{\mathcal{V},f_i}\right] \qquad [4.155]$$

4.3.3.5. *Properties of discrete subspaces*

The main results obtained from incidence matrices are summarized in Table 4.17.

Using relations [4.112], [4.126] and [4.147], it can be shown that the discrete subspaces $W^k(\Omega_d)$, where $k \in \{0, 1, 2, 3\}$, form a sequence of spaces with properties similar (see Figure 4.7) to those met in the continuous domain (see section 2.4) and that depend on the topology of the domain, as shown in Figure 2.10.

Element type	Discrete space	Discrete operator
Node	$W^0(\Omega_d)$	$\left[\mathbf{G}_{\mathcal{A},\mathcal{N}}\right]$
Edge	$W^1(\Omega_d)$	$\left[\mathbf{R}_{\mathcal{F},\mathcal{A}}\right]$
Facet	$W^2(\Omega_d)$	$\left[\mathrm{D}_{\mathcal{V},\mathcal{F}}\right]$
Volume	$W^3(\Omega_d)$	/

Table 4.17. *Discrete domain, function spaces and vector operators*

$$W^0(\Omega_d) \xrightarrow{\mathbf{grad}} W^1(\Omega_d) \xrightarrow{\mathbf{curl}} W^2(\Omega_d) \xrightarrow{\mathrm{div}} W^3(\Omega_d)$$

Figure 4.7. *Sequence of discrete function spaces*

Based on equations [4.130] and [4.151], two properties can be found:

$$\mathrm{Im}(\mathbf{grad}W^0(\Omega_d)) \subset \ker(\mathbf{curl}W^1(\Omega_d)) \qquad [4.156]$$

$$\mathrm{Im}(\mathbf{curl}W^1(\Omega_d)) \subset \ker(\mathrm{div}W^2(\Omega_d)) \qquad [4.157]$$

As indicated in section 2.3.4, the first property becomes an equality if the domain is simply connected. It is also the case for the second if the boundary of the domain is connected. If the domain is contractible, then we have:

$$\mathrm{Im}(\mathbf{grad}W^0(\Omega_d)) = \ker(\mathbf{curl}W^1(\Omega_d)) \qquad [4.158]$$

$$\mathrm{Im}(\mathbf{curl}W^1(\Omega_d)) = \ker(\mathrm{div}W^2(\Omega_d)) \qquad [4.159]$$

Similar to the continuous domain, we can also introduce the homogeneous boundary conditions and define the discrete function subspaces. Consider the subspace $W^0_{\Gamma k}(\Omega_d)$ of nodal elements defined by:

$$W^0_{\Gamma_k}(\Omega_d) = \left\{ u_d \in W^0(\Omega_d), u_d = 0|_{\Gamma_k} \right\} \qquad [4.160]$$

and $W^1_{\Gamma n}(\Omega_d)$, which corresponds to the space of edge elements such that:

$$W^1_{\Gamma_n}(\Omega_d) = \left\{ \mathbf{v}_d \in W^1(\Omega_d), \mathbf{v}_d \wedge \mathbf{n}|_{\Gamma_n} = 0 \right\} \qquad [4.161]$$

and $W^2_{\Gamma m}(\Omega_d)$, which corresponds to the space of facet elements such that:

$$W^2_{\Gamma_m}(\Omega_d) = \left\{ \mathbf{w}_d \in W^2(\Omega_d), \mathbf{w}_d . \mathbf{n}|_{\Gamma_m} = 0 \right\} \qquad [4.162]$$

It should be noted that these various subspaces are similar to those introduced in the continuous domain. As an illustration, Table 4.18 summarizes the correspondence between the continuous and discrete function spaces.

In conclusion, there is a great similarity between the properties of discrete spaces introduced in this section and those encountered in the continuous domain (see Chapter 2). All of these results will be used for the discretization of electromagnetic fields and potentials, and also for taking into account source terms (support fields and associated potentials).

Function spaces	
Continuous domain	**Discrete domain**
$H_{\Gamma_n}(\mathbf{grad},\Omega)$	$W^0_{\Gamma_n}(\Omega_d)$
$H_{\Gamma_n}(\mathbf{curl},\Omega)$	$W^1_{\Gamma_n}(\Omega_d)$
$H_{\Gamma_m}(\mathrm{div},\Omega)$	$W^2_{\Gamma_m}(\Omega_d)$
$L^2(\Omega)$	$W^3(\Omega_d)$

Table 4.18. *Correspondence between the function spaces in the continuous domain and in the discrete domain*

4.3.4. *Discretization of physical quantities and associated fields*

Maxwell's equations, as presented in Chapter 1, introduce the vector fields **E**, **H**, **B**, **J** and **D**, as well as the volume density of charges ρ. In Chapter 3, these fields were associated with function spaces in the continuous domain. Then, the scalar and vector potentials V, **P**, **T**, **A** and φ were introduced for electrostatics, electrokinetics, magnetostatics and also magnetodynamics. On the other hand, in order to take into account the sources of electromagnetic fields, we introduced source fields, defined by their support fields **β** and **λ** or their associated potentials, α and **χ**, respectively. These source fields, support fields and potentials have been defined in function spaces in relation to the physical quantities to which they are associated.

As already seen in section 4.3.3, some properties of the function spaces encountered in the continuous domain can be transposed in the discrete domain. We defined in the continuous domain the spaces to which the fields and potentials belong (see the Tonti diagram, Figure 3.25). As illustrated in Table 4.18, for each function space in the continuous domain, there is an equivalent space in the discrete domain. Therefore, it seems natural to use this correspondence to define the discretization spaces. As an illustration, Table 4.19 presents for the fields **E**, **H**, **B**, **J**, **D** and the potentials V, φ, **P**, **T** and **A**, the function spaces to which they belong in the continuous domain and their equivalent in the discrete domain. Concerning the boundary conditions, they are given as an illustration. They obviously depend on the studied problem and on the source terms imposed on the boundary.

For the discrete function spaces of the support fields and associated potentials, the approach is equivalent, relying on the fields they represent.

The function space $W^0(\Omega_d)$ allows for the discretization, by means of nodal elements, of the fields that are defined, in the continuous domain, in the function space H(**grad**, Ω). Consider the example of the scalar potential $V \in H(\mathbf{grad}, \Omega)$; its approximation can then be written in the discrete domain (see Table 4.19) in the following form:

$$V_d(x) = \sum_{i=1}^{\mathcal{N}} \omega_{n_i}(x) V_i = [\omega_{\mathcal{N}}(x)]^t [V_{\mathcal{N}}] \qquad [4.163]$$

where V_i is the value of the scalar potential $V_d(x)$ at node i and $[V_{\mathcal{N}}]$ is the vector with components V_i.

Physical quantities	Continuous domain		Discrete domain	
	Notation	Function space	Notation	Function space
Electric scalar potential	V	$H_{\Gamma_e}(\mathbf{grad}, \Omega)$	V_d	$W^0_{\Gamma_e}(\Omega_d)$
Magnetic scalar potential	φ	$H_{\Gamma_h}(\mathbf{grad}, \Omega)$	φ_d	$W^0_{\Gamma_h}(\Omega_d)$
Electric field	**E**	$H_{\Gamma_e}(\mathbf{curl}, \Omega)$	$\mathbf{E}_d$	$W^1_{\Gamma_e}(\Omega_d)$
Magnetic field	**H**	$H_{\Gamma_h}(\mathbf{curl}, \Omega)$	$\mathbf{H}_d$	$W^1_{\Gamma_h}(\Omega_d)$
Electric sector potential	**P**	$H_{\Gamma_d}(\mathbf{curl}, \Omega)$	$\mathbf{P}_d$	$W^1_{\Gamma_d}(\Omega_d)$
Electric sector potential	**T**	$H_{\Gamma_j}(\mathbf{curl}, \Omega)$	$\mathbf{T}_d$	$W^1_{\Gamma_j}(\Omega_d)$
Magnetic sector potential	**A**	$H_{\Gamma_b}(\mathbf{curl}, \Omega)$	$\mathbf{A}_d$	$W^1_{\Gamma_b}(\Omega_d)$
Magnetic flux density	**B**	$H_{\Gamma_b}(\mathrm{div}0, \Omega)$	$\mathbf{B}_d$	$W^2_{\Gamma_b}(\Omega_d)$
Current density	**J**	$H_{\Gamma_j}(\mathrm{div}0, \Omega)$	$\mathbf{J}_d$	$W^2_{\Gamma_j}(\Omega_d)$
Electric displacement field	**D**	$H_{\Gamma_d}(\mathrm{div}, \Omega)$	$\mathbf{D}_d$	$W^2_{\Gamma_d}(\Omega_d)$
Electric charge density	ρ	$L^2(\Omega)$	ρ_d	$W^3(\Omega_d)$

Table 4.19. *Function spaces of fields or pote ntials in the continuous domain and the discrete domain*

As for the space $W^1(\Omega_d)$, it allows for the discretization with edge elements of quantities defined in the continuous space H(**curl**, Ω). Under these conditions, the approximation of an electric field $\mathbf{E} \in H(\mathbf{curl}, \Omega)$ can be written in the following form:

$$\mathbf{E}_d(x) = \sum_{i=1}^{\mathcal{A}} \boldsymbol{\omega}_{a_i}(x) E_i = [\boldsymbol{\omega}_{\mathcal{A}}(x)]^t [E_{\mathcal{A}}] \quad [4.164]$$

where E_i is the circulation of the electric field $\mathbf{E}_d(x)$ on the edge i of the mesh and $[E_{\mathcal{A}}]$ is the vector with entries E_i.

Likewise, the space $W^2(\Omega_d)$ allows, with facet elements, for the discretization of fields defined in the continuous domain in the space H(div, Ω).

Hence, the approximation of the magnetic flux density **B**, belonging to H(div0, Ω), has the following expression:

$$\mathbf{B}_d(x) = \sum_{i=1}^{\mathcal{F}} \boldsymbol{\omega}_{f_i}(x) B_i = [\boldsymbol{\omega}_{\mathcal{F}}(x)]^t [B_{\mathcal{F}}] \quad [4.165]$$

where B_i is the magnetic flux density $\mathbf{B}_d(x)$ through the facet i of the mesh and $[B_{\mathcal{F}}]$ is the vector composed of the entries B_i.

Finally, the discrete space $W^3(\Omega_d)$, corresponding to the space $L^2(\Omega)$ in the continuous domain, allows the electric charge density ρ to be expressed with the volume elements in the following form:

$$\rho_d(x) = \sum_{i=1}^{\mathcal{V}} \omega_{v_i}(x) \rho_i = [\omega_{\mathcal{V}}(x)]^t [\rho_{\mathcal{V}}] \quad [4.166]$$

where ρ_i is the charge contained in the element i of the mesh and $[\rho_{\mathcal{V}}]$ is the vector whose entries are ρ_i.

Based on Table 4.19 and the previous formulas, the expressions of the other vector fields or support fields are very easily deduced.

4.3.5. *Taking into account homogeneous boundary conditions*

As already seen in section 4.3.4, the fields are naturally discretized in the space of the node, edge and facet elements. This section shows how to impose, on a part of the boundary of the domain, homogeneous boundary conditions (see equations

[4.160], [4.161] and [4.162]). These conditions depend on the space in which the considered quantity is discretized and it will be shown that they are imposed by acting on the degrees of freedom.

4.3.5.1. *Case of node elements*

The space of the nodal elements $W^0(\Omega_d)$ allows for the discretization of scalar functions continuous on the domain (see Table 4.13). These functions can be associated with homogeneous boundary conditions on one part of the boundary, as indicated by relation [4.160].

Let us consider, as an example, the expression of the electric scalar potential (see equation [4.163]) and denote by Γ_e the boundary on which $V = 0$ should be imposed. Consider now a facet f_k belonging to Γ_e and n_1, n_2 and n_3 the three associated nodes. The value of V on the facet f_k depends only on the node values V_{n1}, V_{n2} and V_{n3} (see section 4.3.2.1). Consequently, if it is expected to have $V = 0$ on f_k, it is sufficient to impose the values $V_{n1} = V_{n2} = V_{n3} = 0$. This reasoning can be generalized to the set of facets covering the surface Γ_e, and it can thus be shown that to impose $V = 0$ on Γ_e, it is sufficient to have the value of V_{ni} zero on all the nodes of this surface.

Let us denote by $\mathcal{N}$ the number of nodes of the mesh and by $\mathcal{N}_0$ the number of nodes located on Γ_e. Let us assume there is a layout of nodes so that those belonging to the boundary Γ_e are the last ones in the order of numbering. Since the $\mathcal{N}_0$ node values are set to zero, the number of remaining nodes is written as $\mathcal{N}_r = \mathcal{N} - \mathcal{N}_0$. Under these conditions, the discretized form of the electric scalar potential, considering the boundary conditions homogeneous on the boundary, is written as:

$$V_d(x) = \sum_{i=1}^{\mathcal{N}_r} \omega_{n_i}(x) V_i = \left[\omega_{\mathcal{N}_r}(x)\right]^t \left[V_{\mathcal{N}_r}\right] \qquad [4.167]$$

4.3.5.2. *Case of edge elements*

As already seen previously, the fields defined in the function space H(**curl**, Ω) are discretized in the space of the edge elements $W^1(\Omega_d)$. As noted in section 4.3.2.2, the tangential component of fields of $W^1(\Omega_d)$ is conserved when passing from one element to another (see Table 4.13). For these fields, the boundary conditions homogeneous on one part of the boundary of the domain are defined by relation [4.161]. As an example, let us consider the electric field, whose discrete form $\mathbf{E}_d(x)$ is given by equation [4.164]. Let us assume that on one part of the boundary, denoted by Γ_e, we impose $\mathbf{E}_d(x) \wedge \mathbf{n} = 0$. Consider now a facet f_k, belonging to the boundary Γ_e, composed of edges a_1, a_2 and a_3. The expression of the tangential component of $\mathbf{E}_d(x)$, on the facet f_k, depends only on the circulations E_{a1}, E_{a2} and E_{a3} along the edges a_1, a_2 and a_3, respectively. Under these conditions, in order to impose to zero the tangential component of $\mathbf{E}_d(x)$ on f_k, we fix the circulations E_{a1}, E_{a2} and E_{a3} equal to zero. This reasoning can be generalized to the set of facets composing surface Γ_e and it can be concluded that in order to impose $\mathbf{E}_d(x) \wedge \mathbf{n} \,|_{\Gamma e} = 0$, the circulation of E_{ak} should be zero on all the edges a_k of Γ_e.

Let us consider now that the mesh is composed of $\mathcal{A}$ edges and that the boundary Γ_e, on which the condition ($\mathbf{E}_d(x) \wedge \mathbf{n} = 0$) is imposed, has $\mathcal{A}_0$ edges. Similar to section 4.3.5.1, let us assume that the layout of edges is such that those belonging to the boundary Γ_e are the last ones in the order of numbering. Since the values of $\mathcal{A}_0$ edges are fixed, the number of remaining unknown edges is written as $\mathcal{A}_r = \mathcal{A} - \mathcal{A}_0$. Then, the expression of the field is:

$$\mathbf{E}_d(x) = \sum_{i=1}^{\mathcal{A}_r} \boldsymbol{\omega}_{a_i}(x) E_i = \left[\boldsymbol{\omega}_{\mathcal{A}_r}(x)\right]^t \left[E_{\mathcal{A}_r}\right] \qquad [4.168]$$

4.3.5.3. *Case of facet elements*

The subspace of the facet elements $W^2(\Omega_d)$ allows for the discretization of fields defined in the function space H(div, Ω). For these fields, the boundary conditions on the boundary are given by relation [4.162]. As an example of discretized magnetic quantity in the space of facet elements, consider the magnetic flux density $\mathbf{B}$ with boundary conditions of type Γ_b (see Table 4.19). If a facet "f_i" belongs to the boundary Γ_b, the flux through this facet is equal to the component B_i associated with the facet function $\boldsymbol{\omega}_{fi}(x)$. Consequently, imposing the facet flux $B_{fi} = 0$, we impose that the normal component of $\mathbf{B}$ is zero on the facet "f_i". This can be generalized to the set of facets belonging to the boundary Γ_b.

Consider now the mesh $\mathcal{M}$, composed of $\mathcal{F}$ facets, among which $\mathcal{F}_0$ facets belong to the boundary Γ_b. Assume that the $\mathcal{F}_0$ facets, belonging to the boundary Γ_b, are organized such that they are the last in the order of numbering. To impose $\mathbf{B}.\mathbf{n}\,|_{\Gamma b} = 0$, we set to zero the flux B_{fi} on the $\mathcal{F}_0$ facets. The number of remaining unknowns of the problem is then $\mathcal{F}_r = \mathcal{F} - \mathcal{F}_0$. Under these conditions, the discrete form $\mathbf{B}_d(x)$ is:

$$\mathbf{B}_d(x) = \sum_{i=1}^{\mathcal{F}_r} \boldsymbol{\omega}_{f_i}(x) B_i = \left[\boldsymbol{\omega}_{\mathcal{F}_r}(x)\right]^t \left[B_{\mathcal{F}_r}\right] \qquad [4.169]$$

4.3.5.4. *Synthesis of properties*

It can be noted that the use of the node, edge and facet elements makes it possible to naturally impose homogeneous conditions on a boundary. For this, it is sufficient to have zero degrees of freedom of the approximation functions associated with the geometric elements belonging to this boundary.

4.3.6. *Gauge conditions in the discrete domain*

As already noted, there are an infinite number of fields verifying a condition carried by a **grad**, **curl** or div operator (see section 2.5). This is the case with scalar and vector potentials introduced in Chapter 3, as well as support fields and associated potentials. To obtain a unique solution, a gauge condition must be imposed (see section 2.5.2). For the fields defined by a gradient, a curl or a divergence, the following section explores the numerical process to be used to impose a gauge in the discrete domain.

For the developments, we assume that the domain Ω_d is contractible.

4.3.6.1. *Case of the gradient operator*

Given a known field $\mathbf{v}_d(x) \in W^1(\Omega_d)$, the objective is to find a scalar field $u_d(x)$ such that $\mathbf{v}_d(x) = \mathbf{grad}u_d(x)$ with $u_d(x) \in W^0(\Omega_d)$. In order to have a good formulation of the problem and to satisfy the property [4.158], $\mathbf{v}_d(x)$ must verify the relation $\mathbf{curl}\mathbf{v}_d(x) = 0$. In a matricial form, searching for $u_d(x)$ means finding a vector $[u_{\mathcal{N}}]$ such that $[v_{\mathcal{A}}] = [\mathbf{G}_{\mathcal{A},\mathcal{N}}][u_{\mathcal{N}}]$ where $[v_{\mathcal{A}}]$ is a known vector that verifies $[\mathbf{R}_{\mathcal{F},\mathcal{A}}][v_{\mathcal{A}}] = 0$. Since the rank of the $[\mathbf{G}_{\mathcal{A},\mathcal{N}}]$ matrix is $\mathcal{N}-1$ (see equation [4.114]), the value of an entry of vector $[u_{\mathcal{N}}]$ must be fixed in order to have a unique solution. This makes it necessary to fix the value of the function $u_d(x)$ at one node of the mesh.

In the general case, to fix a node value of $[u_{\mathcal{N}}]$, an arbitrary node of the mesh can be chosen. Nevertheless, in some cases, the entries of the problem may guide the choice. As an example, let us consider the magnetic scalar potential φ formulation. It is not rare to have at least one surface of the boundary, of type Γ_h, on which the magnetic field has a zero tangential component. Under these conditions, this is an equipotential surface with $\varphi \in H_{\Gamma h}(\mathbf{grad}, \Omega)$ (see equation [3.217]). To impose $\varphi|_{\Gamma h} = 0$, as indicated in section 4.3.5.1, we set at zero the entries of vector $[\varphi_{\mathcal{N}}]$ which corresponds to the nodes located on Γ_h, which equally allows for imposing the gauge condition. The same approach can be taken in the case of the electric scalar potential V formulation, imposing to zero the entries associated with the nodes located on the boundary Γ_e (see relation [3.159]).

For a given mesh, in order to obtain the components u_i $(1 \leq i \leq \mathcal{N})$ of the vector $[u_{\mathcal{N}}]$, knowing the components v_k $(1 \leq k \leq \mathcal{A})$ of the vector $[v_{\mathcal{A}}]$, the following approach can be used: given an edge k linking two nodes i and j, with the circulation v_k being known, as well as the value u_i, it is possible to calculate the value u_j.

Indeed, using the raw k of the matrix relation $[v_{\mathcal{A}}] = [\mathbf{G}_{\mathcal{A},\mathcal{N}}][u_{\mathcal{N}}]$ we obtain the following formula (see section 4.3.3.2):

$$v_k = i(k,j)u_j + i(k,i)u_i \qquad [4.170]$$

Using this relation, an iterative process can be developed for the calculation of $\mathcal{N}$ values u_j of the vector $[u_{\mathcal{N}}]$. It starts at the node i where the component u_i was imposed. Then, we consider the set of m edges connected to node i. For all the nodes j (except the node i), extremity of the m edges, the value of u_j is calculated using relation [4.170]. The above process is then repeated for all the nodes i whose value is known. This process aims to determine the node values of all the nodes in the mesh. During this process, the value at a node j can be calculated several times, but the value of u_j at this node will always be the same. If this is not the case, then the field $\mathbf{v}_d(x)$ is not curl free and does not derive from a gradient.

In order to illustrate this process, we work with an elementary mesh $\mathcal{M}$ composed of two adjacent tetrahedra (see Figure 4.8). The properties of the mesh are: $\mathcal{N} = 5$, $\mathcal{A} = 9$, $\mathcal{F} = 7$ and $\mathcal{V} = 2$. The element $\mathcal{V}_1$ is composed of nodes 1, 2, 3 and 4 and $\mathcal{V}_2$ of nodes 1, 2, 3 and 5. Figure 4.8 shows the orientation of edges. For the sake of a lighter figure, the orientation of facets is not indicated. In fact, except for the internal facet "4", oriented from element $\mathcal{V}_1$ to $\mathcal{V}_2$, the other facets are outwardly oriented.

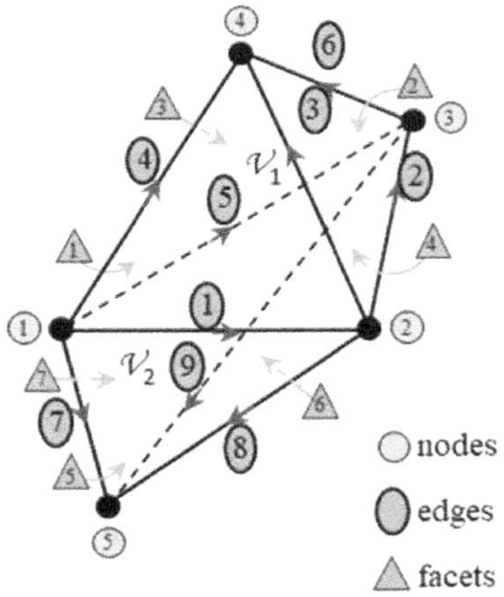

Figure 4.8. *Mesh composed of two adjacent tetrahedrons with numbering of nodes, edges and facets*

As an example, let us fix the circulations of the field $\mathbf{v}_d(x)$ along the nine edges, i.e. $[v_{\mathcal{A}}] = [-1, -1, -2, -3, -2, -1, 4, 5, 6]^t$. To impose the uniqueness of $u_d(x)$, the value at node 4 is set, i.e. $u_4 = 0$. According to the approach proposed above, the first step is to determine the matrix$[\mathbf{G}_{\mathcal{AN}}]$, reproduced in Table 4.20.

i(a, n)	1	2	3	4	5
1	– 1	1	0	0	0
2	0	– 1	1	0	0
3	0	– 1	0	1	0
4	– 1	0	0	1	0
5	– 1	0	1	0	0
6	0	0	– 1	1	0
7	– 1	0	0	0	1
8	0	– 1	0	0	1
9	0	0	– 1	0	1

Table 4.20. *Matrix $[G_{\mathcal{AN}}]$, example of Figure 4.8*

Second, we determine the set of nodes connected to the edges linked with node 4. Figure 4.8 shows that nodes 1, 2 and 3 are linked to node 4 via, respectively, the edges 3, 4 and 6. It should be noted that it also includes the list of edges connected via column 4 of matrix $[\mathbf{G}_{\mathcal{AN}}]$ (see Table 4.20). The value of u_j, in these nodes, can be calculated using relation [4.170], which yields:

$$u_1 = 3 \quad u_2 = 2 \quad u_3 = 1 \qquad [4.171]$$

Nodes 1, 2 and 3 should now be successively considered to calculate the values at the nodes to which they are connected. Node 1 is connected to nodes 2–5. As the values at nodes 2–4 have already been calculated, only node 5 is considered. Applying again relation [4.125], we obtain:

$$u_5 = 7 \qquad [4.172]$$

Since the present case is extremely simple, the process stops here, as all the values at nodes have been determined. Then, we have the vector $[u_{\mathcal{N}}] = [3, 2, 1, 0, 7]^t$.

4.3.6.2. *Case of a curl operator*

Let us now consider a known field $\mathbf{w}_d(x) \in W^2(\Omega_d)$ and search for a field $\mathbf{v}_d(x)$ such that $\mathbf{w}_d(x) = \mathbf{curl}\mathbf{v}_d(x)$ with $\mathbf{v}_d(x) \in W^1(\Omega_d)$. To have a well posed problem and in agreement with property [4.159], $\mathbf{w}_d(x)$ must verify relation div$\mathbf{w}_d(x) = 0$. When a field is uniquely defined by a curl, in order to have a unique solution, a gauge condition must be imposed, either in the continuous or in the discrete domain.

As already seen in section 2.5.2.2, in the continuous domain, there are several gauges, among which is the gauge $\mathbf{v}.\boldsymbol{\eta} = 0$ (see equation [2.51]). Its equivalent, in the discrete domain, relies on the construction of an edge tree (Albanese and Rubinacci 1990), whose process will be developed below.

A tree is a set $\mathcal{T}_\mathcal{A}$ of edges, connecting all the nodes of the mesh without creating loops. This means that all the nodes of the mesh can be linked by a unique path using the edges of the tree. The tree $\mathcal{T}_\mathcal{A}$ contains $\mathcal{N} - 1$ edges. Assigning an arbitrary value to the circulations of $\mathbf{v}_d(x)$ on this set of edges is equivalent to imposing a gauge condition and therefore its uniqueness (see section 2.5.2.2).

Indeed, let us consider two fields $\mathbf{v}_{d1}(x)$ and $\mathbf{v}_{d2}(x)$ belonging to $W^1(\Omega_d)$ so that the circulations on all the edges of tree $\mathcal{T}_\mathcal{A}$ are equal. As the curls of the two fields $\mathbf{v}_{d1}(x)$ and $\mathbf{v}_{d2}(x)$ are equal to $\mathbf{w}_d(x)$, the field $\Delta\mathbf{v}_d(x) = \mathbf{v}_{d1}(x) - \mathbf{v}_{d2}(x)$ is curl free and can be written as the gradient of a scalar function $\Delta u_d(x)$ (i.e. $\Delta\mathbf{v}_d(x) = \mathbf{grad}\Delta u_d(x)$). Along the edges of $\mathcal{T}_\mathcal{A}$, since the circulations of $\mathbf{v}_{d1}(x)$ and $\mathbf{v}_{d2}(x)$ being imposed are equal, the circulations of $\mathbf{grad}\Delta u_d(x)$ are zero. Considering now two nodes A and B of the mesh, since $\mathcal{T}_\mathcal{A}$ is a spanning tree, there is a unique path C made of edges of $\mathcal{T}_\mathcal{A}$ connecting A and B. Since the circulation of $\mathbf{grad}\Delta u_d(x)$ is zero on the edges of $\mathcal{T}_\mathcal{A}$, the circulation of $\mathbf{grad}\Delta u_d(x)$ between A and B is also zero. The values of $\Delta u_d(x)$, at the nodes A and B, are therefore equal. This reasoning can be applied to any pair of nodes of the mesh. Consequently, all the values at nodes of the function $\Delta u_d(x)$ are equal, leading to the function $u_d(x)$, which is constant throughout the domain. The function $\mathbf{grad}\Delta u_d(x)$ is therefore zero; thus, we have $\mathbf{v}_{d1}(x) = \mathbf{v}_{d2}(x)$. Consequently, imposing the circulation of $\mathbf{v}_d(x)$ on the edges of the tree leads to the uniqueness of this function. In practice, these circulations are equal to zero. The objective is now to determine the circulations on the edges of the cotree, i.e. not belonging to the edge tree. An iterative process is used for this purpose. Since the curl of $\mathbf{v}_d(x)$ is equal to $\mathbf{w}_d(x)$, the circulation of $\mathbf{v}_d(x)$ around a facet f is equal to the flux of $\mathbf{w}_d(x)$ through this facet. Using the raw "f" of the incidence matrix $[\mathbf{R}_{\mathcal{F}\mathcal{A}}]$ (see equation [4.127]), we then obtain in the case of a triangular facet:

$$w_f = i(f,i)v_i + i(f,j)v_j + i(f,k)v_k \qquad [4.173]$$

where i, j and k are the indices of the edges belonging to the facet f and i(f,i) is the incidence of the edge i on the facet f of the matrix $[\mathbf{R}_{\mathcal{F}\mathcal{A}}]$ (see section 4.3.3.3). As for the terms v_i, v_j and v_k, they represent the components of vector $[v_\mathcal{A}]$ of the circulations of $v_d(x)$ on the edges and w_f the fth component of vector $[w_{\mathcal{F}}]$ of the fluxes of $\mathbf{w}_d(x)$ through the facets of the mesh.

To calculate the circulations of $\mathbf{v}_d(x)$, therefore of v_k, on the edges of the cotree, all of the facets of the mesh are scanned one after the other. When a facet is met, for which the circulations of $\mathbf{v}_d(x)$ along the two edges are known, the circulation on the third edge is determined using relation [4.173]. This process is repeated until all the circulations on the edges of the cotree are calculated.

NOTE.– Consider relation [4.127], which connects the vector $[v_{\mathcal{A}}]$ of $\mathcal{A}$ values of the circulation of $\mathbf{v}_d(x)$ on the edges with the vector $[w_{\mathcal{F}}]$ of $\mathcal{F}$ values of the flux of facets of $\mathbf{w}_d(x)$ via the discrete curl operator $[\mathbf{R}_{\mathcal{FA}}]$. According to relation [4.128], the rank of the matrix $[\mathbf{R}_{\mathcal{FA}}]$ is equal to $\mathcal{A} - (\mathcal{N} - 1)$, which is in agreement with the previous, where, in order to fix the uniqueness of $\mathbf{v}_d(x)$, $(\mathcal{N} - 1)$ values of vector $[v_{\mathcal{A}}]$ are imposed through an edge tree.

As a simple illustration, let us consider the two adjacent tetrahedra in Figure 4.8. It should be recalled that, except for the internal facet "4", oriented from the element $\mathcal{V}_1$ to $\mathcal{V}_2$, the other facets are oriented outwardly. Based on this information and on the orientation of the edges in the figure, it is easy to build the incidence matrix $[\mathbf{R}_{\mathcal{FA}}]$ reproduced in Table 4.21.

i(f,a)	1	2	3	4	5	6	7	8	9
1	1	0	1	– 1	0	0	0	0	0
2	0	1	– 1	0	0	1	0	0	0
3	0	0	0	1	– 1	– 1	0	0	0
4	– 1	– 1	0	0	1	0	0	0	0
5	– 1	0	0	0	0	0	1	– 1	0
6	0	– 1	0	0	0	0	0	1	– 1
7	0	0	0	0	1	0	– 1	0	1

Table 4.21. *Matrix $[R_{\mathcal{FA}}]$; example of Figure 4.8*

For this mesh, let us now assume the known field $\mathbf{w}_d(x)$ of $W^2(\Omega_d)$, defined by the fluxes through the seven facets, i.e. $[w_{\mathcal{F}}] = [0, 0, -1, 1, 1, 0, 0]^t$. Taking into account the orientation of facets, it can be readily verified that $\mathbf{w}_d(x)$ is divergence free. The objective is to find a field $\mathbf{v}_d(x)$, belonging to $W^1(\Omega_d)$, so that $\mathbf{v}_d(x) = \mathbf{curl}\mathbf{w}_d(x)$. This is equivalent to finding the set of circulations $[v_{\mathcal{A}}] = [v_1, v_2, v_3, v_4, v_5, v_6, v_7, v_8, v_9]^t$.

To calculate the vector $[v_{\mathcal{A}}]$, we build an edge tree. Knowing the number of nodes of the mesh ($\mathcal{N} = 5$), the tree is composed of four edges ($\mathcal{N} - 1$). Let us choose, for example (see Figure 4.9), the edges 1, 3, 6 and 9 on which zero circulation is imposed, hence $v_1 = v_3 = v_6 = v_9 = 0$. To calculate the remaining circulations, we use the incidence matrix to find a facet for which two circulations are known and the third one is calculated using relation [4.173]. This process can be repeated until all the circulations are obtained. Then, for vector $[v_{\mathcal{A}}]$, we have the following values:

$$[v_{\mathcal{A}}] = [0, 0, 0, 0, 1, 0, 1, 0, 0]^t \qquad [4.174]$$

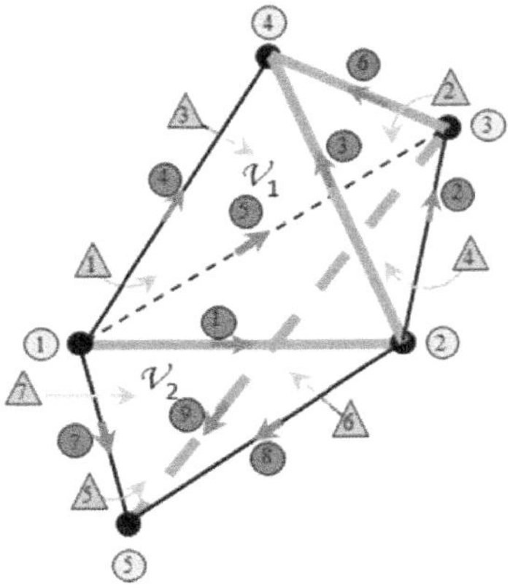

Figure 4.9. *Edge tree in the thick line, composed of edges 1, 3, 6 and 9*

4.3.6.3. *Case of a divergence operator*

The objective is now to calculate a field $\mathbf{w}_d(x) \in W^2(\Omega_d)$ so that its divergence is equal to a known source term, $q_d(x) \in W^3(\Omega_d)$. The field $\mathbf{w}_d(x)$ is not unique, given that if $\mathbf{w}_d(x)$ is a solution, any field written in the form $\mathbf{w}_d(x) + \mathbf{curl}\mathbf{v}_d(x)$ with $\mathbf{v}_d(x) \in W^1(\Omega_d)$ is also a solution. A gauge should therefore be introduced to impose the uniqueness of $\mathbf{w}_d(x)$. A solution is to again use a tree technique (Le Ménach et al. 1998). In fact, this involves building an edge tree denoted by $\mathcal{T}_f$. The analogy existing between "nodes and elements" and "edges and facets" will be used for this purpose (Bossavit 1997). Indeed, an edge joins two nodes and a facet "joins" two elements. Based on this principle, a graph $\mathcal{G}$ can be built. In this context, the elements of the mesh are represented by nodes and the facets by edges. The orientation of the facet f of the mesh defines the orientation of the corresponding edge of graph $\mathcal{G}$. An additional node is required to represent the external environment of the domain Ω_d. All the edges representing the facets located on the boundary of Ω_d are linked to this additional node.

To illustrate this principle, let us consider again the example in Figure 4.8, namely two adjacent tetrahedra. For this example, it should be recalled that, with the exception of the internal facet "4", oriented from the element $\mathcal{V}_1$ to $\mathcal{V}_2$, the other facets are outwardly oriented. Figure 4.10(a) reproduces the two tetrahedra, but for the sake of readability, the notations related to nodes and facets are not indicated. On the other hand, the orientation of the seven facets via their normal component has been introduced. A graph of facets, corresponding to this elementary mesh, is reproduced in Figure 4.10(b), which shows the additional node, denoted by $\mathcal{V}_{ext}$, representing the external domain. The graph $\mathcal{G}$ is therefore composed of $\mathcal{F}$ edges and $\mathcal{V} + 1$ nodes, hence seven edges and three nodes.

A tree is then built on the graph $\mathcal{G}$. The facets of tree $\mathcal{T}_f$ correspond to the edges of the cotree of $\mathcal{G}$. The tree $\mathcal{T}_f$ then contains $\mathcal{F} - \mathcal{V}$ facets.

NOTE.– Let us consider relation [4.148]; it links the vector $[w_{\mathcal{F}}]$ of $\mathcal{F}$ values of the flux of $\mathbf{w}_d(x)$ on the facets with the vector $[p_{\mathcal{V}}]$ of $\mathcal{V}$ values of charge $p_d(x)$ contained in each element via the discrete divergence operator $[D_{VF}]$. According to relation [4.149], the rank of matrix $[D_{\mathcal{VF}}]$ is equal to $\mathcal{F} - (\mathcal{A} - (\mathcal{N} - 1))$. For the studied example, the domain is assumed to be contractible; therefore, according to the Euler–Poincaré formula, we have: $\mathcal{V} - \mathcal{F} + \mathcal{A} - \mathcal{N} = -1$. Therefore, the rank of matrix $[D_{\mathcal{VF}}]$ is equal to $\mathcal{V}$. $\mathcal{F} - \mathcal{V}$ values of vector $[w_{\mathcal{F}}]$ should be fixed to have a unique vector $\mathbf{w}_d(x)$. This leads to a number of facets of tree $\mathcal{T}_f$ equal to $\mathcal{F} - \mathcal{V}$.

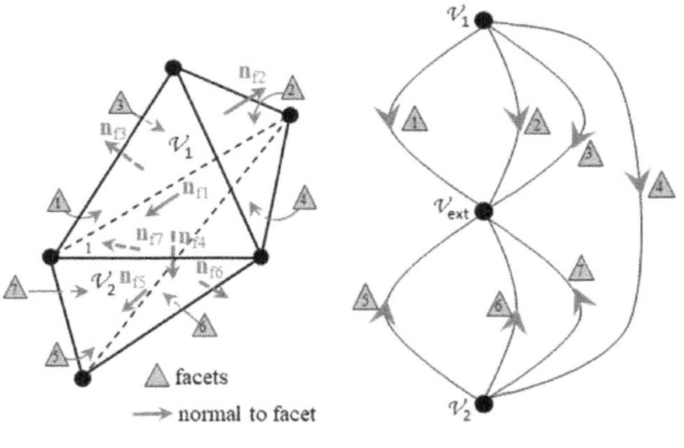

a) b)

Figure 4.10. *a) Two adjacent tetrahedrons with facet orientation; b) corresponding facet graph* $\mathcal{G}$

Let us consider again the calculation of the field $w_d(x)$, defined by its divergence and the way in which the gauge condition is imposed by building a facet tree. However, depending on the boundary conditions on the boundary, some precautions must be taken.

NOTE.– All of the facets located on the boundary on which the normal component of field $\mathbf{w}_d(x)$ is imposed must belong to the tree. Nevertheless, if the normal component is imposed on the entire domain, to avoid creating a closed surface, one facet must be excluded from the tree.

To gauge the solution, the values of the fluxes are then fixed through the facets of $\mathcal{T}_f$. Then, there are three possible cases.

– Case 1: the objective is to find the support field $\boldsymbol{\eta}_{sl}$ of a source term imposed inside the domain and defined by a density (see section 3.2.2.2). This is the case with the electric charge density for electrostatic problems. Arbitrary values of flux on the facets of tree $\mathcal{T}_f$ are then imposed.

– Case 2: the objective is to find the support field $\boldsymbol{\lambda}_s$ of a source term imposed on a part of the domain boundary (see section 3.2.1.2). In this case, the values of the fluxes for the facets of the tree belonging to the boundary are fixed by the boundary conditions. On the other hand, arbitrary values can be imposed on the facets of the tree, located inside the domain.

– Case 3: the objective is to build an approximation of a known source field defined inside the domain and divergence free. Then, it can be defined by its support field $\boldsymbol{\lambda}_{sl}$ (see equation [3.38]). An example is the current density $\mathbf{J}_0$ in the case of a magnetostatics problem (see Figure 3.20). Indeed, if the mesh does not perfectly fit the shapes of the inductor (case of a curved inductor in the case of a tetrahedral mesh), $\boldsymbol{\lambda}_{sl}$ cannot be correctly expressed in the space $W^2(\Omega_d)$. In this case, an approximation $\boldsymbol{\lambda}_{sld}$ of $\boldsymbol{\lambda}_{sl}$ is determined, which is divergence free. For this purpose, a possible approach (Le Ménach et al. 1998) is to use the proposed gauge, but this time by imposing a flux on the facets of the tree:

$$w_f = \iint_f \boldsymbol{\lambda}_{sl}.\mathbf{n}dS \qquad [4.175]$$

In these three cases, the values of fluxes on the facet tree $\mathcal{T}_f$ are fixed. The fluxes through the other facets of the mesh $\mathcal{M}$ are calculated according to an iterative process similar to that presented in section 4.3.6.2 using, for the case of a tetrahedron, the following relation:

$$q_v = i(v,i)w_i + i(v,j)w_j + i(v,k)w_k + i(v,l)w_l \qquad [4.176]$$

where i, j, k and l are the indices of the facets of the element v, i(v, i) is the incidence of the facet i with respect to the element v (see the matrix $[\mathbf{D}_{\mathcal{VF}}]$ in section 4.3.3.4). As for w_i, w_j, w_k and w_l, they represent the components of the vector $[w_{\mathcal{F}}]$ of fluxes of $\mathbf{w}_d(x)$ through the facets of the element v. Finally, q_v represents the vth component of the vector $[q_{\mathcal{V}}]$ of charges related to the density of charges $q_d(x)$.

As an illustration, let us consider again the two adjacent tetrahedra in Figure 4.10(a). Based on the orientation of edges in the figure, the incidence matrix $[\mathbf{D}_{\mathcal{VF}}]$, reproduced in Table 4.22, can be easily built.

i(v,f)	1	2	3	4	5	6	7
1	1	1	1	1	0	0	0
2	0	0	0	– 1	1	1	1

Table 4.22. *Matrix $[D_{\mathcal{VF}}]$; example of Figure 4.10(a)*

For this mesh $\mathcal{M}$, composed of two elements and seven facets, let us now consider a field $q_d(x) \in W3(\Omega_d)$ such that $q_d(x) = 0$, therefore vector $[q_{\mathcal{V}}] = [0, 0]^t$. The objective is to find a field $\mathbf{w}_d(x) \in W^2(\Omega_d)$ such that:

– $\text{div}\mathbf{w}_d(x) = q_d(x) = 0$;

– fluxes entering facet 1 and exiting facet 7 are equal to 1;

– fluxes exiting the other facets of the boundary, i.e. facets 2, 3, 5 and 6 are equal to zero.

All the facets located on the surface have imposed fluxes, but given the fact that $\mathbf{w}_d(x)$ is divergence free, one of these six fluxes is a linear combination of the other five. Consequently, five of these facets must belong to the facet tree $\mathcal{T}_f$ and hence to the cotree of graph $\mathcal{G}$ represented in Figure 4.10(b). Let us fix the edge corresponding to the facet 7 in the tree of $\mathcal{G}$ and complete it by the edge corresponding to facet 4. The tree of $\mathcal{G}$ is therefore composed of facets 4 and 7. The facet tree corresponding to the cotree of $\mathcal{G}$ therefore contains all the facets 1, 2, 3, 5 and 6. The fluxes on these facets are fixed by the boundary conditions. Therefore, we have:

$$w_1 = -1, \ \ w_2 = 0 \ \ w_3 = 0, \ \ w_5 = 0, \ \ w_6 = 0 \qquad [4.177]$$

In order to obtain the missing fluxes, we use the approach involving a loop on the elements. Since three of the four fluxes of the facets of element 1 are known, considering again the first line of the matrix $[\mathbf{D}_{\mathcal{VF}}]$, with $q_1 = 0$, it can be deduced from equation [4.176] that $w_4 = 1$. The same approach applied to element 2 yields $w_7 = 1$. Under these conditions, the vector of fluxes of the facet $[w_{\mathcal{F}}]$ is given by:

$$[w_{\mathcal{F}}] = [-1, 0, 0, 1, 0, 0, 1, 0, 1]^t \qquad [4.178]$$

4.3.7. *Discretization of support fields and associated potentials*

As already seen in section 1.5, source terms, encountered in the low-frequency electromagnetism, can be local quantities (ρ, σ_s, $\mathbf{J}_s$) or global quantities (f_s, e, f_m, I, ϕ). In section 3.2, to facilitate the derivation of the formulations, we introduced source fields and, for their representation, we used support fields ($\boldsymbol{\beta}_s$, $\boldsymbol{\lambda}_s$, $\boldsymbol{\eta}_{sl}$, $\boldsymbol{\xi}_{sl}$), which can be associated with the potentials (α_s, $\boldsymbol{\chi}_s$). The following section will develop, in the discrete domain, the expression of support fields and associated potentials depending on their properties. The notations are the same, namely the mesh $\mathcal{M}$ contains $\mathcal{N}$ nodes, $\mathcal{A}$ edges, $\mathcal{F}$ facets and $\mathcal{V}$ volumes.

4.3.7.1. *Discretization of the source terms α_s and β_s*

For the discretization of the source terms α_s and $\boldsymbol{\beta}_s$, we refer again to the general case considered in section 3.2.1.1. The properties of the scalar potential α_s, given by relations [3.13] and [3.14], show that it is defined in the space H(**grad**, Ω) with boundary conditions on Γ_{nk} with $k \in \{1, 2\}$. The potential α_s must therefore be discretized in the space of node elements $W^0(\Omega_d)$, as shown in Table 4.18. It should be recalled that the boundaries Γ_{n1} and Γ_{n2} are equipotential with $\alpha_{s|\Gamma n1}$ and $\alpha_{s|\Gamma n2}$ equal to 1 and 0, respectively.

In the discrete domain, in order to impose the boundary conditions on Γ_{n1} and Γ_{n2}, we fix to 1 the values associated with the nodes on the boundary Γ_{n1} and to zero those associated with the nodes of Γ_{n2} (Henneron et al. 2004). Indeed, it can be verified that if the values of a function are equal on the nodes of a facet, then the value of the function is constant on this facet. Consequently, the value of $\alpha_{sd}(x)$ is constant and equal to 1 on all the facets of surface Γ_{n1} and is zero on those of Γ_{n2}. Moreover, the other values of $\alpha_{sd}(x)$ at nodes can be arbitrarily fixed (in practice, they are often fixed at zero). Indeed, let us note by $\mathcal{N}_1$ the set of nodes belonging to the boundary Γ_{n1}; then, we can write:

$$\alpha_{sd_i} = 1 \text{ if } i \in \mathcal{N}_1 \text{ else } \alpha_{sdi} = 0 \qquad [4.179]$$

where α_{sdi} is the value of $\alpha_{sd}(x)$ at nodes "i". Let us now consider the $\mathcal{N}$ nodes of mesh $\mathcal{M}$. In this case, the expression of the discrete form $\alpha_{sd}(x)$ of the potential α_s is:

$$\alpha_{sd}(x) = \sum_{i=1}^{\mathcal{N}} \omega_{n_i}(x)\,\alpha_{sd_i} \tag{4.180}$$

where, as a reminder, $\omega_{ni}(x)$ represents the nodal interpolation functions (see equation [4.84]).

To represent the nodal values α_{sdi} in expression [4.180], we can introduce an $\mathcal{N}$-dimensional vector denoted by $[\alpha_{\mathcal{N}}]$. Then, we have:

$$\alpha_{sd}(x) = [\omega_{\mathcal{N}}(x)]^t\,[\alpha_{\mathcal{N}}] \tag{4.181}$$

Considering the properties given by equation [3.12], the support field $\boldsymbol{\beta}_s$ is defined in the function space $H_{\Gamma n1 \cup \Gamma n2}(\mathbf{curl}0, \Omega)$. Moreover, it is expressed as a function of potential α_s via the gradient operator (see equation [3.13]). In the discrete domain, it belongs to the space of edge elements $W^1_{\Gamma n1 \cup \Gamma n2}(\Omega_d)$ (see Table 4.18). Therefore, based on equation [4.111], it can be written in the discrete form:

$$\begin{aligned}\boldsymbol{\beta}_{sd}(x) &= -\mathbf{grad}\,\alpha_{sd}(x) = -\sum_{i \in \mathcal{N}} (\mathbf{grad}\ \omega_{n_i}(x))\alpha_{n_i} \\ &= -[\boldsymbol{\omega}_{\mathcal{A}}(x)]^t\,[\mathbf{G}_{\mathcal{A},\mathcal{N}}][\alpha_{\mathcal{N}}] = [\boldsymbol{\omega}_{\mathcal{A}}(x)]^t\,[\beta_{\mathcal{A}}]\end{aligned} \tag{4.182}$$

where $[\beta_{\mathcal{A}}]$ is written as:

$$[\beta_{\mathcal{A}}] = -[\mathbf{G}_{\mathcal{A},\mathcal{N}}][\alpha_{\mathcal{N}}] \tag{4.183}$$

For a given problem, knowing the vector $[\alpha_{\mathcal{N}}]$ of nodal values, as well as the matrix $[\mathbf{G}_{\mathcal{A}\mathcal{N}}]$, we very easily obtain the vector $[\beta_{\mathcal{A}}]$ of $\mathcal{A}$ circulations of $\boldsymbol{\beta}_s$ on the edges of the mesh. It can then be verified that the only non-zero components of vector $[\beta_{\mathcal{A}}]$ are associated with edges with only one extremity on the surface Γ_{n1}.

4.3.7.2. *Discretization of fields λ_s and χ_s*

This section considers again the example presented in Figure 3.1 with the support fields introduced in section 3.2.1.2. The field $\boldsymbol{\lambda}_s$ is entirely defined in the domain Ω of boundary Γ. This boundary is the union of two gates Γ_{n1} and Γ_{n2} and of the wall Γ_m. The support field $\boldsymbol{\lambda}_s$ belongs to the function space $H_{\Gamma m}(\text{div}0, \Omega)$ (see relation [3.23]). Under these conditions, this field must be discretized in the space of facet elements (see Table 4.18). Then, $\boldsymbol{\lambda}_{sd}(x) \in W^2{}_{\Gamma m}(\Omega_d)$, and its expression is written as:

$$\lambda_{sd}(x) = \sum_{i=1}^{\mathcal{F}} \boldsymbol{\omega}_{f_i}(x)\lambda_{sf_i} = [\boldsymbol{\omega}_{\mathcal{F}}(x)]^t [\lambda_{\mathcal{F}}] \quad [4.184]$$

The objective is to determine the fluxes of $\boldsymbol{\lambda}_{sd}(x)$ through the facets of the mesh, i.e. the components of the vector $[\lambda_{\mathcal{F}}]$ such that:

– $\mathrm{div}\boldsymbol{\lambda}_{sd}(x) = 0$ or in an equivalent manner, in the discrete domain, $[D_{\mathcal{VF}}]\,[\lambda_{\mathcal{F}}] = 0$;

– $\boldsymbol{\lambda}_{sd}(x).\mathbf{n} = 0$ on Γ_m, which means that the fluxes through the facets located on Γ_m are zero;

– $\int_{\Gamma_{n2}} \lambda_{sd}(x).\mathbf{n}dS = -\int_{\Gamma_{n1}} \lambda_{sd}(x).\mathbf{n}dS = 1$, which is equivalent to imposing that the sum of fluxes exiting the facets located on Γ_{n2} is equal to 1 and equal to –1 on Γ_{n1}.

There are an infinite number of solutions satisfying these conditions. To build one, the gauge condition proposed in section 4.3.6.3 is applied. This gauge relies on building a facet tree (case 2). It should be recalled that a requirement when building a tree is that all the facets of the domain boundary except one belong to the facet tree (Le Ménach et al. 1998; Henneron et al. 2004).

Concerning the associated vector potential $\boldsymbol{\chi}_{sd}(x)$, it is linked to the support field $\boldsymbol{\lambda}_{sd}(x)$ via the curl operator. In the case of a not simply connected boundary such as Γ_m, some precautions need to be taken in the continuous domain in order to determine it (see section 3.2.1.2.2). Indeed, in this case, as shown by equations [3.28] and [3.29], we have $\boldsymbol{\chi}_s \in H^{\Delta}{}_{\Gamma m}(\mathbf{curl}, \Omega)$. It will be shown how this difficulty can be overcome during its discretization.

In the discrete domain, the associated vector potential $\boldsymbol{\chi}_{sd}(x)$ belongs to the space of the edge elements $W^1{}_{\Gamma m}(\Omega_d)$ and its expression is given by:

$$\chi_{sd}(x) = \sum_{i=1}^{\mathcal{A}} \boldsymbol{\omega}_{a_i}(x)\chi_{sa_i} = [\boldsymbol{\omega}_{\mathcal{A}}(x)]^t [\chi_{\mathcal{A}}] \quad [4.185]$$

The objective is to determine the components of the vector $[\chi_{\mathcal{A}}]$ that represent the $\mathcal{A}$ circulations of $\boldsymbol{\chi}_{sd}(x)$ on the edges of the mesh. The conditions to be verified are:

– $\boldsymbol{\lambda}_{sd}(x) = \mathbf{curl}\boldsymbol{\chi}_{sd}(x)$ or, in an equivalent manner, $[\lambda_{\mathcal{F}}] = [\mathbf{R}_{\mathcal{FA}}][\chi_{\mathcal{A}}]$;

– $\boldsymbol{\chi}_{sd}(x) \wedge \mathbf{n} = 0$ on Γ_m.

Since $\boldsymbol{\lambda}_{sd}(x)$ belongs to $W^2{}_{\Gamma m}(\Omega_d)$ and it is divergence free, there are an infinite number of fields $\boldsymbol{\chi}_{sd}(x)$ satisfying the two above-mentioned conditions. To determine one solution, we can use the gauge condition (see section 4.3.6.2), based on the construction of an edge tree. To impose the boundary conditions on the boundary Γ_m, the construction of the edge tree must begin on this boundary. It can be shown that this algorithm introduces a cut on the boundary Γ_m between the gates Γ_{n1} and Γ_{n2} (Le Ménach et al. 1998). This makes it possible to alleviate the constraints related to the not simply connected boundary Γ_m.

Knowing the $\mathcal{A}$ circulations of the field on the edges of the mesh, hence the vector $[\chi_{\mathcal{A}}]$, we can express $\boldsymbol{\chi}_{sd}(x)$ in the form of equation [4.185].

4.3.7.3. *Discretization of the source terms ξ_{sl} and η_{sl}*

The support fields of the source terms, ξ_{sl} and $\boldsymbol{\eta}_{sl}$, are defined in section 3.2.2.2. The scalar field ξ_{sl} represents a density as shown in equation [3.43]. As for the field $\boldsymbol{\eta}_{sl}$, it is defined based on ξ_{sl} via relation [3.46].

In order to determine the discrete form of these two fields, denoted by ξ_{sld} and $\boldsymbol{\eta}_{sld}$, we consider the example in Figure 3.9 introduced in section 3.2.3.4. In the continuous domain, these two fields are defined in the function spaces, such that $\xi_{sl} \in L^2(\Omega)$ and $\boldsymbol{\eta}_{sl} \in H_{\Gamma m}(\text{div}, \Omega)$. Under these conditions, transposed in the discrete domain (see Table 4.18), we have $\xi_{sld}(x) \in W^3(\Omega_d)$ and $\boldsymbol{\eta}_{sld}(x) \in W^2{}_m(\Omega_d)$, hence, respectively, the space of the volume elements and the facet elements. As a first step, we determine $\xi_{sld}(x)$ by using its properties defined in the continuous domain by equation [3.43]. Then, we calculate $\boldsymbol{\eta}_{sld}(x)$ by transposing relation [3.46] in the discrete domain.

The field $\xi_{sld}(x)$ is an input of the problem defined in the space of the volume elements. Under these conditions (see equation [4.105]), the expression of its discrete form is:

$$\xi_{sld}(x) = \sum_{i=1}^{\mathcal{V}} \omega_{v_i}(x)\xi_{slv_i} = [\omega_{\mathcal{V}}(x)]^t [\xi_{\mathcal{V}}] \quad [4.186]$$

The entries of the vector $[\xi_{\mathcal{V}}]$ represent the contribution of the volume corresponding to the element. This means the volume integral of $\xi_{sld}(x)$ on each element of the mesh. For example, in electrostatics, if the charge density is constant on a subdomain Ω_{sd} of Ω_d, then we can write for an element v_i, based on equation [3.57]:

$$\xi_{slv_i} = \frac{vol(v_i)}{vol(\Omega_{sd})} \text{ if } v_i \in \Omega_{sd} \text{ and } \xi_{slv_i} = 0 \text{ if } v_i \notin \Omega_{sd} \quad [4.187]$$

in this expression, $vol(v_i)$ represents the volume of the element v_i.

Let us now find the support field $\boldsymbol{\eta}_{sld}(x)$, which is written as follows:

$$\boldsymbol{\eta}_{sld}(x) = \sum_{i=1}^{\mathcal{F}} \boldsymbol{\omega}_{f_i}(x)\eta_{slf_i} = [\boldsymbol{\omega}_{\mathcal{F}}(x)][\eta_{\mathcal{F}}] \quad [4.188]$$

In this expression, the vector $[\eta_{\mathcal{F}}]$ represents the $\mathcal{F}$ fluxes of the facets of the field $\boldsymbol{\eta}_{sld}(x)$, which must satisfy the following conditions:

– $\text{div}\boldsymbol{\eta}_{sld}(x) = \xi_{sld}(x)$ or, in an equivalent manner, $[\xi_{\mathcal{V}}] = [D_{\mathcal{VF}}][\eta_{\mathcal{F}}]$;

– $\boldsymbol{\eta}_{sld}(x).\mathbf{n} = 0$ on Γ_m.

There are an infinite number of fields $\boldsymbol{\eta}_{sld}(x)$ verifying these two conditions. To determine the entries of $[\eta_{\mathcal{F}}]$, we then apply a gauge condition such as the one described in section 4.3.6.3 (case 1). Therefore, we have to build a facet tree $\mathcal{T}_f$. To impose the boundary conditions, all the facets belonging to Γ_m in the tree $\mathcal{T}_f$ must be imposed. However, as indicated in section 4.3.6.3, if Γ_m covers the entire boundary, then all the facets of Γ_m, except one, belong to $\mathcal{T}_f$.

4.4. Discretization of weak formulations

4.4.1. *Notations*

In what follows, we apply the finite element method to the set of weak formulations developed in section 4.2. As a first step, the Ritz–Galerkin method will be presented. This method will be applied to construct the discrete model from the weak formulation obtained in section 4.2.

To alleviate the notations, the next section of this book will no longer refer to the dependence on "x" of the interpolation functions. The node, edge, facet and volume functions will therefore be denoted by ω_n, $\boldsymbol{\omega}_a$, $\boldsymbol{\omega}_f$ and ω_v, respectively. Similarly, the vectors corresponding to these interpolation functions will be written as $[\omega_{\mathcal{N}}]^t$, $[\boldsymbol{\omega}_{\mathcal{A}}]^t$, $[\boldsymbol{\omega}_{\mathcal{F}}]^t$ and $[\omega_{\mathcal{V}}]^t$, where $\mathcal{N}$, $\mathcal{A}$, $\mathcal{F}$ and $\mathcal{V}$ represent their dimension, namely the number of nodes, edges, facets and volume, respectively.

4.4.2. *Ritz–Galerkin method*

Section 4.2 developed the weighted residual method that led, in the continuous domain, to searching for a weak form of the solution to the initial problem.

Consider the example of a scalar function U, defined on a domain Ω. This field must satisfy a gate-type boundary condition on a part Γ_n of the boundary Γ on which we impose U = 0. Consider that the equation to be solved can be written as follows:

$$\mathrm{div}(\mathcal{K}\,\mathbf{grad}\mathrm{U}) + \mathrm{q} = 0 \qquad [4.189]$$

where $\mathcal{K}$ is a strictly positive scalar function depending on the position, $\mathrm{q} \in L^2(\Omega)$ is a volume source term that can be expressed using a global quantity $\mathcal{Q}$ (see equation [3.43]), such that $\mathrm{q} = \mathcal{Q}\,\xi_{sl}$, and $\mathrm{U} \in \mathrm{H}_{\Gamma n}(\mathbf{grad}, \Omega)$ is the unknown function. On the remaining part Γ_m of Γ, we impose Neumann's boundary conditions in the form "$\mathcal{K}$ **grad**U.**n** = 0".

Considering the boundary conditions, the weak form of this equation can be written, after development, in the form:

$$\iiint_{\Omega} \mathcal{K}(\mathbf{grad}\mathrm{U}.\mathbf{grad}\psi)d\tau = \iiint_{\Omega} \mathcal{Q}\xi_{sl}\psi d\tau \;\; \forall \; \psi \in \mathrm{H}_{\Gamma_n}(\mathbf{grad}, \Omega) \qquad [4.190]$$

To find an approximation to the solution, we will associate equation [4.190] and the finite element method (see section 4.3.1). Let us consider that the mesh $\mathcal{M}$ contains $\mathcal{N}$ nodes and $\mathcal{V}$ volumes. Considering the domain of definition of the function U, the objective is to find an approximation (see Table 4.18), in the function space $W^0_{\Gamma n}(\Omega_d)$, namely the space of nodal elements. Moreover, to impose the boundary conditions, we fix at zero the $\mathcal{N}_0$ nodes of the mesh belonging to the boundary Γ_n (see section 4.3.6.1). Then, the number of nodes to be determined is $\mathcal{N}_r = \mathcal{N} - \mathcal{N}_0$. To simplify the notations, the nodes are renumbered so that those belonging to the boundary Γ_n have an index higher than $\mathcal{N}_r$. Under these conditions, the discretized form of the unknown function, denoted by $U_d(x)$, is written as:

$$U_d(x) = \sum_{i=1}^{\mathcal{N}_r} \omega_{n_i} U_i = \lfloor \omega_{\mathcal{N}_r} \rfloor^t \lfloor U_{\mathcal{N}_r} \rfloor \qquad [4.191]$$

On the other hand, the source term q is discretized, taking into account these properties in the space $W^3(\Omega_d)$. Then, it can be written using the discrete form of ξ_{sl} (see equation [4.186]), in the following form:

$$q_d(x) = Q\sum_{i=1}^{\mathcal{V}} \omega_{v_i} \xi_{slv_i} = Q[\omega_{\mathcal{V}}]^t [\xi_{\mathcal{V}}] \quad [4.192]$$

As indicated in section 4.2.1, in the continuous domain, the weighting functions ψ were chosen in the adjoint space of the operator. In our case, the operator is the divergence (see equation [4.189]) and therefore the adjoint operator is the gradient, as shown in Table 2.1. Therefore, the weighting functions ψ belong to $H_{\Gamma n}(\mathbf{grad}, \Omega)$. The approach is similar in the discrete domain and the weighting functions are chosen in the space $W^0{}_{\Gamma n}(\Omega_d)$, which is generated by the set of $\mathcal{N}_r$ node functions associated with the nodes of the mesh that are not located on Γ_n. This is known as the Ritz–Galerkin method. In fact, the space of approximation and the space of the weighting functions is the same. Under these conditions, the expression [4.190] can be rewritten by taking for weighting functions $\psi = \omega_{nj}$ with $1 \le j \le \mathcal{N}_r$, i.e.:

$$\sum_{i=1}^{\mathcal{N}_r} \iiint_{\Omega} \mathcal{K}(\mathbf{grad}\,\omega_{n_i}).(\mathbf{grad}\,\omega_{n_j}) U_i d\tau = Q\sum_{k=1}^{\mathcal{V}} \iiint_{\Omega} \omega_{v_k} \xi_{slv_k} \omega_{n_j} d\tau \quad [4.193]$$
$$\text{with } 1 \le j \le \mathcal{N}_r$$

This equation must be verified for the set of the $\mathcal{N}_r$ functions ω_{nj}. Then, we obtain a system of $\mathcal{N}_r$ equations with $\mathcal{N}_r$ unknowns U_i, contained in the vector $[U_{\mathcal{N}r}]$. The system of equations can then be written in the following matrix form:

$$\iiint_{\Omega} \mathcal{K}(\mathbf{grad}[\omega_{\mathcal{N}_r}]^t)^t (\mathbf{grad}[\omega_{\mathcal{N}_r}])[U_{\mathcal{N}_r}] d\tau$$
$$= Q\iiint_{\Omega} ([\omega_{\mathcal{N}_r}][\omega_{\mathcal{V}}]^t [\xi_{\mathcal{V}}]) d\tau \quad [4.194]$$

Relation [4.110] allows us to rewrite this system by introducing the matrices $\left[\mathbf{G}_{\mathcal{A}\mathcal{N}_r}\right]$ and the vector $[\boldsymbol{\omega}_{\mathcal{A}}]$ of the edge weighting functions:

$$\iiint_{\Omega} \mathcal{K} [\mathbf{G}_{\mathcal{A}\mathcal{N}_r}]^t [\boldsymbol{\omega}_{\mathcal{A}}][\boldsymbol{\omega}_{\mathcal{A}}]^t [\mathbf{G}_{\mathcal{A}\mathcal{N}_r}][U_{\mathcal{N}_r}] d\tau =$$
$$\iiint_{\Omega} Q([\omega_{\mathcal{N}_r}][\omega_{\mathcal{V}}]^t [\xi_{\mathcal{V}}]) d\tau \quad [4.195]$$

The incidence matrix $\left[\mathbf{G}_{\mathcal{A}\mathcal{N}_r}\right]$ and also vectors $\left[\mathrm{U}_{\mathcal{N}_r}\right]$ and $\left[\xi_{\mathcal{V}}\right]$ do not depend on the position "x". Under these conditions, equation [4.195] is written as:

$$\left[\mathbf{G}_{\mathcal{A}\mathcal{N}_r}\right]^t\left[\mathrm{M}^{\mathcal{K}}_{\mathcal{A}\mathcal{A}}\right]\left[\mathbf{G}_{\mathcal{A}\mathcal{N}_r}\right]\left[\mathrm{U}_{\mathcal{N}_r}\right]=\left[\mathrm{M}^{\mathrm{s}}_{\mathcal{N}_r\mathcal{V}}\right]\left[\xi_{\mathcal{V}}\right] \qquad [4.196]$$

The elementary term of the matrix $\left[\mathrm{M}^{\mathcal{K}}_{\mathcal{A}\mathcal{A}}\right]$ is written as:

$$\mathrm{M}^{\mathcal{K}}_{\mathrm{a_i a_j}}=\iiint_{\Omega}\mathcal{K}\,\boldsymbol{\omega}_{\mathrm{a_i}}\cdot\boldsymbol{\omega}_{\mathrm{a_j}}\,\mathrm{d}\tau \qquad [4.197]$$

As for the matrix $\left[\mathrm{M}^{\mathrm{S}}_{\mathcal{N}_r\mathcal{V}}\right]$, it represents the contribution of the source term whose elementary term has the following expression:

$$\mathrm{M}^{\mathrm{s}}_{\mathrm{n_i v_j}}=\mathcal{Q}\iiint_{\Omega}\omega_{\mathrm{n_i}}\cdot\omega_{\mathrm{v_j}}\,\mathrm{d}\tau \qquad [4.198]$$

NOTE.– Two different types of matrices can be noted in the obtained matrix system:

– Incidence matrix $\left[\mathbf{G}_{\mathcal{A}\mathcal{N}_r}\right]$ whose expression depends only on the manner in which the edges and the nodes are interconnected (topology of the mesh). On the other hand, it does not depend on the position of the nodes (the metric associated with domain Ω_d).

– Matrices $\left[\mathrm{M}^{\mathcal{K}}_{\mathcal{A}\mathcal{A}}\right]$ and $\left[\mathrm{M}^{\mathrm{S}}_{\mathcal{N}_r\mathcal{V}}\right]$ whose expressions strongly depend on the metric with the integration of functions depending on the position and on the behavior law of the materials.

This dissociation between topology and metric occurs naturally by using the tools proposed by differential geometry (Bossavit 1997).

Finally, posing:

$$\left[\mathrm{S}_{\mathcal{N}_r\mathcal{N}_r}\right]=\left[\mathrm{G}_{\mathcal{A}\mathcal{N}_r}\right]^t\left[\mathrm{M}^{\mathcal{K}}_{\mathcal{A}\mathcal{A}}\right]\left[\mathrm{G}_{\mathcal{A}\mathcal{N}_r}\right] \text{ and } \left[\mathrm{F}_{\mathcal{N}_r}\right]=\left[\mathrm{M}^{\mathrm{s}}_{\mathcal{N}_r\mathcal{V}}\right]\left[\xi_{\mathcal{V}}\right] \qquad [4.199]$$

equation [4.196] is written as:

$$\left[S_{\mathcal{N}_r \mathcal{N}_r}\right]\left[U_{\mathcal{N}_r}\right]=\left[F_{\mathcal{N}_r}\right] \qquad [4.200]$$

The solution to this system of equations makes it possible to obtain, in this case, the $\mathcal{N}_r$ values of the vector $\left[U_{\mathcal{N}_r}\right]$. Then, using relation [4.85], we deduce the expression $U_d(x)$, namely the approximation to the solution of this problem.

4.4.3. *Electrostatics*

For electrostatics, let us consider the previously studied example in Figure 3.11 (see sections 3.3.2 and 4.2.2). It should be recalled that, for this example, there is a domain Ω, composed of two electrodes in contact with the external environment, denoted by $\mathcal{E}_1$ and $\mathcal{E}_2$, and an internal electrode $\mathcal{E}_3$. The presence of the internal electrode limits the studied domain to $\Omega' = \Omega - \Omega_{\mathcal{E}3}$. The source terms can be the circulations f_{ij} between the electrodes of the electric field (see equation [3.108]) or the charges Q_k on the electrodes (see equation [3.110]). The boundary Γ of the domain is composed of gates Γ_{ek} with $k \in \{1, 2, 3\}$ and a wall Γ_d (see equation [3.106]). On the domain Ω', a mesh $\mathcal{M}$ is built and the number of nodes, edges, facets and volumes is denoted by $\mathcal{N}$, $\mathcal{A}$, $\mathcal{F}$ and $\mathcal{V}$, respectively.

In this section, the finite element method is applied to the two weak formulations (scalar potential and vector potential) that have been developed in section 4.2.2.

4.4.3.1. *Scalar potential V formulation*

The weak scalar potential formulation (see equation [4.15]) is recalled as follows:

$$\iiint_{\Omega'} \varepsilon \,\mathbf{grad}V.\mathbf{grad}\psi d\tau = \iiint_{\Omega'} \varepsilon(f_{13}\boldsymbol{\beta}_{13} + f_{23}\boldsymbol{\beta}_{23}).\mathbf{grad}\psi d\tau \qquad [4.201]$$

where V and ψ belong to the space $H_{\Gamma e1 \cup \Gamma e2 \cup \Gamma e3}(\mathbf{grad}, \Omega')$ and the support fields $\boldsymbol{\beta}_{ij}$ of the source terms defined in [3.113].

To solve this equation with the finite element method, the first step is to discretize the scalar potential V and the support fields $\boldsymbol{\beta}_{ij}$.

4.4.3.1.1. Discretization of the scalar potential V

The scalar potential V belongs to $H_{\Gamma e1 \cup \Gamma e2 \cup \Gamma e3}(\mathbf{curl}0, \Omega')$ and, as shown in Table 4.18, it must be discretized in the space of the node elements

$W^0{}_{\Gamma e1 \cup \Gamma e2 \cup \Gamma e3}(\Omega'_d)$. Its expression is then given by relation [4.163]. The boundaries Γ_{ek} with k = {1, 2, 3} are equipotential surfaces on which the potential is equal to zero (see equation [3.115]). To impose this condition in the discrete domain, we fix at zero the set of nodal values V_i belonging to the boundary $\Gamma_{e1} \cup \Gamma_{e2} \cup \Gamma_{e3}$. It should be noted that, as indicated in section 4.3.6.1, this condition also allows for the gauging of the scalar potential. Given $\mathcal{N}$ the number of nodes of the mesh and $\mathcal{N}_{ek}$ the number of nodes on the boundaries Γ_{ek} with k = {1, 2, 3}, the number of remaining unknowns of the problem, denoted by $\mathcal{N}_r$, is written as:

$$\mathcal{N}_r = \mathcal{N} - \sum_{k=1}^{3} \mathcal{N}_{ek} \qquad [4.202]$$

At this stage of our study, the nodes are renumbered so that those belonging to the boundary $\Gamma_{e1} \cup \Gamma_{e2} \cup \Gamma_{e3}$ have the highest indices. In other terms, the nodes with an index higher than $\mathcal{N}_r$ are located on the boundary $\Gamma_{e1} \cup \Gamma_{e2} \cup \Gamma_{e3}$. Under these conditions, given that the node values of the scalar potential are equal to zero for the nodes located on the boundary $\Gamma_{e1} \cup \Gamma_{e2} \cup \Gamma_{e3}$, the discrete form of the scalar potential $V_d(x)$ is written as:

$$V_d(x) = \sum_{i=1}^{\mathcal{N}_r} \omega_{n_i} V_i = \left[\omega_{\mathcal{N}_r}\right]^t \left[V_{\mathcal{N}_r}\right] \qquad [4.203]$$

where $\left[V_{\mathcal{N}_r}\right]$ represents the vector of dimension $\mathcal{N}_r$ of the nodal values of the scalar potential.

4.4.3.1.2. Discretization of the support fields α_{ij} and β_{ij}

To determine the discrete form of the support fields $\boldsymbol{\beta}_{ij}$, we will use the approach proposed in section 4.3.7.1. The first step is to build in the space of nodal elements the associated scalar potentials α_{ij}, and then deduce from them the support fields in the space of edge elements.

The scalar potentials α_{ij} are defined by the relations given in equation [3.114]. To build their discrete forms, we use the approach developed in section 4.3.7.1. In fact, for $\alpha_{13d}(x)$ and $\alpha_{23d}(x)$, we fix to 1 their node values belonging, respectively, to the boundaries Γ_{e1} and Γ_{e2}, and to zero the other node values (Henneron et al. 2004). Then, we have, for α_{13i} (with $1 \leq i \leq \mathcal{N}$):

$$\alpha_{13i} = 1 \text{ if } i \in \mathcal{N}_{\Gamma_{e1}} \text{ else } \alpha_{13i} = 0 \qquad [4.204]$$

Similarly, for the components α_{23i} (with $1 \leq i \leq \mathcal{N}$) associated with the function $\alpha_{23d}(x)$, we have:

$$\alpha_{23i} = 1 \text{ if } i \in \mathcal{N}_{\Gamma_{e2}} \text{ else } \alpha_{23i} = 0 \qquad [4.205]$$

As we know the expression of nodal values, the discrete form of $\alpha_{13d}(x)$ is written as:

$$\alpha_{13d}(x) = \sum_{i=1}^{\mathcal{N}} \omega_{n_i} \alpha_{13_i} = [\omega_{\mathcal{N}}]^t [\alpha_{13\mathcal{N}}] \qquad [4.206]$$

where $[\alpha_{13\mathcal{N}}]$ represents the $\mathcal{N}$-dimensional vector of nodal values α_{13i}. Similarly, the expression of the discrete form of the associated potential $\alpha_{23\,d}(x)$ is:

$$\alpha_{23d}(x) = \sum_{i=1}^{\mathcal{N}} \omega_{n_i} \alpha_{23_i} = [\omega_{\mathcal{N}}]^t [\alpha_{23\mathcal{N}}] \qquad [4.207]$$

where $[\alpha_{23\mathcal{N}}]$ represents the $\mathcal{N}$-dimensional vector of nodal values α_{23i}.

Knowing the expression of the associated potentials $\alpha_{13d}(x)$ and $\alpha_{23d}(x)$, the discrete form of the support fields $\boldsymbol{\beta}_{13d}(x)$ and $\boldsymbol{\beta}_{23d}(x)$ can be immediately calculated (see equation [3.114]). Relying on equation [4.182], as well as on equation [4.92] and posing $[\beta_{13\mathcal{A}}] = -[\mathbf{G}_{\mathcal{A},\mathcal{N}}]^t [\alpha_{13\mathcal{N}}]$, we deduce the following equation:

$$\boldsymbol{\beta}_{13d}(x) = \sum_{i=1}^{\mathcal{A}} \boldsymbol{\omega}_{a_i} \beta_{13_i} = [\boldsymbol{\omega}_{\mathcal{A}}]^t [\beta_{13\mathcal{A}}] \qquad [4.208]$$

Similarly, the support field $\boldsymbol{\beta}_{23}(x)$ is written as:

$$\boldsymbol{\beta}_{23d}(x) = \sum_{i=1}^{\mathcal{A}} \boldsymbol{\omega}_{a_i} \beta_{23_i} = [\boldsymbol{\omega}_{\mathcal{A}}]^t [\beta_{23\mathcal{A}}] \qquad [4.209]$$

with $[\beta_{23\mathcal{A}}] = -[G_{\mathcal{A},\mathcal{N}}]^t [\alpha_{23\mathcal{N}}]$.

4.4.3.1.3. System of equations to be solved

In order to solve equation [4.201] using the finite element method, we replace the scalar potential and the support fields by their discrete forms and we apply the

Ritz–Galerkin method (see section 4.4.2). In this case, the weighting functions are in the discrete space corresponding to $H_{\Gamma e1\cup\Gamma e2\cup\Gamma e3}(\mathbf{grad}, \Omega')$, namely $W^0_{\Gamma e1\cup\Gamma e2\cup\Gamma e3}(\Omega'_d)$, which corresponds to $\mathcal{N}_r$ nodal approximation functions ω_{nj}. Then, we can write:

$$\sum_{i=1}^{\mathcal{N}_r} \iiint_{\Omega'_d} \varepsilon(\mathbf{grad}\omega_{n_j}.\mathbf{grad}\omega_{n_i})V_i d\tau =$$
$$\sum_{k=1}^{\mathcal{A}} \iiint_{\Omega'_d} \varepsilon f_{13}\mathbf{grad}\omega_{n_j}.\boldsymbol{\omega}_{a_k}\beta_{13_k} d\tau \qquad [4.210]$$
$$+\sum_{l=1}^{\mathcal{A}} \iiint_{\Omega'_d} \varepsilon f_{23}\mathbf{grad}\omega_{n_j}.\boldsymbol{\omega}_{a_l}\beta_{23_l} d\tau \quad \text{avec } 1 \le j \le \mathcal{N}_r$$

Using relation [4.108] and, given that the equation must be verified for the $\mathcal{N}_r$ weighting functions, this is reflected by the matrix equation:

$$\iiint_{\Omega'_d} \varepsilon\,(\mathbf{grad}[\omega_{\mathcal{N}_r}]^t)^t(\mathbf{grad}[\omega_{\mathcal{N}_r}]^t)[V_{\mathcal{N}_r}]d\tau =$$
$$\iiint_{\Omega'_d} \varepsilon f_{13}(\mathbf{grad}[\omega_{\mathcal{N}_r}]^t)^t[\boldsymbol{\omega}_{\mathcal{A}}]^t[\beta_{13\mathcal{A}}]d\tau \qquad [4.211]$$
$$+\iiint_{\Omega'_d} \varepsilon f_{23}(\mathbf{grad}[\omega_{\mathcal{N}_r}]^t)^t[\boldsymbol{\omega}_{\mathcal{A}}]^t[\beta_{23\mathcal{A}}]d\tau$$

Relation [4.110], related to the gradient operator, allows us to rewrite this expression in the following form:

$$\iiint_{\Omega'} \varepsilon[\mathbf{G}_{\mathcal{A}\mathcal{N}_r}]^t[\boldsymbol{\omega}_{\mathcal{A}}][\boldsymbol{\omega}_{\mathcal{A}}]^t[\mathbf{G}_{\mathcal{A}\mathcal{N}_r}][V_{\mathcal{N}_r}]d\tau =$$
$$\iiint_{\Omega'} \varepsilon f_{13}[\mathbf{G}_{\mathcal{A}\mathcal{N}_r}]^t[\boldsymbol{\omega}_{\mathcal{A}}][\boldsymbol{\omega}_{\mathcal{A}}]^t[\beta_{13\mathcal{A}}]d\tau \qquad [4.212]$$
$$+\iiint_{\Omega'} \varepsilon f_{23}[\mathbf{G}_{\mathcal{A}\mathcal{N}_r}]^t[\boldsymbol{\omega}_{\mathcal{A}}][\boldsymbol{\omega}_{\mathcal{A}}]^t[\beta_{23\mathcal{A}}]d\tau$$

Knowing that, besides the circulations f_{13} and f_{23}, the matrix $[\mathbf{G}_{\mathcal{A}\mathcal{N}_r}]$ and the vectors $[V_{\mathcal{N}r}]$, $[\beta_{13\mathcal{A}}]$ and $[\beta_{23\mathcal{A}}]$ do not depend on the position (see equation [4.196]), we can modify this equation and write it as follows:

$$[\mathbf{G}_{\mathcal{A}\mathcal{N}_r}]^t[M^{\varepsilon}_{\mathcal{A}\mathcal{A}}][\mathbf{G}_{\mathcal{A}\mathcal{N}_r}][V_{\mathcal{N}_r}] = f_{13}[\mathbf{G}_{\mathcal{A}\mathcal{N}_r}]^t[M^{\varepsilon}_{\mathcal{A}\mathcal{A}}][\beta_{13\mathcal{A}}] \qquad [4.213]$$
$$+ f_{23}[\mathbf{G}_{\mathcal{A}\mathcal{N}_r}]^t[M^{\varepsilon}_{\mathcal{A}\mathcal{A}}][\beta_{23\mathcal{A}}]$$

where the elementary terms of the matrix $\left[M^{\varepsilon}_{\mathcal{AA}}\right]$ are:

$$M^{\varepsilon}_{a_i a_j} = \iiint_{\Omega_d'} \varepsilon \boldsymbol{\omega}_{a_i} . \boldsymbol{\omega}_{a_j} d\tau \qquad [4.214]$$

Finally, posing:

$$\begin{aligned} &\left[S^{\varepsilon}_{\mathcal{N}_r\mathcal{N}_r}\right] = \left[G_{\mathcal{AN}_r}\right]^t \left[M^{\varepsilon}_{\mathcal{AA}}\right] \left[G_{\mathcal{AN}_r}\right] \text{ and} \\ &\left[F^{\varepsilon}_{\mathcal{N}_r}\right] = f_{13}\left[G_{\mathcal{AN}_r}\right]^t \left[M^{\varepsilon}_{\mathcal{AA}}\right]\left[\beta_{13\mathcal{A}}\right] + f_{23}\left[G_{\mathcal{AN}_r}\right]^t \left[M^{\varepsilon}_{\mathcal{AA}}\right]\left[\beta_{23\mathcal{A}}\right] \end{aligned} \qquad [4.215]$$

Equation [4.213] to be solved can be written in the generic form as follows:

$$\left[S^{\varepsilon}_{\mathcal{N}_r\mathcal{N}_r}\right] \left[V_{\mathcal{N}_r}\right] = \left[F^{\varepsilon}_{\mathcal{N}_r}\right] \qquad [4.216]$$

The solution to this system of equations allows us to obtain the nodal values of the scalar potential. Then, using expression [4.203], we can express the scalar potential $V_d(x)$ in any point of the domain Ω_d'.

4.4.3.1.4. Imposed charges

Sections 3.3.2.1 and 4.2.2.1 considered that the source terms may be either the total charges Q_k or a combination of the circulation f_{ij} and charges Q_k. In this case, we need to add equations [3.125] or [3.126] depending on whether we impose the charges Q_1 or Q_2 or both. These relations are recalled below, where the total charges Q_1 are written as:

$$Q_1 = \iiint_{\Omega'} \varepsilon \boldsymbol{\beta}_{13} .(f_{13} \boldsymbol{\beta}_{13} + f_{23} \boldsymbol{\beta}_{23} - \textbf{grad } V) d\tau \qquad [4.217]$$

Similarly, for the total charges Q_2, we have:

$$Q_2 = \iiint_{\Omega'} \varepsilon \boldsymbol{\beta}_{23} .(f_{13} \boldsymbol{\beta}_{13} + f_{23} \boldsymbol{\beta}_{23} - \textbf{grad } V) d\tau \qquad [4.218]$$

These equations can be made compatible with the matrix equation [4.213] if we introduce the discrete form of the scalar potential V and that of the support fields $\boldsymbol{\beta}_{13}$

and $\boldsymbol{\beta}_{23}$. For this purpose, we can use expressions [4.203], [4.208] and [4.209]. The expression of the total charges Q_1 is then:

$$\begin{aligned} Q_1 = &-\iiint_{\Omega'_d} \varepsilon([\boldsymbol{\omega}_{\mathcal{A}}]^t [\beta_{13\mathcal{A}}])^t \mathbf{grad}[\omega_{\mathcal{N}_r}]^t [V_{\mathcal{N}_r}] d\tau \\ &+ f_{13}\iiint_{\Omega'_d} \varepsilon([\boldsymbol{\omega}_{\mathcal{A}}]^t [\beta_{13\mathcal{A}}])^t [\boldsymbol{\omega}_{\mathcal{A}}]^t [\beta_{13\mathcal{A}}] d\tau \\ &+ f_{23}\iiint_{\Omega'_d} \varepsilon([\boldsymbol{\omega}_{\mathcal{A}}]^t [\beta_{13\mathcal{A}}])^t [\boldsymbol{\omega}_{\mathcal{A}}]^t [\beta_{23\mathcal{A}}] d\tau \end{aligned} \quad [4.219]$$

Introducing the incidence matrix $\left[\mathbf{G}_{\mathcal{A}\mathcal{N}_r}\right]$ (see equation [4.110]) and after rearranging the equation, we can write:

$$\begin{aligned} Q_1 = &-\iiint_{\Omega'_d} \varepsilon[\beta_{13\mathcal{A}}]^t [\boldsymbol{\omega}_{\mathcal{A}}][\boldsymbol{\omega}_{\mathcal{A}}]^t [\mathbf{G}_{\mathcal{A}\mathcal{N}_r}][V_{\mathcal{N}_r}] d\tau \\ &+ f_{13}\iiint_{\Omega'_d} \varepsilon[\beta_{13\mathcal{A}}]^t [\boldsymbol{\omega}_{\mathcal{A}}][\boldsymbol{\omega}_{\mathcal{A}}]^t [\beta_{13\mathcal{A}}] d\tau \\ &+ f_{23}\iiint_{\Omega'_d} \varepsilon[\beta_{13\mathcal{A}}]^t [\boldsymbol{\omega}_{\mathcal{A}}][\boldsymbol{\omega}_{\mathcal{A}}]^t [\beta_{23\mathcal{A}}] d\tau \end{aligned} \quad [4.220]$$

On the other hand, the matrix $\left[\mathbf{G}_{\mathcal{A}\mathcal{N}_r}\right]$ and the vectors $\left[V_{\mathcal{N}_r}\right]$, $[\beta_{13\mathcal{A}}]$, $[\beta_{23\mathcal{A}}]$ do not depend on the position. Hence, we can introduce the mass matrix $\left[M^{\varepsilon}_{\mathcal{A}\mathcal{A}}\right]$, whose elementary term is given by relation [4.214]. Then, we have:

$$\begin{aligned} Q_1 = &-[\beta_{13\mathcal{A}}]^t [M^{\varepsilon}_{\mathcal{A}\mathcal{A}}][\mathbf{G}_{\mathcal{A}\mathcal{N}_r}][V_{\mathcal{N}_r}] \\ &+ f_{13}[\beta_{13\mathcal{A}}]^t [M^{\varepsilon}_{\mathcal{A}\mathcal{A}}][\beta_{13\mathcal{A}}] + f_{23}[\beta_{13\mathcal{A}}]^t [M^{\varepsilon}_{\mathcal{A}\mathcal{A}}][\beta_{23\mathcal{A}}] \end{aligned} \quad [4.221]$$

The same developments, applied to expression [4.218], can be used to write the total charges Q_2:

$$\begin{aligned} Q_2 = &-[\beta_{23\mathcal{A}}]^t [M^{\varepsilon}_{\mathcal{A}\mathcal{A}}][\mathbf{G}_{\mathcal{A}\mathcal{N}_r}][V_{\mathcal{N}_r}] \\ &+ f_{13}[\beta_{23\mathcal{A}}]^t [M^{\varepsilon}_{\mathcal{A}\mathcal{A}}][\beta_{13\mathcal{A}}] + f_{23}[\beta_{23\mathcal{A}}]^t [M^{\varepsilon}_{\mathcal{A}\mathcal{A}}][\beta_{23\mathcal{A}}] \end{aligned} \quad [4.222]$$

In order to build a complete matrix system, the following notations are introduced:

$$\begin{aligned}
&\left[C^{\varepsilon}_{13\mathcal{N}_r}\right] = -\left[G_{\mathcal{AN}_r}\right]^t\left[M^{\varepsilon}_{\mathcal{AA}}\right]\left[\beta_{13\mathcal{A}}\right], \quad \left[C^{\varepsilon}_{23\mathcal{N}_r}\right] = -\left[G_{\mathcal{AN}_r}\right]^t\left[M^{\varepsilon}_{\mathcal{AA}}\right]\left[\beta_{23\mathcal{A}}\right] \\
&B^{\varepsilon}_{\beta_{13}} = \left[\beta_{13\mathcal{A}}\right]^t\left[M^{\varepsilon}_{\mathcal{AA}}\right]\left[\beta_{13\mathcal{A}}\right], \quad B^{\varepsilon}_{\beta_{23}} = \left[\beta_{23\mathcal{A}}\right]^t\left[M^{\varepsilon}_{\mathcal{AA}}\right]\left[\beta_{23\mathcal{A}}\right] \\
&H^{\varepsilon} = \left[\beta_{13\mathcal{A}}\right]^t\left[M^{\varepsilon}_{\mathcal{AA}}\right]\left[\beta_{23\mathcal{A}}\right]
\end{aligned} \qquad [4.223]$$

where $\left[C^{\varepsilon}_{13\mathcal{N}_r}\right]$ and $\left[C^{\varepsilon}_{23\mathcal{N}_r}\right]$ are the notations for vectors of dimension $\mathcal{N}_r$ and $B^{\varepsilon}_{\beta_{13}}$, $B^{\varepsilon}_{\beta_{23}}$ and H^{ε} for scalars.

Let us now gather equations [4.213], [4.221] and [4.222] with the notations defined in equations [4.215] and [4.223]. Then, for an electrostatic problem in the case of the scalar potential formulation when the source terms are total charges Q_1 and Q_2, we have the following system to solve:

$$\begin{bmatrix} \left[S^{\varepsilon}_{\mathcal{N}_r\mathcal{N}_r}\right] & \left[C^{\varepsilon}_{13\mathcal{N}_r}\right] & \left[C^{\varepsilon}_{23\mathcal{N}_r}\right] \\ \left[C^{\varepsilon}_{13\mathcal{N}_r}\right]^t & B^{\varepsilon}_{\beta_{13}} & H^{\varepsilon} \\ \left[C^{\varepsilon}_{23\mathcal{N}_r}\right]^t & H^{\varepsilon} & B^{\varepsilon}_{\beta_{23}} \end{bmatrix} \begin{bmatrix} \left[V_{\mathcal{N}_r}\right] \\ f_{13} \\ f_{23} \end{bmatrix} = \begin{bmatrix} [0] \\ Q_1 \\ Q_2 \end{bmatrix} \qquad [4.224]$$

In this system of equations, the unknowns are the nodal values of the scalar potential and the circulations f_{13} and f_{23}. As for the source terms, they are the total charges Q_1 and Q_2. It should also be noted that the system is symmetric. If we now impose a circulation and a value of charges, the new system to be solved is deduced from equation [4.224].

4.4.3.2. *Vector potential P formulation*

For the vector potential formulation, the weak form of our electrostatic problem is given by relation [4.24], which is recalled as follows:

$$\begin{aligned}
&\iiint_{\Omega'} \varepsilon^{-1}\mathbf{curlP.curl\Psi}d\tau = -\iiint_{\Omega'} \varepsilon^{-1}(Q_1\boldsymbol{\lambda}_{13} + Q_2\boldsymbol{\lambda}_{23}).\mathbf{curl\Psi}d\tau, \\
&\mathbf{P} \text{ and } \mathbf{\Psi} \in H_{\Gamma_d}(\mathbf{curl}, \Omega')
\end{aligned} \qquad [4.225]$$

To solve this equation with the finite element method, besides the vector potential **P**, we have to discretize the support fields $\boldsymbol{\lambda}_{12}$ and $\boldsymbol{\lambda}_{23}$. These support fields are defined by equations [3.131] and [3.132] and they belong, respectively, to $H_{\Gamma e2 \cup \Gamma d}(\text{div}, \Omega')$ and $H_{\Gamma e1 \cup \Gamma d}(\text{div}, \Omega')$.

4.4.3.2.1. Discretization of the vector potential P

In the continuous domain, the vector potential **P** is defined in the function space $H_{\Gamma d}(\textbf{curl}, \Omega')$. In the discrete domain, this potential therefore belongs (see Table 4.18) to the function space $W^1_{\Gamma d}(\Omega'_d)$. Moreover, a gauge condition and the boundary conditions should be imposed on Γ_d (see equation [3.138]). To impose the gauge condition, we can use the process presented in section 4.3.6.2, leading to the construction of an edge tree. It should be recalled that, in order to impose the boundary conditions and the gauge, we first build the edge tree on the boundary Γ_d, then we extend it to the set of edges of the mesh. Then, we fix the circulations of **P** to zero on the $\mathcal{N} - 1$ edges of the tree. For the boundary conditions, we denote by $\mathcal{A}_d$ the number of edges belonging to the boundary Γ_d and by $\mathcal{A}_{da}$ the number of edges of the tree also belonging to the boundary Γ_d. To impose the boundary conditions, besides the edges belonging to the tree, we also fix to zero the $\mathcal{A}_d - \mathcal{A}_{da}$ edges of the boundary. The number of unknowns of the problem, namely the number $\mathcal{A}_r$ of circulations of the vector potential on the edges, has the following expression:

$$\mathcal{A}_r = \mathcal{A} - (\mathcal{N} - 1) - (\mathcal{A}_d - \mathcal{A}_{da}) \qquad [4.226]$$

To facilitate the implementation of the gauge and the boundary conditions in the following, as already noted in sections 4.4.2 and 4.4.3.1, we renumber the edges so that the unknown circulations of the vector **P** are numbered from 1 to $\mathcal{A}_r$. Hence, the indices of the edges belonging to the boundary Γ_d and to the edge tree are higher than $\mathcal{A}_r$. Under these conditions, the discrete form of the vector potential $\mathbf{P}_d(x)$ is expressed using the vector $[\boldsymbol{\omega}_{\mathcal{A}r}]$ of interpolation functions $\boldsymbol{\omega}_{ai}$ and the circulations of the vector potential (denoted by P_i) on the $\mathcal{A}_r$ edges as follows:

$$\mathbf{P}_d(x) = \sum_{i=1}^{\mathcal{A}_r} \boldsymbol{\omega}_{a_i} P_i = \left[\boldsymbol{\omega}_{\mathcal{A}_r}\right]^t \left[P_{\mathcal{A}_r}\right] \qquad [4.227]$$

In this expression, $[P_{\mathcal{A}r}]$ represents the vector of P_i circulations of the vector potential on the $\mathcal{A}_r$ edges of the mesh associated with the unknowns of the problem.

4.4.3.2.2. Discretization of the support fields $\boldsymbol{\lambda}_{ij}$

For the discretization of the support fields $\boldsymbol{\lambda}_{13}$ and $\boldsymbol{\lambda}_{23}$, we rely on the properties defined by relations [3.131] and [3.132]. These two fields belong to the function spaces $H_{\Gamma e2\cup\Gamma d}(\text{div}, \Omega')$ and $H_{\Gamma e1\cup\Gamma d}(\text{div}, \Omega')$. Therefore, they are discretized (see Table 4.18) in the spaces $W^2_{\Gamma e2\cup\Gamma d}(\Omega'_d)$ and $W^2_{\Gamma e1\cup\Gamma d}(\Omega'_d)$, respectively, namely the space of facet elements. The expression of the support field $\boldsymbol{\lambda}_{13}$ is:

$$\boldsymbol{\lambda}_{13d}(x) = \sum_{i=1}^{\mathcal{F}} \boldsymbol{\omega}_{f_i} \lambda_{13f_i} = [\boldsymbol{\omega}_{\mathcal{F}}]^t [\lambda_{13_{\mathcal{F}}}] \quad [4.228]$$

and $\boldsymbol{\lambda}_{23}$:

$$\boldsymbol{\lambda}_{23d}(x) = \sum_{i=1}^{\mathcal{F}} \boldsymbol{\omega}_{f_i} \lambda_{23f_i} = [\boldsymbol{\omega}_{\mathcal{F}}]^t [\lambda_{23_{\mathcal{F}}}] \quad [4.229]$$

For the calculation of the vectors $[\boldsymbol{\lambda}_{13}]$ and $[\boldsymbol{\lambda}_{23}]$, we use the process developed in section 4.3.7.2. For the construction of $[\boldsymbol{\lambda}_{13}]$, the two gates are Γ_{e1} and Γ_{e3} and the wall boundary Γ_m is the union of Γ_{e2} and Γ_d. In a complementary manner, $[\boldsymbol{\lambda}_{23}]$ is built with the surfaces Γ_{e2} and Γ_{e3} as gates, and the union of Γ_{e1} and Γ_d as wall Γ_m.

4.4.3.2.3. System of equations to be solved

In order to solve equation [4.225] with the finite element method, we replace the source fields and the vector potential with their discretized expressions given, respectively, by relations [4.227], [4.228] and [4.229].

The Ritz–Galerkin method is applied, consisting of having the $\mathcal{A}_r$ edge functions $\boldsymbol{\omega}_{aj}$ as weighting functions. We then obtain:

$$\begin{aligned}
&\sum_{i=1}^{\mathcal{A}_r} \iiint_{\Omega'_d} \varepsilon^{-1} \mathbf{curl}\boldsymbol{\omega}_{a_j} . \mathbf{curl}\,\boldsymbol{\omega}_{a_i}\, d\tau\, P_i = \\
&\quad -Q_1 \sum_{k=1}^{\mathcal{F}} \iiint_{\Omega'_d} \varepsilon^{-1} \mathbf{curl}\boldsymbol{\omega}_{a_j} . \boldsymbol{\omega}_{f_k} \lambda_{13f_k}\, d\tau \\
&\quad -Q_2 \sum_{k=1}^{\mathcal{F}} \iiint_{\Omega'_d} \varepsilon^{-1} \mathbf{curl}\boldsymbol{\omega}_{a_j} . \boldsymbol{\omega}_{f_k} \lambda_{23f_k}\, d\tau, \text{ with } 1 \le j \le \mathcal{A}_r
\end{aligned} \quad [4.230]$$

This equation must be verified for the set of $\mathcal{A}_r$ weighting functions $\boldsymbol{\omega}_{aj}$ of the function space $W^1_{\Gamma d}(\Omega'_d)$.

It can therefore be rewritten in the matrix form as:

$$\begin{aligned}&\iiint_{\Omega_d'} \varepsilon^{-1}(\mathbf{curl}\left[\omega_{\mathcal{A}_r}\right]^t)^t \mathbf{curl}\left[\omega_{\mathcal{A}_r}\right]^t\left[P_{\mathcal{A}_r}\right]d\tau = \\ &\quad -Q_1\iiint_{\Omega_d'} \varepsilon^{-1}(\mathbf{curl}\left[\omega_{\mathcal{A}_r}\right]^t)^t\left[\omega_{\mathcal{F}}\right]^t\left[\lambda_{13\mathcal{F}}\right]d\tau \\ &\qquad -Q_2\iiint_{\Omega_d'} \varepsilon^{-1}(\mathbf{curl}\left[\omega_{\mathcal{A}_r}\right]^t)^t\left[\omega_{\mathcal{F}}\right]^t\left[\lambda_{23\mathcal{F}}\right]d\tau\end{aligned} \quad [4.231]$$

Using relation [4.124], we can write the system of equations by introducing the matrices $[\mathbf{R}_{\mathcal{F}\mathcal{A}}]$ and $[\boldsymbol{\omega}_{\mathcal{F}}]$ as follows:

$$\begin{aligned}&\iiint_{\Omega} \varepsilon^{-1}\left[\mathbf{R}_{\mathcal{F}\mathcal{A}_r}\right]^t\left[\omega_{\mathcal{F}}\right]\left[\omega_{\mathcal{F}}\right]^t\left[\mathbf{R}_{\mathcal{F}\mathcal{A}_r}\right]\left[P_{\mathcal{A}_r}\right]d\tau = \\ &\quad -Q_1\iiint_{\Omega} \varepsilon^{-1}\left[\mathbf{R}_{\mathcal{F}\mathcal{A}_r}\right]^t\left[\omega_{\mathcal{F}}\right]\left[\omega_{\mathcal{F}}\right]^t\left[\lambda_{13\mathcal{F}}\right]d\tau \\ &\qquad -Q_2\iiint_{\Omega} \varepsilon^{-1}\left[\mathbf{R}_{\mathcal{F}\mathcal{A}_r}\right]^t\left[\omega_{\mathcal{F}}\right]\left[\omega_{\mathcal{F}}\right]^t\left[\lambda_{23\mathcal{F}}\right]d\tau\end{aligned} \quad [4.232]$$

The incidence matrices $\left[\mathbf{R}_{\mathcal{F}\mathcal{A}_r}\right]$ as well as the vectors $\left[P_{\mathcal{A}_r}\right]$, $[\lambda_{13\mathcal{F}}]$ and $[\lambda_{23\mathcal{F}}]$ are vectors whose components do not depend on the position. Under these conditions, equation [4.232] can be written in the following form:

$$\begin{aligned}&\left[\mathbf{R}_{\mathcal{F}\mathcal{A}_r}\right]^t\left[M_{\mathcal{F}\mathcal{F}}^{\varepsilon^{-1}}\right]\left[\mathbf{R}_{\mathcal{F}\mathcal{A}_r}\right]\left[P_{\mathcal{A}_r}\right] = -Q_1\left[\mathbf{R}_{\mathcal{F}\mathcal{A}_r}\right]^t\left[M_{\mathcal{F}\mathcal{F}}^{\varepsilon^{-1}}\right]\left[\lambda_{13\mathcal{F}}\right] \\ &\quad -Q_2\left[\mathbf{R}_{\mathcal{F}\mathcal{A}_r}\right]^t\left[M_{\mathcal{F}\mathcal{F}}^{\varepsilon^{-1}}\right]\left[\lambda_{13\mathcal{F}}\right]\end{aligned} \quad [4.233]$$

The elementary term of the matrix $\left[M_{\mathcal{F}\mathcal{F}}^{\varepsilon^{-1}}\right]$ is written as:

$$M_{f_i f_j}^{\varepsilon^{-1}} = \iiint_{\Omega_d'} \varepsilon^{-1}\boldsymbol{\omega}_{f_i}.\boldsymbol{\omega}_{f_j}\,d\tau \quad [4.234]$$

Let us now pose:

$$\begin{aligned}&\left[S_{\mathcal{A}_r\mathcal{A}_r}^{\varepsilon^{-1}}\right] = \left[\mathbf{R}_{\mathcal{F}\mathcal{A}_r}\right]^t\left[M_{\mathcal{F}\mathcal{F}}^{\varepsilon^{-1}}\right]\left[\mathbf{R}_{\mathcal{F}\mathcal{A}_r}\right] \text{ and} \\ &\left[F_{\mathcal{A}_r}^{\varepsilon^{-1}}\right] = -Q_1\left[\mathbf{R}_{\mathcal{F}\mathcal{A}_r}\right]^t\left[M_{\mathcal{F}\mathcal{F}}^{\varepsilon^{-1}}\right]\left[\lambda_{13\mathcal{F}}\right] - Q_2\left[\mathbf{R}_{\mathcal{F}\mathcal{A}_r}\right]^t\left[M_{\mathcal{F}\mathcal{F}}^{\varepsilon^{-1}}\right]\left[\lambda_{23\mathcal{F}}\right]\end{aligned} \quad [4.235]$$

The system of equations is then written as:

$$\left[S^{\varepsilon^{-1}}_{\mathcal{A}_r\mathcal{A}_r}\right]\left[P_{\mathcal{A}_r}\right]=\left[F^{\varepsilon^{-1}}_{\mathcal{A}_r}\right] \quad [4.236]$$

The solution of this matrix equation can be used to obtain the circulations of the vector potential on the edges of the mesh. Then, using expression [4.227], we express the vector potential $\mathbf{P}_d(x)$ in any point of the domain Ω'_d.

4.4.3.2.4. Imposed circulation of the electric field

According to sections 3.3.2.2 and 4.2.2.2, the source terms could be the circulations f_{ij} or a combination of a circulation and a value of charges imposed on an electrode. Under these conditions, to obtain a full equation system, the system [4.233] must be completed with equation [3.146] and/or, depending on the case, equation [3.147]. These two equations must therefore be discretized. Therefore, equation [3.146] is recalled below:

$$f_{13}=\iiint_{\Omega'}\boldsymbol{\lambda}_{13}.\varepsilon^{-1}(Q_1\boldsymbol{\lambda}_{13}+Q_2\boldsymbol{\lambda}_{23}+\mathbf{curlP})d\tau \quad [4.237]$$

Replacing the vector potential $\mathbf{P}$ and the support fields $\boldsymbol{\lambda}_{13}$ and $\boldsymbol{\lambda}_{23}$ by their expressions given, respectively, by equations [4.227], [4.228] and [4.229], we obtain:

$$\begin{aligned} f_{13}=&\iiint_{\Omega'_d}\varepsilon^{-1}([\boldsymbol{\omega}_{\mathcal{F}}]^t[\lambda_{13\mathcal{F}}])^t(\mathbf{curl}[\boldsymbol{\omega}_{\mathcal{A}_r}])^t[P_{\mathcal{A}_r}]d\tau \\ &+Q_1\iiint_{\Omega'_d}\varepsilon^{-1}([\boldsymbol{\omega}_{\mathcal{F}}]^t[\lambda_{13\mathcal{F}}])^t[\boldsymbol{\omega}_{\mathcal{F}}]^t[\lambda_{13\mathcal{F}}]d\tau \\ &+Q_2\iiint_{\Omega'_d}\varepsilon^{-1}([\boldsymbol{\omega}_{\mathcal{F}}]^t[\lambda_{13\mathcal{F}}])^t[\boldsymbol{\omega}_{\mathcal{F}}]^t[\lambda_{23\mathcal{F}}]d\tau \end{aligned} \quad [4.238]$$

Using relation [4.124] related to the discrete curl operator and then introducing matrix $\left[M^{\varepsilon^{-1}}_{\mathcal{F}\mathcal{F}}\right]$ defined in equation [4.234], equation [4.238] takes the following form:

$$\begin{aligned} f_{13}=&[\lambda_{13\mathcal{F}}]^t\left[M^{\varepsilon^{-1}}_{\mathcal{F}\mathcal{F}}\right][\mathbf{R}_{\mathcal{F}\mathcal{A}_r}][P_{\mathcal{A}_r}] \\ &+Q_1[\lambda_{13\mathcal{F}}]^t\left[M^{\varepsilon^{-1}}_{\mathcal{F}\mathcal{F}}\right][\lambda_{13\mathcal{F}}]+Q_2[\lambda_{13\mathcal{F}}]^t\left[M^{\varepsilon^{-1}}_{\mathcal{F}\mathcal{F}}\right][\lambda_{23\mathcal{F}}] \end{aligned} \quad [4.239]$$

The discretization of equation [3.147] can be very easily deduced from this expression and we obtain:

$$\begin{aligned} f_{23} &= [\lambda_{23\mathcal{F}}]^t \left[M_{\mathcal{FF}}^{\varepsilon^{-1}}\right] [\mathbf{R}_{\mathcal{F}\mathcal{A}_r}] [P_{\mathcal{A}_r}] \\ &+ Q_1 [\lambda_{23\mathcal{F}}]^t \left[M_{\mathcal{FF}}^{\varepsilon^{-1}}\right] [\lambda_{13\mathcal{F}}] + Q_2 [\lambda_{23\mathcal{F}}]^t \left[M_{\mathcal{FF}}^{\varepsilon^{-1}}\right] [\lambda_{23\mathcal{F}}] \end{aligned} \quad [4.240]$$

Similar to the case of the scalar potential formulation, a full system of equations is built. The first step is to introduce the following notations:

$$\begin{aligned} &\left[C_{13\mathcal{A}_r}^{\varepsilon^{-1}}\right] = [\mathbf{R}_{\mathcal{F}\mathcal{A}_r}]^t \left[M_{\mathcal{FF}}^{\varepsilon^{-1}}\right] [\lambda_{13\mathcal{F}}], \quad \left[C_{23\mathcal{A}_r}^{\varepsilon^{-1}}\right] = [R_{\mathcal{F}\mathcal{A}_r}]^t \left[M_{\mathcal{FF}}^{\varepsilon^{-1}}\right] [\lambda_{23\mathcal{F}}] \\ &B_{\lambda_{13}}^{\varepsilon^{-1}} = [\lambda_{13\mathcal{F}}]^t \left[M_{\mathcal{FF}}^{\varepsilon^{-1}}\right] [\lambda_{13\mathcal{F}}], \quad B_{\lambda_{23}}^{\varepsilon^{-1}} = [\lambda_{23\mathcal{F}}]^t \left[M_{\mathcal{FF}}^{\varepsilon^{-1}}\right] [\lambda_{23\mathcal{F}}] \\ &H^{\varepsilon^{-1}} = [\lambda_{13\mathcal{F}}]^t \left[M_{\mathcal{FF}}^{\varepsilon^{-1}}\right] [\lambda_{23\mathcal{F}}] \end{aligned} \quad [4.241]$$

In this expression, $\left[C_{13\mathcal{A}_r}^{\varepsilon^{-1}}\right]$ and $\left[C_{23\mathcal{A}_r}^{\varepsilon^{-1}}\right]$ denote vectors of dimension $\mathcal{A}_r$. As for the terms $B_{\lambda_{13}}^{\varepsilon^{-1}}$, $B_{\lambda_{23}}^{\varepsilon^{-1}}$ and $H^{\varepsilon^{-1}}$, they denote scalars.

Gathering equations [4.233], [4.239] and [4.240] using the previous notations, the system of equations to be solved is written as:

$$\begin{bmatrix} \left[S_{\mathcal{A}_r\mathcal{A}_r}^{\varepsilon^{-1}}\right] & \left[C_{13\mathcal{A}_r}^{\varepsilon^{-1}}\right] & \left[C_{23\mathcal{A}_r}^{\varepsilon^{-1}}\right] \\ \left[C_{13\mathcal{A}_r}^{\varepsilon^{-1}}\right]^t & B_{\lambda_{13}}^{\varepsilon^{-1}} & H^{\varepsilon^{-1}} \\ \left[C_{23\mathcal{A}_r}^{\varepsilon^{-1}}\right]^t & H^{\varepsilon^{-1}} & B_{\lambda_{23}}^{\varepsilon^{-1}} \end{bmatrix} \begin{bmatrix} [P_{\mathcal{A}_r}] \\ Q_1 \\ Q_2 \end{bmatrix} = \begin{bmatrix} [0] \\ f_{13} \\ f_{23} \end{bmatrix} \quad [4.242]$$

In this system of equations, the unknowns are the circulations of the vector potential on the edges of the mesh and the values of charges Q_1 and Q_2. The source terms are then the circulations f_{13} and f_{23}. Moreover, similar to the scalar potential formulation, the matrix of equations to be solved is symmetric.

Let us now assume that the source terms are the circulation f_{13} and the value of charges Q_2, the unknowns of the problem are the circulations of the vector potential on the edges of the mesh and the value of charges Q_1. The system of equations to be solved can be readily obtained from equation [4.242].

4.4.4. *Electrokinetics*

For electrokinetics, we consider again the example presented in section 3.4.2 and illustrated in Figure 3.14. In this case, we must solve equations [3.149] and [3.150]. For the boundary conditions (see equation [3.191]), we recall that the boundary Γ is decomposed into three gates Γ_{e1}, Γ_{e2} and Γ_{e3} for the electric field and a wall Γ_j for the current density. For this example, we studied several combinations of source terms that can be imposed on the three gates. We have considered two electromotive forces e_{13} and e_{23}, two current density fluxes I_1 and I_2 and the possibility of combining an electromotive force and a current. We developed the scalar potential formulation and the vector potential formulation. Finally, section 4.2.2 introduces the weak forms of the formulations. This section develops these various weak formulations in the case of the finite element method.

To solve these equations, we consider a mesh $\mathcal{M}$ composed of $\mathcal{N}$ nodes, $\mathcal{A}$ edges, $\mathcal{F}$ facets and $\mathcal{V}$ elements.

4.4.4.1. *Scalar potential V formulation*

When electromotive forces e_{13} and e_{23} are imposed, the weak scalar potential formulation is given by relation [4.29], with V and ψ belonging to $H_{\Gamma e1 \cup \Gamma e2 \cup \Gamma e3}(\mathbf{grad}, \Omega)$. To solve this equation, we need to first discretize the potential V and the support fields $\boldsymbol{\beta}_{13}$ and $\boldsymbol{\beta}_{23}$.

4.4.4.1.1. Discretization of the scalar potential V

Concerning the scalar potential V, as already seen in section 4.4.3.1, it is discretized in the space of nodal elements, with $V_d(x) \in W^0_{\Gamma e1 \cup \Gamma e2 \cup \Gamma e3}(\Omega_d)$. The three boundaries Γ_{ek}, $k \in \{1, 2, 3\}$, are equipotential surfaces on which (see equation [3.159]) the unknown electric scalar potential is imposed to zero. Imposing this condition in the discrete domain means fixing to zero the nodal values (denoted by V_i) of the scalar potential. Proceeding this way, it can be noted that the scalar potential V is automatically gauged (see section 4.3.6.1). Under these conditions, the number of unknown nodal values of the problem (denoted by $\mathcal{N}_r$) is written as:

$$\mathcal{N}_r = \mathcal{N} - (\mathcal{N}_1 + \mathcal{N}_2 + \mathcal{N}_3) \qquad [4.243]$$

where $\mathcal{N}_1$, $\mathcal{N}_2$ and $\mathcal{N}_3$ represent the number of nodes on the boundaries Γ_{e1}, Γ_{e2} and Γ_{e3}, respectively.

Using a renumbering, the indices of the nodes belonging to the boundaries Γ_{e1}, Γ_{e2} and Γ_{e3} are imposed higher than $\mathcal{N}_r$. Under these conditions, considering the boundary conditions on the boundary $\Gamma_{e1} \cup \Gamma_{e2} \cup \Gamma_{e3}$, the scalar potential $V_d(x)$ is written as:

$$V_d(x) = \sum_{i=1}^{\mathcal{N}_r} \omega_{n_i} V_i = \left[\omega_{\mathcal{N}_r}\right]^t \left[V_{\mathcal{N}_r}\right] \qquad [4.244]$$

4.4.4.1.2. Discretization of the support fields α_{ij} and β_{ij}

The properties of the support fields $\boldsymbol{\beta}_{ij}$ and α_{ij} are given by relations [3.192] and [3.193]. Since these properties are the same as in the case of electrostatics, an approach similar to that in section 4.4.3.1.2 will be used for their construction. As a first step, we build the associated potentials α_{ij} in the space of node elements. Then, we deduce the fields $\boldsymbol{\beta}_{ij}$, which belong to the space of edge elements, using relations [4.182] and [4.183].

As already seen in section 4.4.3.1.2, a possible solution for expressing the potentials α_{ij} involves fixing to 1 the values of $\alpha_{ij}(x)$ of the set of nodes of Γ_{ei} and to zero the other ones (see relations [4.204] and [4.205]). The expressions of $\alpha_{13d}(x)$ and $\alpha_{23d}(x)$ are then given by equations [4.206] and [4.207], respectively.

Based on the associated potentials α_{ij}, the support fields $\boldsymbol{\beta}_{ij}$ are obtained by applying the discrete operator $[\mathbf{G}_{\mathcal{AN}}]$ of the gradient.

Using equation [4.208], presented in the case of electrostatics, the discrete form of the support field $\boldsymbol{\beta}_{13d}(x)$ is written as:

$$\begin{aligned} &\boldsymbol{\beta}_{13d}(x) = \sum_{i=1}^{\mathcal{A}} \boldsymbol{\omega}_{a_i} \beta_{13_i} = \left[\boldsymbol{\omega}_{\mathcal{A}}\right]^t \left[\beta_{13\mathcal{A}}\right], \\ &\text{with}: \left[\beta_{13\mathcal{A}}\right] = -\left[G_{\mathcal{AN}}\right]\left[\alpha_{13\mathcal{N}}\right] \end{aligned} \qquad [4.245]$$

and, for the field $\boldsymbol{\beta}_{23d}(x)$, we have:

$$\begin{aligned} &\boldsymbol{\beta}_{23d}(x) = \sum_{i=1}^{\mathcal{A}} \boldsymbol{\omega}_{a_i} \beta_{23_i} \left[\boldsymbol{\omega}_{\mathcal{A}}\right]^t \left[\beta_{23\mathcal{A}}\right], \\ &\text{with}: \left[\beta_{23\mathcal{A}}\right] = -\left[\mathbf{G}_{\mathcal{AN}}\right]\left[\alpha_{23\mathcal{N}}\right] \end{aligned} \qquad [4.246]$$

4.4.4.1.3. System of equations to be solved

Applying the finite element method means discretizing the weak formulation [4.29]. To this end, the scalar potential V and the support fields $\boldsymbol{\beta}_{13}$ and $\boldsymbol{\beta}_{23}$ are replaced by their discrete form and the Ritz–Galerkin method is applied (see section 4.4.2) by taking as weighting functions the $\mathcal{N}_r$ node functions ω_{nj}. Based on relations [4.244], [4.245] and [4.246], we can write:

$$\begin{aligned}&\sum_{i=1}^{\mathcal{N}_r}\iiint_{\Omega_d}\sigma(\mathbf{grad}\omega_{n_j}.\mathbf{grad}\omega_{n_i})V_i d\tau =\\&\quad\sum_{k=1}^{\mathcal{A}}\iiint_{\Omega_d}\sigma e_{13}\mathbf{grad}\omega_{n_j}.\omega_{a_k}\beta_{13_k}d\tau\\&\quad+\sum_{l=1}^{\mathcal{A}}\iiint_{\Omega_d}\sigma e_{23}\mathbf{grad}\omega_{n_j}.\omega_{al_k}\beta_{23_l}d\tau \text{ with } 1\le j\le\mathcal{N}_r\end{aligned}\qquad[4.247]$$

Let us now consider the set of $\mathcal{N}_r$ weighting functions, which leads to the matrix system:

$$\begin{aligned}&\iiint_{\Omega}\sigma(\mathbf{grad}[\omega_{\mathcal{N}_r}]^t)^t(\mathbf{grad}[\omega_{\mathcal{N}_r}]^t)[V_{\mathcal{N}_r}]d\tau =\\&\quad\iiint_{\Omega}\sigma e_{13}(\mathbf{grad}[\omega_{\mathcal{N}_r}]^t)^t[\boldsymbol{\omega}_{\mathcal{A}}]^t[\beta_{13_{\mathcal{A}}}]d\tau\\&\quad+\iiint_{\Omega}\sigma e_{23}(\mathbf{grad}[\omega_{\mathcal{N}_r}]^t)^t[\boldsymbol{\omega}_{\mathcal{A}}]^t[\beta_{23_{\mathcal{A}}}]d\tau\end{aligned}\qquad[4.248]$$

Using relation [4.110], we can rewrite the matrix equation in the following form:

$$\begin{aligned}&\iiint_{\Omega_d}\sigma[\mathbf{G}_{\mathcal{A}\mathcal{N}_r}]^t[\boldsymbol{\omega}_{\mathcal{A}}][\boldsymbol{\omega}_{\mathcal{A}}]^t[\mathbf{G}_{\mathcal{A}\mathcal{N}_r}][V_{\mathcal{N}_r}]d\tau\\&\quad=\iiint_{\Omega_d}\sigma e_{13}[\mathbf{G}_{\mathcal{A}\mathcal{N}_r}]^t[\boldsymbol{\omega}_{\mathcal{A}}][\boldsymbol{\omega}_{\mathcal{A}}]^t[\beta_{13_{\mathcal{A}}}]d\tau\\&\quad+\iiint_{\Omega_d}\sigma e_{23}[\mathbf{G}_{\mathcal{A}\mathcal{N}_r}]^t[\boldsymbol{\omega}_{\mathcal{A}}][\boldsymbol{\omega}_{\mathcal{A}}]^t[\beta_{23_{\mathcal{A}}}]d\tau\end{aligned}\qquad[4.249]$$

Given that the electromotive forces e_{13} and e_{23}, the matrix $\left[\mathbf{G}_{\mathcal{A}\mathcal{N}_r}\right]$ and also vectors $\left[V_{\mathcal{N}_r}\right]$, $[\beta_{13\mathcal{A}}]$ and $[\beta_{23\mathcal{A}}]$ are independent of position, we can write:

$$\begin{aligned}&[\mathbf{G}_{\mathcal{A}\mathcal{N}_r}]^t[M^{\sigma}_{\mathcal{A}\mathcal{A}}][\mathbf{G}_{\mathcal{A}\mathcal{N}_r}][V_{\mathcal{N}_r}]=e_{13}[\mathbf{G}_{\mathcal{A}\mathcal{N}_r}]^t[M^{\sigma}_{\mathcal{A}\mathcal{A}}][\beta_{13_{\mathcal{A}}}]\\&\quad+e_{23}[\mathbf{G}_{\mathcal{A}\mathcal{N}_r}]^t[M^{\sigma}_{\mathcal{A}\mathcal{A}}][\beta_{23_{\mathcal{A}}}]\end{aligned}\qquad[4.250]$$

where the elementary term of the matrix $\left[M^{\sigma}_{\mathcal{AA}}\right]$ is written as:

$$M^{\sigma}_{a_i a_j} = \iiint_{\Omega_d} \sigma \boldsymbol{\omega}_{a_i} . \boldsymbol{\omega}_{a_j} d\tau \qquad [4.251]$$

In expression [4.250], the following can be defined:

$$\left[S^{\sigma}_{\mathcal{N}_r\mathcal{N}_r}\right] = \left[\mathbf{G}_{\mathcal{AN}_r}\right]^t \left[M^{\sigma}_{\mathcal{AA}}\right] \left[\mathbf{G}_{\mathcal{AN}_r}\right] \text{ et}$$
$$\left[F^{\sigma}_{\mathcal{N}_r}\right] = e_{13}\left[\mathbf{G}_{\mathcal{AN}_r}\right]^t \left[M^{\sigma}_{\mathcal{AA}}\right]\left[\beta_{13\,\mathcal{A}}\right] + e_{23}\left[\mathbf{G}_{\mathcal{AN}_r}\right]^t \left[M^{\sigma}_{\mathcal{AA}}\right]\left[\beta_{23\,\mathcal{A}}\right] \qquad [4.252]$$

Equation [4.250] can then be written as follows:

$$\left[S^{\sigma}_{\mathcal{N}_r\mathcal{N}_r}\right]\left[V_{\mathcal{N}_r}\right] = \left[F^{\sigma}_{\mathcal{N}_r}\right] \qquad [4.253]$$

The solution to this matrix equation enables us to calculate in this case the $\mathcal{N}_r$ node values of the vector potential, namely the vector $\left[V_{\mathcal{N}_r}\right]$. Then, using relation [4.244], we deduce expression $V_d(x)$ which represents the approximation to the solution to the problem.

4.4.4.1.4. Imposed current density flux

As noted in section 3.4.2.1, when the source terms are the current density flux, the electromotive forces become unknowns. System [4.253] should be added to the discretized form of relations [3.198] and [3.199], which are recalled as follows:

Given the expression of current I_1:

$$I_1 = \iiint_{\Omega} \sigma \boldsymbol{\beta}_{13} . (e_{13}\boldsymbol{\beta}_{13} + e_{23}\boldsymbol{\beta}_{23} - \mathbf{grad}\ V) d\tau \qquad [4.254]$$

Similarly, for current I_2, we have:

$$I_2 = \iiint_{\Omega} \sigma \boldsymbol{\beta}_{23} . (e_{13}\boldsymbol{\beta}_{13} + e_{23}\boldsymbol{\beta}_{23} - \mathbf{grad}\ V) d\tau \qquad [4.255]$$

In order to make these equations compatible with the finite element method, we have to introduce the discrete form of the scalar potential V and the support fields $\boldsymbol{\beta}_{13}$ and $\boldsymbol{\beta}_{23}$. For this, we can use expressions [4.244], [4.245] and [4.246].

The resulting expression for current I_1 is then:

$$\begin{aligned} I_1 = & -\iiint_\Omega \sigma([\omega_{\mathcal{A}}]^t [\beta_{13\mathcal{A}}])^t (\mathbf{grad}[\omega_{\mathcal{N}_r}]^t)[V_{\mathcal{N}_r}]d\tau \\ & + e_{13}\iiint_\Omega \sigma([\omega_{\mathcal{A}}]^t [\beta_{13\mathcal{A}}])^t [\omega_{\mathcal{A}}]^t [\beta_{13\mathcal{A}}]d\tau \\ & + e_{23}\iiint_\Omega \sigma([\omega_{\mathcal{A}}]^t [\beta_{13\mathcal{A}}])^t [\omega_{\mathcal{A}}]^t [\beta_{23\mathcal{A}}]d\tau \end{aligned} \qquad [4.256]$$

Using the properties of the gradient operator in the discrete domain, by means of equation [4.110], we can write:

$$\begin{aligned} I_1 = & -\iiint_\Omega \sigma[\beta_{13\mathcal{A}}]^t [\omega_{\mathcal{A}}][\omega_{\mathcal{A}}]^t [\mathbf{G}_{\mathcal{A}\mathcal{N}_r}][V_{\mathcal{N}_r}]d\tau \\ & + e_{13}\iiint_\Omega \sigma([\omega_{\mathcal{A}}]^t [\beta_{13\mathcal{A}}])^t [\omega_{\mathcal{A}}]^t [\beta_{13\mathcal{A}}]d\tau \\ & + e_{23}\iiint_\Omega \sigma([\omega_{\mathcal{A}}]^t [\beta_{13\mathcal{A}}])^t [\omega_{\mathcal{A}}]^t [\beta_{23\mathcal{A}}]d\tau \end{aligned} \qquad [4.257]$$

Then, taking into account that the vectors $[V_{\mathcal{N}_r}]$, $[\beta_{13\mathcal{A}}]$, $[\beta_{23\mathcal{A}}]$ and the matrix $[\mathbf{G}_{\mathcal{A}\mathcal{N}_r}]$ are independent of the position, we can introduce the mass matrix $[M^\sigma_{\mathcal{A}\mathcal{A}}]$, whose elementary term is given by relation [4.251].

We can then write:

$$\begin{aligned} I_1 = & -[\beta_{13\mathcal{A}}]^t [M^\sigma_{\mathcal{A}\mathcal{A}}][\mathbf{G}_{\mathcal{A}\mathcal{N}_r}][V_{\mathcal{N}_r}] \\ & + e_{13}[\beta_{13\mathcal{A}}]^t [M^\sigma_{\mathcal{A}\mathcal{A}}][\beta_{13\mathcal{A}}] + e_{23}[\beta_{13\mathcal{A}}]^t [M^\sigma_{\mathcal{A}\mathcal{A}}][\beta_{23\mathcal{A}}] \end{aligned} \qquad [4.258]$$

The same developments, applied to expression [4.255], yield the following expression for current I_2:

$$\begin{aligned} I_2 = & -[\beta_{23\mathcal{A}}]^t [M^\sigma_{\mathcal{A}\mathcal{A}}][\mathbf{G}_{\mathcal{A}\mathcal{N}_r}][V_{\mathcal{N}_r}] \\ & + e_{13}[\beta_{23\mathcal{A}}]^t [M^\sigma_{\mathcal{A}\mathcal{A}}][\beta_{13\mathcal{A}}] + e_{23}[\beta_{23\mathcal{A}}]^t [M^\sigma_{\mathcal{A}\mathcal{A}}][\beta_{23\mathcal{A}}] \end{aligned} \qquad [4.259]$$

In order to simplify the expressions to build the matrix system, the following notations are introduced:

$$\begin{aligned}&\left[C^{\sigma}_{13\mathcal{N}_r}\right] = -\left[\mathbf{G}_{\mathcal{A}\mathcal{N}_r}\right]^t\left[M^{\sigma}_{\mathcal{A}\mathcal{A}}\right]\left[\beta_{13\mathcal{A}}\right],\ \left[C^{\sigma}_{23\mathcal{N}_r}\right] = -\left[\mathbf{G}_{\mathcal{A}\mathcal{N}_r}\right]^t\left[M^{\sigma}_{\mathcal{A}\mathcal{A}}\right]\left[\beta_{23\mathcal{A}}\right]\\&B^{\sigma}_{\beta_{13}} = \left[\beta_{13\mathcal{A}}\right]^t\left[M^{\sigma}_{\mathcal{A}\mathcal{A}}\right]\left[\beta_{13\mathcal{A}}\right],\quad B^{\sigma}_{\beta_{23}} = \left[\beta_{23\mathcal{A}}\right]^t\left[M^{\sigma}_{\mathcal{A}\mathcal{A}}\right]\left[\beta_{23\mathcal{A}}\right]\\&H^{\sigma} = \left[\beta_{13\mathcal{A}}\right]^t\left[M^{\sigma}_{\mathcal{A}\mathcal{A}}\right]\left[\beta_{23\mathcal{A}}\right]\end{aligned} \qquad [4.260]$$

where $\left[C^{\sigma}_{13\mathcal{N}_r}\right]$ and $\left[C^{\sigma}_{23\mathcal{N}_r}\right]$ denote vectors of dimension $\mathcal{N}_r$ and $B^{\sigma}_{\beta_{13}}$, $B^{\sigma}_{\beta_{23}}$ and H^{σ} scalar terms.

Let us now gather equations [4.250], [4.258] and [4.259] with the notations defined in equations [4.252] and [4.260]. For an electrokinetics problem, with the vector potential formulation where the source terms are the currents I_1 and I_2, the system below should be solved:

$$\begin{bmatrix}\left[S^{\sigma}{}_{\mathcal{N}_r\mathcal{N}_r}\right] & \left[C^{\sigma}_{13\mathcal{N}_r}\right] & \left[C^{\sigma}_{23\mathcal{N}_r}\right]\\ \left[C^{\sigma}_{13\mathcal{N}_r}\right]^t & B^{\sigma}_{\beta_{13}} & H^{\sigma}\\ \left[C^{\sigma}_{23\mathcal{N}_r}\right]^t & H^{\sigma} & B^{\sigma}_{\beta_{23}}\end{bmatrix}\begin{bmatrix}\left[V_{\mathcal{N}_r}\right]\\ e_{13}\\ e_{23}\end{bmatrix} = \begin{bmatrix}[0]\\ I_1\\ I_2\end{bmatrix} \qquad [4.261]$$

The matrix equation thus obtained is symmetric. It can also be noted that the structure is equivalent to that of the electrostatic problem. If an electromotive force and a current are imposed, the system to be solved can readily be obtained from equation [4.261].

4.4.4.2. *Vector potential T formulation*

It should be recalled that the studied problem is presented in Figure 3.14 and the equations to be solved are developed in section 3.4.1.2. The weak form of the vector potential **T** formulation, where the current density fluxes I_1 and I_2 are imposed, is given by relation [4.34], which is recalled below:

$$\begin{aligned}&\iiint_{\Omega}\sigma^{-1}\mathbf{curl}\mathbf{T}.\mathbf{curl}\boldsymbol{\Psi}d\tau = -\iiint_{\Omega}\sigma^{-1}(I_1\boldsymbol{\lambda}_{13} + I_2\boldsymbol{\lambda}_{23}).\mathbf{curl}\boldsymbol{\Psi}dS\\&\quad\text{with } \mathbf{T} \text{ and } \boldsymbol{\Psi} \in H_{\Gamma_j}(\mathbf{curl},\Omega)\end{aligned} \qquad [4.262]$$

To solve this equation with the finite element method, **T** and the fields $\boldsymbol{\lambda}_{13}$ and $\boldsymbol{\lambda}_{23}$ must be discretized. It should be recalled that $\boldsymbol{\lambda}_{13} \in H_{\Gamma e2 \cup \Gamma j}(\mathrm{div}0,\Omega)$ and $\boldsymbol{\lambda}_{23} \in H_{\Gamma e1 \cup \Gamma j}(\mathrm{div}0,\Omega)$ (see equations [3.200] and [3.201]).

4.4.4.2.1. Discretization of vector potential T

The vector potential **T** belonging to $H_{\Gamma j}(\mathbf{curl}, \Omega)$ must be discretized (see Table 4.18) in the space of edge elements $W^1_{\Gamma j}(\Omega_d)$. Moreover, a gauge condition must be imposed and the boundary conditions must be taken into account.

To gauge the vector potential, as indicated in section 4.3.6.2, an edge tree is built. To make it easier to consider the boundary conditions, we start by building the tree considering the edges of boundary Γ_j. Then, we extend it to the entire domain. We then fix the $\mathcal{N} - 1$ edges of the tree to zero. We denote by $\mathcal{A}_j$ the number of edges of the boundary Γ_j and by $\mathcal{A}_{ja}$ those of Γ_j belonging to the tree. To impose the boundary condition on Γ_j, besides the $\mathcal{A}_{ja}$ edges of the boundary belonging to the tree, zero must be imposed to the remaining $\mathcal{A}_j - \mathcal{A}_{ja}$ edges. Then, for the number of unknown edges of the problem denoted by $\mathcal{A}_r$, we have:

$$\mathcal{A}_r = \mathcal{A} - (\mathcal{N} - 1) - (\mathcal{A}_j - \mathcal{A}_{ja}) \qquad [4.263]$$

The order of the indices of the edges of the mesh is such that those belonging to the tree and the boundary Γ_j are the last ones in the order of numbering. Under these conditions, the discretized electric vector potential $\mathbf{T}_d(x)$ has the following expression:

$$\mathbf{T}_d(x) = \sum_{i=1}^{\mathcal{A}_r} \boldsymbol{\omega}_{a_i} T_i = \left[\boldsymbol{\omega}_{\mathcal{A}_r}\right]^t \left[T_{\mathcal{A}_r}\right] \qquad [4.264]$$

4.4.4.2.2. Discretization of the support fields λ_{ij}

Concerning the support fields, $\boldsymbol{\lambda}_{13}$ and $\boldsymbol{\lambda}_{23}$, they belong, respectively, to the function spaces $H_{\Gamma e2 \cup \Gamma j}(\mathrm{div}0, \Omega)$ and $H_{\Gamma e1 \cup \Gamma j}(\mathrm{div}0, \Omega)$. They are therefore discretized in the space of facet elements (see Table 4.18), i.e. $\boldsymbol{\lambda}_{13d}(x) \in W^2_{\Gamma e2 \cup \Gamma j}(\Omega_d)$ and $\boldsymbol{\lambda}_{23d}(x) \in W^2_{\Gamma e1 \cup \Gamma j}(\Omega_d)$.

To calculate their discrete form, we use the process presented in section 4.3.7.2 (see equation [4.184]). It can be written in the following form for $\boldsymbol{\lambda}_{13d}(x)$:

$$\boldsymbol{\lambda}_{13d}(x) = \sum_{i=1}^{\mathcal{F}} \boldsymbol{\omega}_{f_i} \lambda_{13f_i} = \left[\boldsymbol{\omega}_{\mathcal{F}}\right]^t \left[\lambda_{13\mathcal{F}}\right] \qquad [4.265]$$

Similarly, the expression of $\boldsymbol{\lambda}_{23d}(x)$ is:

$$\boldsymbol{\lambda}_{23d}(x) = \sum_{i=1}^{\mathcal{F}} \boldsymbol{\omega}_{f_i} \lambda_{23f_i} = [\boldsymbol{\omega}_{\mathcal{F}}]^t [\lambda_{23\mathcal{F}}] \quad [4.266]$$

4.4.4.2.3. System of equations to be solved

The system of equations to be solved is obtained by replacing in equation [4.262] the vector potential and the support fields by their discrete form and by applying the Ritz–Galerkin method. Then, considering as the weighting function the $\mathcal{A}_r$ edge functions $\boldsymbol{\omega}_{aj}$ belonging to $W^1_{\Gamma j}(\Omega_d)$, we have:

$$\begin{aligned}
&\sum_{i=1}^{\mathcal{A}_r} \iiint_{\Omega_d} \sigma^{-1} (\mathbf{curl}\boldsymbol{\omega}_{a_j} . \mathbf{curl}\,\boldsymbol{\omega}_{a_i}) T_i d\tau = \\
&\quad -\sum_{k=1}^{\mathcal{F}} I_1 \iiint_{\Omega_d} \sigma^{-1} \mathbf{curl}\boldsymbol{\omega}_{a_j} . \boldsymbol{\omega}_{f_k} \lambda_{13f_k} d\tau \\
&\quad -\sum_{l=1}^{\mathcal{F}} I_2 \iiint_{\Omega_d} \sigma^{-1} \mathbf{curl}\boldsymbol{\omega}_{a_j} . \boldsymbol{\omega}_{f_l} \lambda_{23f_l} d\tau \text{ with } 1 \le j \le \mathcal{A}_r
\end{aligned} \quad [4.267]$$

Then, using the same approach as in section 4.4.3.2.3, we obtain the following expression:

$$\begin{aligned}
&[\mathbf{R}_{\mathcal{F}\mathcal{A}_r}]^t \left[M_{\mathcal{F}\mathcal{F}}^{\sigma^{-1}}\right] [\mathbf{R}_{\mathcal{F}\mathcal{A}_r}][T_{\mathcal{A}_r}] = -I_1 [\mathbf{R}_{\mathcal{F}\mathcal{A}_r}]^t \left[M_{\mathcal{F}\mathcal{F}}^{\sigma^{-1}}\right] [\lambda_{13\mathcal{F}}] \\
&\quad - I_2 [\mathbf{R}_{\mathcal{F}\mathcal{A}_r}]^t \left[M_{\mathcal{F}\mathcal{F}}^{\sigma^{-1}}\right] [\lambda_{23\mathcal{F}}]
\end{aligned} \quad [4.268]$$

It can be easily found that the elementary term of the matrix $\left[M_{\mathcal{F}\mathcal{F}}^{\sigma^{-1}}\right]$ is written as:

$$M_{f_i f_j}^{\sigma^{-1}} = \iiint_{\Omega_d} \sigma^{-1} \boldsymbol{\omega}_{f_i} . \boldsymbol{\omega}_{f_j} d\tau \quad [4.269]$$

Posing:

$$\begin{aligned}
&\left[S_{\mathcal{A}_r \mathcal{A}_r}^{\sigma^{-1}}\right] = \left[R_{\mathcal{F}\mathcal{A}_r}\right]^t \left[M_{\mathcal{F}\mathcal{F}}^{\sigma^{-1}}\right] \left[R_{\mathcal{F}\mathcal{A}_r}\right] \text{ and} \\
&\left[F_{\mathcal{A}_r}^{\sigma^{-1}}\right] = -I_1 \left[R_{\mathcal{F}\mathcal{A}_r}\right]^t \left[M_{\mathcal{F}\mathcal{F}}^{\sigma^{-1}}\right] [\lambda_{13\mathcal{F}}] - I_2 \left[R_{\mathcal{F}\mathcal{A}_r}\right]^t \left[M_{\mathcal{F}\mathcal{F}}^{\sigma^{-1}}\right] [\lambda_{23\mathcal{F}}]
\end{aligned} \quad [4.270]$$

The matrix equation to be solved can then be written as follows:

$$\left[S^{\sigma^{-1}}_{\mathcal{A}_r\mathcal{A}_r}\right]\left[T_{\mathcal{A}_r}\right] = \left[F^{\sigma^{-1}}_{\mathcal{A}_r}\right] \qquad [4.271]$$

The solution to the matrix equation given in equation [4.271] allows us to obtain the circulations of the vector potential on the edges of the mesh when the source terms are the currents I_1 and I_2 imposed on the boundaries Γ_{e1} and Γ_{e2}. To obtain the expression of the vector potential $\mathbf{T}_d(x)$ at any point of the domain Ω_d, we can use equation [4.264].

4.4.4.2.4. Imposed electromotive forces

In section 3.4.2.2, we considered the case in which source terms are the electromotive forces e_{13} and e_{23} or a combination of an electromotive force and a current. Under these conditions, to obtain a full system, we need to add equation [3.205] and/or, as applicable, equation [3.206]. In order to integrate them into system [4.270], they must be discretized. As an example, let us consider equation [3.205]. Following the same process as the one that led to expression [4.239] in electrostatics, we obtain, after development:

$$\begin{aligned} e_{13} = &\left[\lambda_{13\mathcal{F}}\right]^t\left[M^{\sigma^{-1}}_{\mathcal{F}\mathcal{F}}\right]\left[\mathbf{R}_{\mathcal{F}\mathcal{A}_r}\right]\left[T_{\mathcal{A}_r}\right] \\ &+ I_1\left[\lambda_{13\mathcal{F}}\right]^t\left[M^{\sigma^{-1}}_{\mathcal{F}\mathcal{F}}\right]\left[\lambda_{13\mathcal{F}}\right] + I_2\left[\lambda_{13\mathcal{F}}\right]^t\left[M^{\sigma^{-1}}_{\mathcal{F}\mathcal{F}}\right]\left[\lambda_{23\mathcal{F}}\right] \end{aligned} \qquad [4.272]$$

The elementary terms of the matrix $\left[M^{\sigma^{-1}}_{\mathcal{F}\mathcal{F}}\right]$ are given by equation [4.269]. The discretization of equation [3.308] can also be deduced very easily:

$$\begin{aligned} e_{23} = &\left[\lambda_{23\mathcal{F}}\right]^t\left[M^{\sigma^{-1}}_{\mathcal{F}\mathcal{F}}\right]\left[\mathbf{R}_{\mathcal{F}\mathcal{A}_r}\right]\left[T_{\mathcal{A}_r}\right] \\ &+ I_1\left[\lambda_{23\mathcal{F}}\right]^t\left[M^{\sigma^{-1}}_{\mathcal{F}\mathcal{F}}\right]\left[\lambda_{13\mathcal{F}}\right] + I_2\left[\lambda_{23\mathcal{F}}\right]^t\left[M^{\sigma^{-1}}_{\mathcal{F}\mathcal{F}}\right]\left[\lambda_{23\mathcal{F}}\right] \end{aligned} \qquad [4.273]$$

Using the same approach as in section 4.4.3.2.4, in order to build a full equation system, we introduce the following notations:

$$\begin{aligned}
&\left[C^{\sigma^{-1}}_{13\mathcal{A}_r}\right] = \left[\mathbf{R}_{\mathcal{F}\mathcal{A}_r}\right]^t\left[M^{\sigma^{-1}}_{\mathcal{F}\mathcal{F}}\right]\left[\lambda_{13\mathcal{F}}\right], \left[C^{\sigma^{-1}}_{23\mathcal{A}_r}\right] = \left[\mathbf{R}_{\mathcal{F}\mathcal{A}_r}\right]^t\left[M^{\sigma^{-1}}_{\mathcal{F}\mathcal{F}}\right]\left[\lambda_{23\mathcal{F}}\right]\\
&B^{\sigma^{-1}}_{\lambda_{13}} = \left[\lambda_{13\mathcal{F}}\right]^t\left[M^{\sigma^{-1}}_{\mathcal{F}\mathcal{F}}\right]\left[\lambda_{13\mathcal{F}}\right], \quad B^{\sigma^{-1}}_{\lambda_{23}} = \left[\lambda_{23\mathcal{F}}\right]^t\left[M^{\sigma^{-1}}_{\mathcal{F}\mathcal{F}}\right]\left[\lambda_{23\mathcal{F}}\right]\\
&H^{\sigma^{-1}} = \left[\lambda_{13\mathcal{F}}\right]^t\left[M^{\sigma^{-1}}_{\mathcal{F}\mathcal{F}}\right]\left[\lambda_{23\mathcal{F}}\right]
\end{aligned} \qquad [4.274]$$

where $\left[C^{\sigma^{-1}}_{13\mathcal{A}_r}\right]$ and $\left[C^{\sigma^{-1}}_{23\mathcal{A}_r}\right]$ denote vectors of dimension $\mathcal{A}_r$ and $B^{\sigma^{-1}}_{\lambda_{13}}$, $B^{\sigma^{-1}}_{\lambda_{23}}$ and $H^{\sigma^{-1}}$ are scalars.

Gathering equations [4.268], [4.272] and [4.273] and relying on the notations introduced in equations [4.269] and [4.274], we obtain the following system to be solved:

$$\begin{bmatrix}
\left[S^{\sigma^{-1}}_{\mathcal{A}_r\mathcal{A}_r}\right] & \left[C^{\sigma^{-1}}_{13\mathcal{A}_r}\right] & \left[C^{\sigma^{-1}}_{23\mathcal{A}_r}\right]\\
\left[C^{\sigma^{-1}}_{13\mathcal{A}_r}\right]^t & B^{\sigma^{-1}}_{\lambda_{13}} & H^{\sigma^{-1}}\\
\left[C^{\sigma^{-1}}_{23\mathcal{N}_r}\right]^t & H^{\sigma^{-1}} & B^{\sigma^{-1}}_{\lambda_{23}}
\end{bmatrix}
\begin{bmatrix}\left[T_{\mathcal{A}_r}\right]\\ I_1\\ I_2\end{bmatrix} = \begin{bmatrix}[0]\\ e_{13}\\ e_{23}\end{bmatrix} \qquad [4.275]$$

It should be noted that the matrix of this matrix equation is symmetric. The unknowns are the vector $\left[T_{\mathcal{A}_r}\right]$ of the circulations of the vector potential and the currents I_1 and I_2. If the source terms are, for example, the electromotive force e_{13} and the current I_2, the system of equations to be solved can be readily obtained from system [4.275].

4.4.5. *Magnetostatics*

In the case of magnetostatic problems, the approach used for introducing the finite element method relies on the formulation of the equation in sections 3.5.2.5 and 3.5.3.5. This is related to the problem presented in Figure 3.20. The domain Ω contains two source terms, a permanent magnet and an inductor. As for the boundary Γ, it is composed of two gates Γ_{h1} and Γ_{h2} and a wall Γ_b. A magnetomotive force f_m between the gates Γ_{h1} and Γ_{h2} or the magnetic flux ϕ through these gates can be imposed.

In section 4.2.4, we have developed the weak forms of the φ and **A** potential formulations. In this section, we use the finite element method for solving these weak forms.

The domain Ω is discretized using a mesh $\mathcal{M}$ composed of $\mathcal{N}$ elements, $\mathcal{A}$ edges, $\mathcal{F}$ facets and $\mathcal{V}$ volumes. It will then be denoted by Ω_d.

4.4.5.1. *Magnetic scalar potential φ formulation*

As a first step in the scalar potential formulation, we will study the case in which the source terms are a magnetomotive force f_m, a current intensity I in the inductor and a permanent magnet. The second step will be to replace, on the boundaries Γ_{h1} and Γ_{h2}, the boundary condition by imposing the magnetic flux ϕ.

However, the first step is to discretize the magnetic scalar potential φ. In the continuous domain, it belongs to the function space $H_{\Gamma h1 \cup \Gamma h2}(\mathbf{grad}, \Omega)$ (see equation [4.35]). Consequently, in the discrete domain (see Table 4.18), the scalar potential, denoted by $\varphi_d(x)$, belongs to $W^0_{\Gamma h1 \cup \Gamma h2}(\Omega_d)$. It is written in the form of equation [4.85] or [4.87]. Similar to the approach in sections 4.4.3.1 and 4.4.4.1, we define the number of nodal values, unknowns of the problem, denoted by $\mathcal{N}_r$, and then we write the scalar potential in a discrete form.

The boundaries Γ_{h1} and Γ_{h2} are equipotential surfaces on which the magnetic scalar potential is imposed to zero. Considering now the mesh $\mathcal{M}$, we denote by $\mathcal{N}_1$ and $\mathcal{N}_2$ the number of nodes belonging to boundaries Γ_{h1} and Γ_{h2}, respectively. Consequently, to impose the boundary conditions, we fix at zero (see section 4.3.5.1) the values φ_i of potential at the nodes located on Γ_{h1} and Γ_{h2}. It should be recalled that this makes it possible to engage the gauge condition. The number of unknown nodal values of the problem, denoted by $\mathcal{N}_r$, is then written as:

$$\mathcal{N}_r = \mathcal{N} - (\mathcal{N}_1 + \mathcal{N}_2) \qquad [4.276]$$

Let us now consider a numbering of nodes so that those belonging to the boundaries Γ_{h1} and Γ_{h2} have the highest indices. In fact, the nodes whose index is higher than $\mathcal{N}_r$ are located on the boundaries Γ_{h1} and Γ_{h2}. Under these conditions, the expression $\varphi_d(x)$ of the magnetic scalar potential is written as:

$$\varphi_d(x) = \sum_{i=1}^{\mathcal{N}_r} \omega_{n_i} \varphi_i = \left[\omega_{\mathcal{N}_r}\right]^t \left[\varphi_{\mathcal{N}_r}\right] \qquad [4.277]$$

4.4.5.1.1. Imposed source terms: f_m, I, H_c

When the source terms are f_m, I and the coercive field $\mathbf{H}_c$ of the permanent magnet, the local form of the equation to be solved is given by equation [3.265] with $\varphi \in H_{\Gamma h1 \cup \Gamma h2}(\mathbf{grad},\Omega)$. Applying the weighted residual method leads to equation [4.39], which is recalled below:

$$\iiint_{\Omega} \mu(\mathbf{grad}\varphi.\mathbf{grad}\psi)d\tau = \iiint_{\Omega} (\mu f_m \boldsymbol{\beta}_s + \mu I \chi_I - \mu \mathbf{H}_c).\mathbf{grad}\psi d\tau \quad \text{with } \psi \in H_{\Gamma_{h1} \cup \Gamma_{h2}}(\mathbf{grad},\Omega) \qquad [4.278]$$

To solve this equation, we have to discretize, besides the scalar potential φ, the support field $\boldsymbol{\beta}_s$ and the associated potential $\boldsymbol{\chi}_I$. On the other hand, the coercive field $\mathbf{H}_c$ is a special case as, unlike $\boldsymbol{\beta}_s$ and $\boldsymbol{\chi}_I$, its distribution on the domain Ω_d is an input of the problem and does not require any specific construction.

NOTE.– As already seen in section 3.5.2.4, the coercive field $\mathbf{H}_c$ features discontinuities of its normal and tangential components; therefore, it is not possible to exactly approximate this field in the space of facet or edge elements. However, the distribution of permanent magnets and the direction of their magnetization, given by $\mathbf{H}_c$ or $\mathbf{B}_r$, are perfectly known. The coercive field is therefore a vector function depending on the position and that is zero outside of permanent magnets.

The support field $\boldsymbol{\beta}_s$ belongs to $H_{\Gamma h1 \cup \Gamma h2}(\mathbf{curl}0, \Omega)$ (see equation [3.12]). It is therefore discretized in the space of edge elements $W^1_{\Gamma h1 \cup \Gamma h2}(\Omega_d)$. For its construction, we can use the method proposed in section 4.3.7.1 (see equation [4.182]). This makes it possible to write $\boldsymbol{\beta}_{sd}(x)$ as follows:

$$\boldsymbol{\beta}_{sd}(x) = \sum_{i=1}^{\mathcal{A}} \boldsymbol{\omega}_{a_i} \beta_{s_i} = [\boldsymbol{\omega}_{\mathcal{A}}]^t [\beta_{s\mathcal{A}}] \qquad [4.279]$$

Concerning $\boldsymbol{\chi}_I$, it should be recalled that it is the potential associated with the support field $\boldsymbol{\lambda}_I$ that represents the current density in the inductor located inside the domain (see equations [3.236] and [3.237]). As shown in section 3.5.2.3, $\boldsymbol{\lambda}_I$ and $\boldsymbol{\chi}_I$ are built on the entire domain, thus avoiding the problem of connexity related to the inductor geometry. Then, we have $\boldsymbol{\lambda}_I \in H_0(\mathrm{div}0, \Omega)$ and $\boldsymbol{\chi}_I \in H_0(\mathbf{curl}, \Omega)$. In the discrete domain (see Table 4.18), these two fields belong to the function spaces $W^2_0(\Omega_d)$ and $W^1_0(\Omega_d)$, respectively. These fields are linked by the curl operator, and $\boldsymbol{\chi}_I$ is generally deduced from $\boldsymbol{\lambda}_I$. For the construction of the discrete form $\boldsymbol{\lambda}_{Id}(x)$, we use the process proposed in section 4.3.6.3 referring to case 3. Then, we deduce

$\boldsymbol{\chi}_{Id}(x)$, based on the method proposed in section 4.3.7.2 (see equation [4.185]). This allows us to write $\boldsymbol{\chi}_{Id}(x)$ as follows:

$$\chi_{Id}(x) = \sum_{i=1}^{\mathcal{A}} \boldsymbol{\omega}_{a_i} \chi_{I_i} = [\boldsymbol{\omega}_{\mathcal{A}}]^t [\chi_{I\mathcal{A}}] \qquad [4.280]$$

where $[\chi_{I\mathcal{A}}]$ represents the vector of the circulations of the associated potential $\boldsymbol{\chi}_{Id}(x)$ on the edges of the mesh.

To build the matrix equation to be solved, based on the weak form, recalled in equation [4.278], we use the same approach as the one developed in electrostatics in section 4.4.3.1.3. In equation [4.278], the scalar potential φ and the fields $\boldsymbol{\beta}_s$ and $\boldsymbol{\chi}_I$ are replaced by their given expressions, respectively, by equations [4.277], [4.279] and [4.280], and applying the Ritz–Galerkin method, we can write the equation as follows:

$$\begin{aligned} &\sum_{i=1}^{\mathcal{N}_r} \iiint_{\Omega_d} \mu(\mathbf{grad}\omega_{n_j} . \mathbf{grad}\omega_{n_i}) \varphi_i d\tau = \\ &\quad \sum_{k=1}^{\mathcal{A}} \iiint_{\Omega_d} \mu f_m \mathbf{grad}\omega_{n_j} . \boldsymbol{\omega}_{a_k} \beta_{s_k} d\tau \\ &\qquad + \sum_{l=1}^{\mathcal{A}} \iiint_{\Omega_d} \mu I \mathbf{grad}\omega_{n_j} . \boldsymbol{\omega}_{a_l} \chi_{I_l} \\ &\qquad\quad - \iiint_{\Omega_d} \mu \mathbf{grad}\omega_{n_j} . \mathbf{H}_c d\tau \ \text{ for } 1 \le j \le \mathcal{N}_r \end{aligned} \qquad [4.281]$$

The fact that $\mathbf{H}_c$ is discontinuous on certain surfaces of permanent magnets (see section 3.5.2.4) does not pose any problems, since the difficulty related to the term div$\mathbf{H}_c$ was removed using the weighted residual method (see section 4.2.4.1).

Continuing the developments, similar to section 4.4.3.1.3, the above equation can be written as follows:

$$\begin{aligned} &[\mathbf{G}_{\mathcal{A}\mathcal{N}_r}]^t [M^{\mu}_{\mathcal{A}\mathcal{A}}][\mathbf{G}_{\mathcal{A}\mathcal{N}_r}][\varphi_{\mathcal{N}_r}] = f_m [\mathbf{G}_{\mathcal{A}\mathcal{N}_r}]^t [M^{\mu}_{\mathcal{A}\mathcal{A}}][\beta_{s\mathcal{A}}] \\ &\quad + I[\mathbf{G}_{\mathcal{A}\mathcal{N}_r}]^t [M^{\mu}_{\mathcal{A}\mathcal{A}}][\chi_{I\mathcal{A}}] - [\mathbf{G}_{\mathcal{A}\mathcal{N}_r}]^t [H^{\mu}_{c\mathcal{A}}] \end{aligned} \qquad [4.282]$$

In this expression, the elementary term of the matrix $\left[M^{\mu}_{\mathcal{AA}}\right]$ is written as:

$$M^{\mu}_{a_i a_j} = \iiint_{\Omega_d} \mu \boldsymbol{\omega}_{a_i} . \boldsymbol{\omega}_{a_j} d\tau \qquad [4.283]$$

As for the elementary term of the vector $\left[H^{\mu}_{c\mathcal{A}}\right]$, corresponding to the source term of the permanent magnet, its expression is:

$$H^{\mu}_{ca_i} = \iiint_{\Omega_d} \mu \boldsymbol{\omega}_{a_i} . \mathbf{H}_c d\tau \qquad [4.284]$$

Consider now:

$$\begin{aligned} &\left[S^{\mu}_{\mathcal{N}_r\mathcal{N}_r}\right] = \left[\mathbf{G}_{\mathcal{AN}_r}\right]^t \left[M^{\mu}_{\mathcal{AA}}\right] \left[\mathbf{G}_{\mathcal{AN}_r}\right] \text{ and} \\ &\left[F^{\mu}_{\mathcal{N}_r}\right] = f_m \left[\mathbf{G}_{\mathcal{AN}_r}\right]^t \left[M^{\mu}_{\mathcal{AA}}\right]\left[\beta_{s\mathcal{A}}\right] + I\left[\mathbf{G}_{\mathcal{AN}_r}\right]^t \left[M^{\mu}_{\mathcal{AA}}\right]\left[\chi_{I\mathcal{A}}\right] \\ &\qquad - \left[\mathbf{G}_{\mathcal{AN}_r}\right]^t \left[H^{\mu}_{c\mathcal{A}}\right] \end{aligned} \qquad [4.285]$$

Taking the above notations into account, equation [4.282] to be solved can be written as follows:

$$\left[S^{\mu}_{\mathcal{N}_r\mathcal{N}_r}\right]\left[\varphi_{\mathcal{N}_r}\right] = \left[F^{\mu}_{\mathcal{N}_r}\right] \qquad [4.286]$$

In this system of equations, the unknowns are the nodal values of the scalar potential represented by the vector $[\varphi_{\mathcal{N}r}]$. Neglecting the effects of the saturation of magnetic materials (see equation [1.26]), the system to be solved is linear.

4.4.5.1.2. Imposed source terms ϕ, I, H_c

As already shown in section 3.5.2.5.2, if the magnetic flux is imposed, the magnetomotive force becomes an unknown of the problem.

Then, we have to add equation [3.268], which is recalled below:

$$\phi = \iiint_{\Omega} (-\mu \boldsymbol{\beta}_s . \mathbf{grad}\varphi + f_m \mu \boldsymbol{\beta}_s . \boldsymbol{\beta}_s + \mu I \boldsymbol{\beta}_s . \boldsymbol{\chi}_I - \mu \boldsymbol{\beta}_s . \mathbf{H}_c) d\tau \qquad [4.287]$$

The approach used for discretization is similar to that in section 4.4.5.1.2, with φ, $\boldsymbol{\beta}_s$ and $\boldsymbol{\chi}_I$ being replaced by their discrete form.

It can then be shown that equation [4.287] is written as follows:

$$\begin{aligned}\phi = &-\iiint_{\Omega_d} \mu([\boldsymbol{\omega}_{\mathcal{A}}]^t [\beta_{s\mathcal{A}}])^t (\mathbf{grad}[\omega_{\mathcal{N}_r}]^t [\varphi_{\mathcal{N}_r}]) d\tau \\ &+ f_m \iiint_{\Omega_d} \mu([\boldsymbol{\omega}_{\mathcal{A}}]^t [\beta_{s\mathcal{A}}])^t [\boldsymbol{\omega}_{\mathcal{A}}]^t [\beta_{s\mathcal{A}}] d\tau \\ &+ I \iiint_{\Omega_d} \mu([\boldsymbol{\omega}_{\mathcal{A}}]^t [\beta_{s\mathcal{A}}])^t [\boldsymbol{\omega}_{\mathcal{A}}]^t [\chi_{I\mathcal{A}}] d\tau \\ &- \iiint_{\Omega_d} \mu([\boldsymbol{\omega}_{\mathcal{A}}]^t [\beta_{s\mathcal{A}}])^t . \mathbf{H}_c d\tau\end{aligned} \qquad [4.288]$$

After development and using the matrix [4.283] and the vector [4.284], we can write the expression of ϕ in the following form:

$$\begin{aligned}\phi = &-[\beta_{s\mathcal{A}}]^t [M^{\mu}_{\mathcal{A}\mathcal{A}}][\mathbf{G}_{\mathcal{A}\mathcal{N}_r}][\varphi_{\mathcal{N}_r}] + f_m [\beta_{s\mathcal{A}}]^t [M^{\mu}_{\mathcal{A}\mathcal{A}}][\beta_{s\mathcal{A}}] \\ &+ I[\beta_{s\mathcal{A}}]^t [M^{\mu}_{\mathcal{A}\mathcal{A}}][\chi_{I\mathcal{A}}] - [\beta_{s\mathcal{A}}]^t [H^{\mu}_{c\mathcal{A}}]\end{aligned} \qquad [4.289]$$

The system of equations to be solved is built by gathering relations [4.282] and [4.289]. Nevertheless, to simplify the expressions, we introduce the following notations:

$$\begin{aligned}&\left[C^{\mu}_{\beta\mathcal{N}_r}\right] = -[\mathbf{G}_{\mathcal{A}\mathcal{N}_r}]^t \left[M^{\mu}_{\mathcal{A}\mathcal{A}}\right][\beta_{s\mathcal{A}}], \\ &\left[F^{\mu}_{I\text{-}H_c\mathcal{N}_r}\right] = I[\mathbf{G}_{\mathcal{A}\mathcal{N}_r}]^t \left[M^{\mu}_{\mathcal{A}\mathcal{A}}\right][\chi_{I\mathcal{A}}] - [\mathbf{G}_{\mathcal{A}\mathcal{N}_r}]^t \left[H^{\mu}_{c\mathcal{A}}\right], \\ &B^{\mu}_{\beta} = [\beta_{s\mathcal{A}}]^t \left[M^{\mu}_{\mathcal{A}\mathcal{A}}\right][\beta_{s\mathcal{A}}] \\ &\phi^{\mu}_{I-H_c} = I[\beta_{s\mathcal{A}}]^t \left[M^{\mu}_{\mathcal{A}\mathcal{A}}\right][\chi_{I\mathcal{A}}] - [\beta_{s\mathcal{A}}]^t \left[H^{\mu}_{c\mathcal{A}}\right]\end{aligned} \qquad [4.290]$$

where $\left[C^{\mu}_{\beta\mathcal{N}_r}\right]$ denotes a vector of dimension $\mathcal{N}_r$ and $\left[F^{\mu}_{I\text{-}H_c\mathcal{N}_r}\right]$ denotes a vector of dimension $\mathcal{N}_r$, representing the contributions of the current in the inductor and of the permanent magnet. Moreover, B^{μ}_{β} is a scalar term representing a permeance and $\phi^{\mu}_{I-H_c}$ is a scalar representing the flux created on the boundaries Γ_{h1} and Γ_{h2} by the current in the inductor and the permanent magnet.

Gathering equations [4.286] and [4.289] and relying on the notations introduced in equation [4.290], the system to be solved is written as:

$$\begin{bmatrix} \left[S^{\mu}_{\mathcal{N}_r\mathcal{N}_r} \right] & \left[C^{\mu}_{\beta\mathcal{N}_r} \right] \\ \left[C^{\mu}_{\beta\mathcal{N}_r} \right]^t & B^{\mu}_{\beta} \end{bmatrix} \begin{bmatrix} \left[\varphi_{\mathcal{N}_r} \right] \\ f_m \end{bmatrix} = \begin{bmatrix} \left[F^{\mu}_{I-H_c\mathcal{N}_r} \right] \\ \phi - \phi^{\mu}_{I-H_c} \end{bmatrix} \quad [4.291]$$

It can be noted that the matrix of this system of equations is symmetric. The unknowns are the vector $\left[\varphi_{\mathcal{N}_r} \right]$ of nodal values of the scalar potential and the magnetomotive force f_m. If the characteristic of the magnetic materials is linear (see equation [1.26]), the solution to the problem results from solving the system of equations [4.291].

4.4.5.1.3. Taking nonlinearity into account

If the nonlinearity of ferromagnetic materials is taken into account (see section 1.2.2.4, equation [1.25]), the magnetic permeability depends on **H** and therefore on the unknowns of the problem, namely the vector $\left[\varphi_{\mathcal{N}_r} \right]$ of the node values of the scalar potential and f_m. Under these conditions, considering the behavior law of magnetic materials, the systems of equations [4.286] and [4.291] are nonlinear. Numerical methods are available for solving this type of problem, such as the substitution method or the Newton–Raphson method (Dhatt et al. 2012).

4.4.5.2. *Vector potential A formulation*

For the vector potential formulation, the studied problem is the same (see Figure 3.20). As for writing the equation, it is presented in section 3.5.3.5. The first step is to consider as source terms the magnetic flux ϕ imposed on the boundaries Γ_{h1} and Γ_{h2}, the inductor through which flows a current I and the permanent magnet represented by the remanent induction $\mathbf{B}_r$. The second step is to replace the magnetic flux by a magnetomotive force f_m.

Irrespective of the source terms used, we have to discretize the vector potential **A**. As already shown (see equation [3.275]), we have $\mathbf{A} \in H_{\Gamma b}(\mathbf{curl}, \Omega)$. Under these conditions, as shown in Table 4.18, in the discrete domain, the vector potential belongs to the space of edge elements, i.e. $\mathbf{A}_d(x) \in W^1_{\Gamma b}(\Omega_d)$. It is then written in the form of equations [4.92] and [4.93]. However, to express the discrete form $\mathbf{A}_d(x)$, we have to take into account the number of unknowns of the problem considering the boundary conditions and the gauge condition.

In order to gauge the vector potential, by integrating the boundary conditions, we use the process developed in section 4.3.6.2, which relies on an edge tree. With this process, to take into account the boundary conditions, we start the construction of the tree by the edges of the boundary Γ_b, and then extend it to the set of edges of the mesh. The circulations of the vector potential **A** on the $\mathcal{N} - 1$ edges of the tree are then set to zero. In what follows, we denote by $\mathcal{A}_b$ the number of edges belonging to the boundary Γ_b and by $\mathcal{A}_{ba}$ the number of edges of the boundary Γ_b also belonging to the edge tree. To impose the boundary condition on the boundary Γ_b, besides the $\mathcal{A}_{ba}$ edges of the tree, zero must be imposed on the $\mathcal{A}_b - \mathcal{A}_{ba}$ remaining edges. Then, the number of unknown circulations of the vector potential **A** on the edges denoted by $\mathcal{A}_r$ satisfies:

$$\mathcal{A}_r = \mathcal{A} - (\mathcal{N} - 1) - (\mathcal{A}_b - \mathcal{A}_{ba}) \qquad [4.292]$$

The edges are renumbered so that those on which the circulation was imposed to zero have an index higher than $\mathcal{A}_r$. Under these conditions, the expression $\mathbf{A}_d(x)$ is written as:

$$\mathbf{A}_d(x) = \sum_{i=1}^{\mathcal{A}_r} \boldsymbol{\omega}_{a_i} A_{a_i} = \left[\boldsymbol{\omega}_{\mathcal{A}_r}\right]^t \left[A_{\mathcal{A}_r}\right] \qquad [4.293]$$

4.4.5.2.1. Imposed source terms ϕ, I, B_r

As already seen in section 3.5.3.5.1, we must solve equation [3.298]. Applying the weighted residual method (see section 4.2.4.2.1), the weak form is given by equation [4.44] recalled below:

$$\iiint_\Omega \mu^{-1}\mathbf{curlA}.\mathbf{curl\Psi}d\tau = -\iiint_\Omega \mu^{-1}(\phi\boldsymbol{\lambda}_\phi + \mathbf{B}_r).\mathbf{curl\Psi} + \iiint_\Omega I\boldsymbol{\lambda}_I.\boldsymbol{\Psi}d\tau \qquad [4.294]$$
$$\text{with } \boldsymbol{\Psi} \in H_{\Gamma_b}(\mathbf{curl}, \Omega)$$

Besides the vector potential **A**, the solution to this equation requires the discretization of the support fields $\boldsymbol{\lambda}_\phi$ and $\boldsymbol{\lambda}_I$. Concerning the term $\mathbf{B}_r$, it is considered as perfectly known and its discretization is not needed (on this matter, please see the note in section 4.4.5.1.1).

The support fields $\boldsymbol{\lambda}_\phi$ and $\boldsymbol{\lambda}_I$ (see equations [3.273] and [3.289]) belong to the function spaces $H_{\Gamma b}(\text{div}0, \Omega)$ and $H_0(\text{div}0, \Omega)$, respectively. As shown in Table 4.18, they are discretized with facet elements with $\boldsymbol{\lambda}_{\phi d}(x) \in W^2_{\Gamma b}(\Omega_d)$ and $\boldsymbol{\lambda}_{Id}(x) \in W^2_0(\Omega_d)$.

In order to determine the support field $\boldsymbol{\lambda}_{\phi d}(x)$, we directly use the process presented in section 4.3.7.2, which yields the expression:

$$\lambda_{\phi d}(x) = \sum_{i=1}^{\mathcal{F}} \omega_{f_i} \lambda_{\phi f_i} = [\omega_{\mathcal{F}}]^t [\lambda_{\phi \mathcal{F}}] \qquad [4.295]$$

Concerning the support field $\boldsymbol{\lambda}_{Id}(x)$, its construction is introduced in section 4.4.5.1.1 and its expression is:

$$\lambda_{Id}(x) = \sum_{i=1}^{\mathcal{F}} \omega_{f_i} \lambda_{If_i} = [\omega_{\mathcal{F}}]^t [\lambda_{I\mathcal{F}}] \qquad [4.296]$$

Knowing the discrete form of the various terms used in the weak form [4.294], we will now build the matrix equation to be solved.

For this purpose, after having introduced the discrete forms [4.293], [4.295] and [4.296], we apply the Ritz–Galerkin formula, which allows us to write the equation as follows:

$$\begin{aligned}
&\sum_{i=1}^{\mathcal{A}_r} \iiint_{\Omega_d} \mu^{-1} \mathbf{curl} \omega_{a_j} . \mathbf{curl}\, \omega_{a_i} A_i d\tau = \\
&\quad -\sum_{k=1}^{\mathcal{F}} \phi \iiint_{\Omega_d} \mu^{-1} \mathbf{curl} \omega_{a_j} . \omega_{f_k} \lambda_{\phi f_k} d\tau \\
&\quad +\sum_{l=1}^{\mathcal{F}} I \iiint_{\Omega_d} \mu^{-1} \mathbf{curl} \omega_{a_j} \omega_{f_l} \lambda_{If_l} d\tau \\
&\quad -\iiint_{\Omega_d} \mu^{-1} \mathbf{curl} \omega_{a_j} . \mathbf{B}_r d\tau \quad \text{with } 1 \le j \le \mathcal{A}_r
\end{aligned} \qquad [4.297]$$

This equation, which must be verified for the set of $\mathcal{A}_r$ weighting functions $\boldsymbol{\omega}_{aj}$, leads to a matrix equation.

Using the same approach as that proposed in section 4.4.3.2.3, we can write the matrix form:

$$\begin{aligned}
&\left[\mathbf{R}_{\mathcal{F}\mathcal{A}_r}\right]^t \left[M_{\mathcal{F}\mathcal{F}}^{\mu^{-1}}\right] \left[\mathbf{R}_{\mathcal{F}\mathcal{A}_r}\right] \left[A_{\mathcal{A}_r}\right] = -\phi \left[\mathbf{R}_{\mathcal{F}\mathcal{A}_r}\right]^t \left[M_{\mathcal{F}\mathcal{F}}^{\mu^{-1}}\right] \left[\lambda_{\phi\mathcal{F}}\right] \\
&+I \left[M_{\mathcal{A}\mathcal{F}}\right] \left[\lambda_{I\mathcal{F}}\right] - \left[\mathbf{R}_{\mathcal{F}\mathcal{A}_r}\right]^t \left[B_{r\mathcal{F}}^{\mu^{-1}}\right]
\end{aligned} \qquad [4.298]$$

In this expression, the elementary term of the matrix $\left[M_{\mathcal{FF}}^{\mu^{-1}}\right]$ is written as:

$$M_{f_i f_j}^{\mu^{-1}} = \iiint_{\Omega} \mu^{-1} \boldsymbol{\omega}_{f_i} . \boldsymbol{\omega}_{f_j} d\tau \qquad [4.299]$$

As for the elementary term of the matrix $\left[M_{\mathcal{AF}}\right]$, it is written as:

$$M_{a_i f_j} = \iiint_{\Omega} \omega_{a_i} . \omega_{f_j} d\tau \qquad [4.300]$$

The expression of the elementary term of the vector $\left[B_{r\mathcal{F}}^{\mu^{-1}}\right]$ is:

$$B_{rf_i}^{\mu^{-1}} = \iiint_{\Omega_d} \mu^{-1} \boldsymbol{\omega}_{f_i} . \mathbf{B}_r d\tau \qquad [4.301]$$

Now, we pose:

$$\begin{aligned} &\left[S_{\mathcal{A}_r\mathcal{A}_r}^{\mu^{-1}}\right] = \left[\mathbf{R}_{\mathcal{F}\mathcal{A}_r}\right]^t \left[M_{\mathcal{FF}}^{\mu^{-1}}\right] \left[\mathbf{R}_{\mathcal{F}\mathcal{A}_r}\right], \\ &\left[F_{\mathcal{A}_r}^{\mu^{-1}}\right] = -\phi\left[\mathbf{R}_{\mathcal{F}\mathcal{A}_r}\right]^t \left[M_{\mathcal{FF}}^{\mu^{-1}}\right]\left[\lambda_{\phi\mathcal{F}}\right] + I\left[M_{\mathcal{AF}}\right]\left[\lambda_{I\mathcal{F}}\right] \\ &\qquad -\left[\mathbf{R}_{\mathcal{F}\mathcal{A}_r}\right]^t \left[B_{r\mathcal{F}}^{\mu^{-1}}\right] \end{aligned} \qquad [4.302]$$

The system of equations to be solved can then be written as follows:

$$\left[S_{\mathcal{A}_r\mathcal{A}_r}^{\mu^{-1}}\right]\left[A_{\mathcal{A}_r}\right] = \left[F_{\mathcal{A}_r}^{\mu^{-1}}\right] \qquad [4.303]$$

In this system of equations, the unknowns are the circulations of the magnetic vector potential along the edges of the mesh contained in the vector $\left[A_{\mathcal{A}_r}\right]$. If the saturation of magnetic materials is considered negligible (see equation [1.26]), the system to be solved is linear.

NOTE.– The system of equations [4.303] was built by gauging the vector potential. A similar approach can be used to build a system of equations in which the vector potential is not gauged. The number of solutions of the resulting system of equations is then infinite. It has been shown that iterative methods like the conjugate gradient method converge and lead to a unique solution in field (Ren 1996). Indeed, during the iterative solution process, the divergence of the potential is implicitly fixed in the weak sense and automatically imposes the gauge condition.

4.4.5.2.2. Imposed source terms f_m, I, B_r

The objective is to impose the magnetomotive force f_m rather than the magnetic flux between the boundaries Γ_{h1} and Γ_{h2}. As shown in section 4.2.4.2.2, the weak formulation [4.44] is maintained, but the flux ϕ becomes an unknown.

To obtain a full equation system, we then add equation [3.300], which is recalled as follows:

$$f_m = \iiint_{\Omega} \mu^{-1}(\mathbf{curl A} + \phi\boldsymbol{\lambda}_\phi - \mathbf{B}_r).\boldsymbol{\lambda}_\phi d\tau \qquad [4.304]$$

To discretize this equation, the vector potential $\mathbf{A}$ and the support field $\boldsymbol{\lambda}_\phi$ are replaced by their discrete form, namely equations [4.293] and [4.295]. Then, we obtain:

$$\begin{aligned} f_m = &\iiint_{\Omega_d} \mu^{-1}\left(\left[\boldsymbol{\omega}_{\mathcal{F}}\right]^t \left[\lambda_{\phi\mathcal{F}}\right]\right)^t \left(\mathbf{curl}\left[\boldsymbol{\omega}_{\mathcal{A}_r}\right]^t \left[\mathrm{A}_{\mathcal{A}_r}\right]\right) d\tau \\ &+ \phi\iiint_{\Omega_d} \mu^{-1}\left(\left[\boldsymbol{\omega}_{\mathcal{F}}\right]^t \left[\lambda_{\phi\mathcal{F}}\right]\right)^t \left[\boldsymbol{\omega}_{\mathcal{F}}\right]^t \left[\lambda_{\phi\mathcal{F}}\right] d\tau) \\ &- \iiint_{\Omega_d} \mu^{-1}\left(\left[\boldsymbol{\omega}_{\mathcal{F}}\right]^t \left[\lambda_{\phi\mathcal{F}}\right]\right)^t .\mathbf{B}_r d\tau) \end{aligned} \qquad [4.305]$$

Using the properties of the discrete curl operator [4.124] and introducing the matrix $\left[\mathrm{M}^{\mu^{-1}}_{\mathcal{FF}}\right]$, whose elementary term is given in equation [4.299], as well as the vector $\left[\mathrm{B}^{\mu^{-1}}_{r\mathcal{F}}\right]$, having as an elementary term equation [4.301], we obtain:

$$\begin{aligned} f_m = &\left[\lambda_{\phi\mathcal{F}}\right]^t \left[\mathrm{M}^{\mu^{-1}}_{\mathcal{FF}}\right] \left[\mathbf{R}_{\mathcal{F}\mathcal{A}_r}\right] \left[\mathrm{A}_{\mathcal{A}_r}\right] + \phi\left[\lambda_{\phi\mathcal{F}}\right]^t \left[\mathrm{M}^{\mu^{-1}}_{\mathcal{FF}}\right] \left[\lambda_{\phi\mathcal{F}}\right] \\ &- \left[\lambda_{\phi\mathcal{F}}\right]^t \left[\mathrm{B}^{\mu^{-1}}_{r\mathcal{F}}\right] \end{aligned} \qquad [4.306]$$

In order to build the matrix equation to be solved, we will gather relations [4.298] and [4.306]. To simplify the expression, we first introduce the following notations:

$$\begin{aligned}
&\left[C^{\mu^{-1}}_{\lambda\mathcal{A}_r}\right] = \left[\mathbf{R}_{\mathcal{F}\mathcal{A}_r}\right]^t \left[M^{\mu^{-1}}_{\mathcal{F}\mathcal{F}}\right]\left[\lambda_{\phi\mathcal{F}}\right],\\
&\left[F^{\mu^{-1}}_{I-B_r\mathcal{A}_r}\right] = I\left[M_{\mathcal{A}\mathcal{F}}\right]\left[\lambda_{I\mathcal{F}}\right] - \left[\mathbf{R}_{\mathcal{F}\mathcal{A}_r}\right]^t\left[B^{\mu^{-1}}_{r\mathcal{F}}\right],\\
&B^{\mu^{-1}}_{\lambda} = \left[\lambda_{\phi\mathcal{F}}\right]^t\left[M^{\mu^{-1}}_{\mathcal{F}\mathcal{F}}\right]\left[\lambda_{\phi\mathcal{F}}\right]\\
&f^{\mu^{-1}}_{B_r} = \left[\lambda_{\phi\mathcal{F}}\right]^t\left[B^{\mu^{-1}}_{r\mathcal{F}}\right]
\end{aligned} \qquad [4.307]$$

In this expression, $\left[C^{\mu^{-1}}_{\lambda\mathcal{A}_r}\right]$ denotes a vector of dimension $\mathcal{A}_r$. Similarly, $\left[F^{\mu^{-1}}_{I\text{-}B_r\mathcal{A}_r}\right]$ is a vector, of the same dimension $\mathcal{A}_r$, representing the contribution of the current in the inductor and that of the permanent magnet. On the other hand, the term $B^{\mu^{-1}}_{\lambda}$ is a scalar term corresponding to a reluctance. Finally, the term $f^{\mu^{-1}}_{B_r}$ is a scalar representing the magnetomotive force created between the boundaries Γ_{h1} and Γ_{h2} by the permanent magnet.

Gathering equations [4.298] and [4.306] and relying on the notations introduced in equation [4.307], the matrix equation to be solved is written as:

$$\begin{bmatrix} \left[S^{\mu-1}_{\mathcal{N}_r\mathcal{N}_r}\right] & \left[C^{\mu-1}_{\lambda\mathcal{A}_r}\right] \\ \left[C^{\mu-1}_{\lambda\mathcal{A}_r}\right]^t & B^{\mu-1}_{\lambda} \end{bmatrix} \begin{bmatrix} \left[A_{\mathcal{A}_r}\right] \\ \phi \end{bmatrix} = \begin{bmatrix} \left[F^{\mu^{-1}}_{I-H_c\mathcal{N}_r}\right] \\ f_m + f^{\mu^{-1}}_{B_r} \end{bmatrix} \qquad [4.308]$$

This system of equations allows us to solve a magnetostatics problem with the vector potential formulation when the source terms are a magnetomotive force imposed on the boundary with an inductor and a permanent magnet inside the domain. This is a linear system if the effects of the saturation of magnetic materials are considered negligible.

4.4.5.2.3. Consideration of the magnetic nonlinearity

As noted in section 1.2.2.4, equation [1.25], magnetic permeability could be considered as a nonlinear function of the field **H** or the magnetic flux density **B.** In this case, the elementary terms of matrices $\left[M_{\mathcal{FF}}^{\mu^{-1}}\right]$ (see equation [4.299]) depend on **B** and therefore on the unknowns of the problem, namely on the vector $\left[A_{\mathcal{A}_r}\right]$ of the circulations of the vector potential. Then, taking into account the behavior law of magnetic materials, we have to solve a system of nonlinear equations. As indicated in section 4.4.5.2.3, methods are available to solve this type of problem, for example, the substitution method or the Newton–Raphson method (Dhatt et al. 2012).

4.4.6. *Magnetodynamics*

For magnetodynamics, the focus is on the study of the example presented in Figure 3.23. The domain is denoted by Ω and contains a conductor Ω_c with two gates Γ_{e1} and Γ_{e2} on which an electromotive force "e" or the current density flux "I", two quantities varying with time, can be imposed. One part of the boundary of the conductor is in contact with a subdomain Ω_0 whose boundary with the external environment, denoted by Γ_b, is a wall for the magnetic flux density. The interface between Ω_c and Ω_0 (denoted by Γ_j) is a wall for the current density.

For this example, we will develop the finite element method in the case of electric (**A**-V) and magnetic (**T**-φ) formulations, introduced in section 3.6.1. To this end, we will rely on the weak forms developed in section 4.2.5.1.

In this part, we consider that the magnetic materials have linear characteristics (see equation [1.25]). In the nonlinear case, a substitution or Newton–Raphson iterative method can be introduced after time discretization (Dhatt et al. 2012). We build a mesh $\mathcal{M}$ on the domain Ω, denoted by Ω_d in what follows. This mesh is composed of $\mathcal{N}$ nodes, $\mathcal{A}$ edges, $\mathcal{F}$ facets and $\mathcal{V}$ volumes. We consider the mesh $\mathcal{M}_c$ of the conductor (domain Ω_{cd}), composed of $\mathcal{N}_c$ nodes, $\mathcal{A}_c$ edges, $\mathcal{F}_c$ facets and $\mathcal{V}_c$ volumes. The mesh $\mathcal{M}_c$ is a "sub-mesh" of the mesh $\mathcal{M}$. In what follows, we will develop the potential formulations for the case in which electric quantities are imposed. If magnetic quantities are imposed (see Figure 3.24), the system of equations results from developments similar to those that will be presented.

4.4.6.1. *Electric A-V formulation*

The first step in obtaining this formulation is to impose the electromotive force that appears as a natural source term (see section 3.6.1.1.1). The second step is to study the case in which current intensity I is imposed. This requires the introduction of an additional equation (see section 3.6.1.1.2). Before this, we have to discretize the potentials **A** and φ by taking into account the boundary conditions and the gauge conditions.

The vector potential (see equation [3.316]) belongs to the space $H_0(\mathbf{curl}, \Omega)$. As shown in section 4.4.5.2 and in Table 4.18, it is discretized in the space of edge elements, i.e. $W^1{}_0(\Omega_d)$. To impose the gauge condition, we build an edge tree (see sections 4.3.6.2 and 4.4.5.2). The construction of a tree starts by the edges belonging to the boundary Γ of the domain. For the boundary conditions, we impose a zero circulation on all the edges located on the boundary Γ. Under these conditions, if we denote by $\mathcal{A}_b$ the set of edges belonging to Γ, the number of unknown circulations $\mathcal{A}_r$ of the discretized vector potential $\mathbf{A}_d(x)$ is written based on equation [4.292] in the following form:

$$\mathcal{A}_r = \mathcal{A} - (\mathcal{N} - 1) - (\mathcal{A}_b - \mathcal{A}_{ba}) \qquad [4.309]$$

where $\mathcal{A}_{ba}$ represents the number of edges on the boundary Γ belonging to the edge tree. The edges are renumbered so that those belonging to the tree and to the boundary have an index number higher than $\mathcal{A}_r$. In this case, the expression of the vector potential $\mathbf{A}_d(x)$ is written in the form [4.293], which is recalled below.

$$\mathbf{A}_d(x) = \sum_{i=1}^{\mathcal{A}_r} \boldsymbol{\omega}_{a_i} A_{a_i} = \left[\boldsymbol{\omega}_{\mathcal{A}_r}\right]^t \left[A_{\mathcal{A}_r}\right] \qquad [4.310]$$

The scalar potential V belongs (see equation [3.319]) to the space $H_{\Gamma e1 \cup \Gamma e2}(\mathbf{grad}, \Omega_c)$. It should be recalled that it is calculated only in the conducting domain Ω_c. In the discrete domain, it belongs to the space of nodal elements (see section 4.4.4.1.1 and Table 4.18), i.e. $V_d(x) \in W^0{}_{\Gamma e1 \cup \Gamma e2}(\Omega_{cd})$. To impose the gauge condition and the boundary conditions, we proceed as in electrokinetics, setting to zero the node values "V_i" of the potential on the boundary $\Gamma_{e1} \cup \Gamma_{e2}$ (see section 4.2.4.1.1). Let us denote by $\mathcal{N}_{1c}$ and $\mathcal{N}_{2c}$ the number of nodes belonging to the boundaries Γ_{e1} and Γ_{e2}, respectively. Under these conditions, the number of unknown nodal values of the discrete scalar potential $V_d(x)$ is written as:

$$\mathcal{N}_{rc} = \mathcal{N}_c - (\mathcal{N}_{1c} + \mathcal{N}_{2c}) \qquad [4.311]$$

All the nodes $\mathcal{N}_c$ of the mesh of the conductor are then renumbered so that those located on the boundaries Γ_{e1} and Γ_{e2} have an index higher than $\mathcal{N}_{rc}$. The scalar potential $V_d(x)$ can be written in the form of equation [4.244] as follows:

$$V_d(x) = \sum_{i=1}^{\mathcal{N}_{rc}} \omega_{n_i} V_i = \left[\omega_{\mathcal{N}_{rc}}\right]^t \left[V_{\mathcal{N}_{rc}}\right] \qquad [4.312]$$

4.4.6.1.1. Imposed electromotive force

For the example in Figure 3.23, when the source term is the electromotive force, the weak form of the equations to be solved is given by equations [4.49] and [4.53] that are recalled as follows:

$$\iiint_{\Omega}(\mu^{-1}(\mathbf{curlA.curl\Psi})+\sigma(\frac{\partial \mathbf{A}}{\partial t}+\mathbf{grad}\mathrm{V}-\mathrm{e}\boldsymbol{\beta}_{\mathrm{e}}).\boldsymbol{\Psi})d\tau=0 \qquad [4.313]$$

$$\iiint_{\Omega_c}(\sigma\mathbf{grad}\,\psi\,\mathbf{grad}\,\mathrm{V}+\sigma(\frac{\partial \mathbf{A}}{\partial t}.\mathbf{grad}\,\psi-\mathrm{e}\boldsymbol{\beta}_{\mathrm{e}}).\mathbf{grad}\,\psi)d\tau=0 \qquad [4.314]$$

with $\boldsymbol{\Psi} \in H_0(\mathbf{curl}, \Omega)$ and $\psi \in H_{\Gamma e1 \cup \Gamma e2}(\mathbf{grad}, \Omega_c)$ (see Table 4.7).

In order to solve this system of equations, besides the potentials **A** and V, we have to discretize the support field $\boldsymbol{\beta}_e$. Similar to the approach for electrostatics and electrokinetics, this field can be built from the associated potential α_e. For the studied example, the properties of these two fields ($\boldsymbol{\beta}_e$ and α_e) are defined in equation [3.313]. The support field α_e belongs to the function space $H(\mathbf{grad}, \Omega_c)$ with two equipotential surfaces Γ_{e1} and Γ_{e2}. It is therefore discretized in the space of nodal elements. It belongs to $W^0(\Omega_{cd})$ and is denoted by $\alpha_{ed}(x)$. As for $\boldsymbol{\beta}_e$, it belongs to $H_{\Gamma e1 \cup \Gamma e2}(\mathbf{grad}, \Omega_c)$; it is discretized in the space of edge elements $W^1_{\Gamma e1 \cup \Gamma e2}(\Omega_{cd})$ and is denoted by $\boldsymbol{\beta}_{ed}(x)$. It should be recalled that the fields $\boldsymbol{\beta}_e$ and α_e are uniquely defined on the conducting domain Ω_{cd}. Indeed, in equations [4.303] and [4.304], the contribution of the support field $\boldsymbol{\beta}_e$, on the non-conducting domain $\Omega_d - \Omega_{cd}$, is zero because the conductivity $\sigma = 0$ (see equation [3.307]).

For the construction of $\boldsymbol{\beta}_{ed}(x)$, we use the same procedure as that used in section 4.4.4.1.2 and therefore we first determine $\alpha_{ed}(x)$. To do this, we set to 1 the values "α_{ei}" associated with the nodes located on the boundary Γ_{e1} and to zero the other nodes. Based on equation [4.180], its expression can be written as follows:

$$\alpha_{ed}(x) = \sum_{i=1}^{\mathcal{N}_c} \omega_{n_i}(x)\,\alpha_{e_i} = \left[\omega_{\mathcal{N}_c}\right]^t\left[\alpha_{e\mathcal{N}_c}\right] \qquad [4.315]$$

where $\mathcal{N}_c$ represents the number of nodes of mesh $\mathcal{M}_c$ of Ω_{cd}. Knowing the associated potential $\alpha_{ed}(x)$, we determine the support field $\boldsymbol{\beta}_{ed}(x)$ by applying the gradient operator. In the discrete domain, we calculate the vector of the circulations of the support field $\left[\beta_{e\mathcal{A}_c}\right]$ using the expression [4.245]. Based on vector $\left[\beta_{e\mathcal{A}_c}\right]$, the field $\boldsymbol{\beta}_{ed}(x)$ can be written using equation [4.182]:

$$\boldsymbol{\beta}_{ed}(x) = \sum_{i=1}^{\mathcal{A}_c} \boldsymbol{\omega}_{a_i} \beta_{e_i} = \left[\boldsymbol{\omega}_{\mathcal{A}_c}\right]^t \left[\beta_{e\mathcal{A}_c}\right] \qquad [4.316]$$

It should be recalled that $\mathcal{A}_c$ represents the number of edges of the mesh $\mathcal{M}_c$. In magnetodynamics, we have a system of two differential equations to solve, [4.313] and [4.314]. To build the discrete form of our problem, let us first consider equation [4.313]. In this expression, the vector and scalar potentials, as well as the support field $\boldsymbol{\beta}_e$, are replaced by their discretized form ($\mathbf{A}_d(x)$, $V_d(x)$ and $\boldsymbol{\beta}_{ed}(x)$) and the Ritz–Galerkin method is applied. Then, we obtain:

$$\begin{aligned}&\sum_{i=1}^{\mathcal{A}_r} \iiint_{\Omega_d} \mu^{-1}(\mathbf{curl}\boldsymbol{\omega}_{a_j}.\mathbf{curl}\boldsymbol{\omega}_{a_i} A_i)d\tau + \sum_{k=1}^{\mathcal{A}_r} \iiint_{\Omega_d} \sigma\boldsymbol{\omega}_{a_j}.\boldsymbol{\omega}_{a_k} \frac{dA_k}{dt} d\tau + \\ &\sum_{m=1}^{\mathcal{N}_{rc}} \iiint_{\Omega_d} \sigma\boldsymbol{\omega}_{a_j}.\mathbf{grad}\omega_{n_m} V_{n_m} d\tau - e\sum_{l=1}^{\mathcal{A}_c} \iiint_{\Omega_d} \sigma\boldsymbol{\omega}_{a_j}.\boldsymbol{\omega}_{a_l} \beta_{e_l} d\tau = 0 \\ &\text{for } 1 \le j \le \mathcal{A}_r\end{aligned} \qquad [4.317]$$

The terms related to $V_d(x)$ and $\boldsymbol{\beta}_{ed}(x)$ are only defined on Ω_{cd}, as shown by equations [4.312] and [4.316]. However, since the conductivity σ is zero on $\Omega_d - \Omega_{cd}$ (see equation [3.307]), the last two terms of equation [4.317] are perfectly defined and can be integrated on Ω_d. As this expression must be verified by the set of $\mathcal{A}_r$ weighting functions $\boldsymbol{\omega}_{aj}$, it can be written in the form of a matrix system:

$$\begin{aligned}&\iiint_{\Omega_d} \mu^{-1}(\mathbf{curl}\left[\boldsymbol{\omega}_{\mathcal{A}_r}\right]^t)^t(\mathbf{curl}\left[\boldsymbol{\omega}_{\mathcal{A}_r}\right]^t)\left[A_{\mathcal{A}_r}\right]d\tau + \\ &\iiint_{\Omega_d} \sigma\left[\boldsymbol{\omega}_{\mathcal{A}_r}\right]\left[\boldsymbol{\omega}_{\mathcal{A}_r}\right]^t \frac{d}{dt}\left[A_{\mathcal{A}_r}\right]d\tau + \iiint_{\Omega_d} \sigma\left[\boldsymbol{\omega}_{\mathcal{A}_r}\right](\mathbf{grad}\left[\omega_{\mathcal{N}_{rc}}\right]^t)\left[V_{\mathcal{N}_{rc}}\right]d\tau - \\ &e\iiint_{\Omega_d} \sigma\left[\boldsymbol{\omega}_{\mathcal{A}_r}\right]\left[\boldsymbol{\omega}_{\mathcal{A}_c}\right]^t\left[\beta_{e\mathcal{A}_c}\right]d\tau = 0\end{aligned} \qquad [4.318]$$

Using the properties of the curl [4.124] and gradient [4.110] discrete operators, we have:

$$\begin{aligned}&\iiint_{\Omega_d} \mu^{-1}\left[\mathbf{R}_{\mathcal{F}\mathcal{A}_r}\right]^t\left[\boldsymbol{\omega}_{\mathcal{F}}\right]\left[\boldsymbol{\omega}_{\mathcal{F}}\right]^t\left[\mathbf{R}_{\mathcal{F}\mathcal{A}_r}\right]\left[A_{\mathcal{A}_r}\right]d\tau + \\ &\iiint_{\Omega_d} \sigma\left[\boldsymbol{\omega}_{\mathcal{A}_r}\right]\left[\boldsymbol{\omega}_{\mathcal{A}_r}\right]^t \frac{d}{dt}\left[A_{\mathcal{A}_r}\right]d\tau + \iiint_{\Omega_d} \sigma\left[\boldsymbol{\omega}_{\mathcal{A}_r}\right]\left[\boldsymbol{\omega}_{\mathcal{A}}\right]^t\left[\mathbf{G}_{\mathcal{A}\mathcal{N}_{rc}}\right]\left[V_{\mathcal{N}_{rc}}\right]d\tau - \\ &e\iiint_{\Omega_d} \sigma\left[\boldsymbol{\omega}_{\mathcal{A}_r}\right]\left[\boldsymbol{\omega}_{\mathcal{A}_c}\right]^t\left[\beta_{e\mathcal{A}_c}\right]d\tau = 0\end{aligned} \qquad [4.319]$$

The incidence matrices $\left[\mathbf{R}_{\mathcal{F}\mathcal{A}_r}\right]$ and $\left[\mathbf{G}_{\mathcal{A}\mathcal{N}_{rc}}\right]$ as well as the vectors $\left[\mathrm{A}_{\mathcal{A}_r}\right]$, $\left[\mathrm{V}_{\mathcal{N}_{rc}}\right]$ and $\left[\beta_{e\mathcal{A}_c}\right]$ are vectors with constant terms, therefore independent of the position in space. The equation can then be rewritten as follows:

$$\begin{aligned}&\left[\mathbf{R}_{\mathcal{F}\mathcal{A}_r}\right]^t\left[\mathrm{M}_{\mathcal{F}\mathcal{F}}^{\mu-1}\right]\left[\mathbf{R}_{\mathcal{F}\mathcal{A}_r}\right]\left[\mathrm{A}_{\mathcal{A}_r}\right]+\left[\mathrm{M}_{\mathcal{A}_r\mathcal{A}_r}^{\sigma}\right]\frac{d}{dt}\left[\mathrm{A}_{\mathcal{A}_r}\right]\\&+\left[\mathrm{M}_{\mathcal{A}_r\mathcal{A}}^{\sigma}\right]\left[\mathbf{G}_{\mathcal{A}\mathcal{N}_{rc}}\right]\left[\mathrm{V}_{\mathcal{N}_{rc}}\right]+e\left[\mathrm{M}_{\mathcal{A}_r\mathcal{A}_c}^{\sigma}\right]\left[\beta_{e\mathcal{A}_c}\right]=0\end{aligned}\qquad [4.320]$$

The elementary term of the matrix $\left[\mathrm{M}_{\mathcal{F}\mathcal{F}}^{\mu^{-1}}\right]$ is written as:

$$\mathrm{M}_{f_i f_j}^{\mu^{-1}}=\iiint_{\Omega_d}\mu^{-1}\boldsymbol{\omega}_{f_i}.\boldsymbol{\omega}_{f_j}\,d\tau \qquad [4.321]$$

Concerning the matrices $\left[\mathrm{M}_{\mathcal{A}_r\mathcal{A}_r}^{\sigma}\right]$ and $\left[\mathrm{M}_{\mathcal{A}_r\mathcal{A}}^{\sigma}\right]$, their elementary terms have the following form:

$$\mathrm{M}_{a_i a_j}^{\sigma}=\iiint_{\Omega_d}\sigma\boldsymbol{\omega}_{a_i}.\boldsymbol{\omega}_{a_j}\,d\tau \qquad [4.322]$$

However, as defined in section 3.6 (see equation [3.307]), the conductivity σ is zero in $\Omega_d - \Omega_{cd}$. Consequently, the integrations corresponding to the 2nd, 3rd and 4th terms of equation [4.319] can be limited to domain Ω_{cd}. Under these conditions, the elementary terms of matrices $\left[\mathrm{M}_{\mathcal{A}_r\mathcal{A}_r}^{\sigma}\right]$ and $\left[\mathrm{M}_{\mathcal{A}_r\mathcal{A}}^{\sigma}\right]$ given by equation [4.322] are written as:

$$\mathrm{M}_{a_i a_j}^{\sigma}=\iiint_{\Omega_{cd}}\sigma\boldsymbol{\omega}_{a_i}.\boldsymbol{\omega}_{a_j}\,d\tau \qquad [4.323]$$

It can be noted that, if an edge i or j is located outside the domain Ω_{cd}, then the term $\mathrm{M}^{\sigma}_{aiaj}$ is zero, since the function $\boldsymbol{\omega}_{ai}$ or $\boldsymbol{\omega}_{aj}$ is zero on Ω_{cd}.

Let us now consider equation [4.314], which should also be solved. To do this, we can use a similar approach to the one used for the discretization of relation [4.313]. First, we introduce the discrete forms of the scalar potential, vector potential and support field $\boldsymbol{\beta}_e$, and then we apply the Ritz–Galerkin method. Then, we can write as:

$$\sum_{i=1}^{\mathcal{N}_{rc}} \iiint_{\Omega_{cd}} \sigma \mathbf{grad}\omega_{n_j} . \mathbf{grad}\omega_{n_i} V_{n_i} d\tau + \sum_{k=1}^{\mathcal{A}_r} \iiint_{\Omega_{cd}} \sigma \mathbf{grad}\omega_{n_j} . \boldsymbol{\omega}_{a_k} \frac{d}{dt} A_k \, d\tau - e\sum_{m=1}^{\mathcal{A}_c} \iiint_{\Omega_{cd}} \sigma \mathbf{grad}\omega_{n_j} . \boldsymbol{\omega}_{a_m} \beta_{em} d\tau = 0 \text{ with } 1 \le j \le \mathcal{N}_{rc} \quad [4.324]$$

Based on this expression, the approach used for equation [4.317] readily leads to the following relation:

$$\left[\mathbf{G}_{\mathcal{A}_c\mathcal{N}_{rc}}\right]^t \left[M^{\sigma}_{\mathcal{A}_c\mathcal{A}_c}\right] \left[\mathbf{G}_{\mathcal{A}_c\mathcal{N}_{rc}}\right] \left[V_{\mathcal{N}_{rc}}\right] + \left[\mathbf{G}_{\mathcal{A}_c\mathcal{N}_{rc}}\right]^t \left[M^{\sigma}_{\mathcal{A}_c\mathcal{A}_r}\right] \frac{d}{dt} \left[A_{\mathcal{A}_r}\right] - e\left[\mathbf{G}_{\mathcal{A}_c\mathcal{N}_{rc}}\right]^t \left[M^{\sigma}_{\mathcal{A}_c\mathcal{A}_c}\right] \left[\beta_{e\mathcal{A}_c}\right] = 0 \quad [4.325]$$

For this expression, the elementary terms of matrices $\left[M^{\sigma}_{\mathcal{A}_c\mathcal{A}_c}\right]$ and $\left[M^{\sigma}_{\mathcal{A}_c\mathcal{A}_r}\right]$ are given by expression [4.323].

Gathering the matrix forms [4.320] and [4.325], we obtain the matrix equation to be solved as follows:

$$\begin{bmatrix} \left[\mathbf{R}_{\mathcal{F}\mathcal{A}_r}\right]^t \left[M^{\mu-1}_{\mathcal{F}\mathcal{F}}\right] \left[\mathbf{R}_{\mathcal{F}\mathcal{A}_r}\right] & \left[M^{\sigma}_{\mathcal{A}_r\mathcal{A}}\right] \left[\mathbf{G}_{\mathcal{A}\mathcal{N}_{rc}}\right] \\ [0] & \left[\mathbf{G}_{\mathcal{A}_c\mathcal{N}_{rc}}\right]^t \left[M^{\sigma}_{\mathcal{A}_c\mathcal{A}_c}\right] \left[\mathbf{G}_{\mathcal{A}_c\mathcal{N}_{rc}}\right] \end{bmatrix} \begin{bmatrix} \left[A_{\mathcal{A}_r}\right] \\ \left[V_{\mathcal{N}_{rc}}\right] \end{bmatrix} + \begin{bmatrix} \left[M^{\sigma}_{\mathcal{A}_r\mathcal{A}_r}\right] & [0] \\ \left[\mathbf{G}_{\mathcal{A}_c\mathcal{N}_{rc}}\right]^t \left[M^{\sigma}_{\mathcal{A}_c\mathcal{A}_r}\right] & [0] \end{bmatrix} \frac{d}{dt} \begin{bmatrix} \left[A_{\mathcal{A}_r}\right] \\ \left[V_{\mathcal{N}_r}\right] \end{bmatrix} = e \begin{bmatrix} \left[M^{\sigma}_{\mathcal{A}_r\mathcal{A}_c}\right] \left[\beta_{e\mathcal{A}_c}\right] \\ \left[\mathbf{G}_{\mathcal{A}_c\mathcal{N}_{rc}}\right]^t \left[M^{\sigma}_{\mathcal{A}_c\mathcal{A}_c}\right] \left[\beta_{e\mathcal{A}_c}\right] \end{bmatrix} \quad [4.326]$$

In order to take into account the time derivative of the unknown vectors $\left[A_{\mathcal{A}_r}\right]$ and $\left[V_{\mathcal{N}_r}\right]$, two possibilities can be considered. The objective is to find the solution in the steady state when the source term is periodic. In this case, when the magnetic behavior laws are linear, the complex notation can be introduced (Alonso Rodriguez and Valli 2010) leading to a complex matrix equation. Conversely, if an arbitrary state is considered or if, in the presence of a sinusoidal source, we want to consider a transient state, then the differential operator d/dt should be discretized using a time discretization scheme like, for example, Euler or Crank–Nicholson methods.

NOTE.– Similar to the magnetostatics case, we can take into account a nonlinear magnetic behavior law. This leads to a matrix $\left[M^{\mu^{-1}}_{\mathcal{FF}}\right]$ that depends on the solution. The system of equations to be solved is then nonlinear, and fixed point iteration methods or the Newton–Raphson method (Dhatt et al. 2012) should then be implemented. On the other hand, precautions should be taken for the time discretization.

4.4.6.1.2. Imposed current intensity

When the current intensity is imposed through the boundaries Γ_{e1} and Γ_{e2}, the electromotive force becomes an unknown of the problem. As shown in section 3.6.1.1.2, an additional equation is then required. In what follows, we use equation [3.339], which is recalled below:

$$I = \iiint_{\Omega_c} \boldsymbol{\beta}_e.\sigma(-(\frac{\partial \mathbf{A}}{\partial t} + \mathbf{grad}\, V) + e\,\boldsymbol{\beta}_e)d\tau \qquad [4.327]$$

Then, we need to consider the discrete forms of the potentials **A** and V as well as those of the support field $\boldsymbol{\beta}_e$. Using expressions [4.310], [4.312] and [4.316], we obtain:

$$\begin{aligned} I = &-\iiint_{\Omega_{cd}} \sigma(\left[\boldsymbol{\omega}_{\mathcal{A}_c}\right]^t\left[\beta_{e\mathcal{A}_c}\right])^t\left[\boldsymbol{\omega}_{\mathcal{A}_r}\right]^t\frac{d}{dt}\left[A_{\mathcal{A}_r}\right]d\tau - \\ &\iiint_{\Omega_{cd}} \sigma(\left[\boldsymbol{\omega}_{\mathcal{A}_c}\right]^t\left[\beta_{e\mathcal{A}_c}\right])^t\mathbf{grad}\left[\omega_{\mathcal{N}_{rc}}\right]^t\left[V_{\mathcal{N}_{rc}}\right]d\tau + \\ &e\iiint_{\Omega_{cd}} \sigma(\left[\boldsymbol{\omega}_{\mathcal{A}_c}\right]^t\left[\beta_{e\mathcal{A}_c}\right])^t\left[\boldsymbol{\omega}_{\mathcal{A}_c}\right]^t\left[\beta_{e\mathcal{A}_c}\right]d\tau \end{aligned} \qquad [4.328]$$

After rearrangement, and using the expression of the elementary term [4.323], this equation is written as:

$$\begin{aligned} I = &-\left[\beta_{e\mathcal{A}_c}\right]^t\left[M^{\sigma}_{\mathcal{A}_c\mathcal{A}_r}\right]\frac{d}{dt}\left[A_{\mathcal{A}_r}\right] - \left[\beta_{e\mathcal{A}_c}\right]^t\left[M^{\sigma}_{\mathcal{A}_c\mathcal{A}_c}\right]\left[\mathbf{G}_{\mathcal{A}_c\mathcal{N}_{rc}}\right]\left[V_{\mathcal{N}_{rc}}\right] + \\ &e\left[\beta_{e\mathcal{A}_c}\right]^t\left[M^{\sigma}_{\mathcal{A}_c\mathcal{A}_c}\right]\left[\beta_{e\mathcal{A}_c}\right] \end{aligned} \qquad [4.329]$$

whose elementary terms of matrices $\left[M^{\sigma}_{\mathcal{A}_c\mathcal{A}_r}\right]$ and $\left[M^{\sigma}_{\mathcal{A}_c\mathcal{A}_c}\right]$ are given by equation [4.323].

In conclusion, when the current I through the boundaries Γ_{e1} and Γ_{e2} is the source term, we have to solve the matrix equation composed of expressions [4.326] and [4.329]. The unknowns are then the circulations of the magnetic vector potential on the edges of the mesh, the nodal values of the electric scalar potential in the conductor and the electromotive force imposed between the boundaries Γ_{e1} and Γ_{e2}.

4.4.6.2. *Magnetic formulation T-φ*

The studied example is still the one in Figure 3.23, having the electromotive force e(t) and the current I(t) as source terms. For this configuration, the magnetic formulation **T**-φ is developed in section 3.6.1.2. As shown in this section, when the current intensity I is imposed, the source term appears naturally in the developments. On the other hand, if the source term is the electromotive force, then an additional equation is required. The first to be studied is the case in which a current I is imposed through the boundaries Γ_{e1} and Γ_{e2}. Then, the electromotive force "e" will be considered a source term.

Before writing the equation with the finite element method, we will discretize the magnetic scalar potential φ and the electric vector potential **T**.

For the studied example, the scalar potential φ (see equation [3.354]) belongs to the function space H(**grad**, Ω). In the discrete domain, it belongs to the space of nodal elements (see Table 4.18), i.e. $\varphi_d(x) \in W^0(\Omega_d)$. To impose the gauge condition, we then set the potential to zero in one node of the mesh. In our case, the node with the highest index will be chosen. The number of unknown nodal values of the potential, denoted by $\mathcal{N}_r$, is then equal to $\mathcal{N} - 1$. The discrete form of the magnetic scalar potential $\varphi_d(x)$ can be written as follows:

$$\varphi_d(x) = \sum_{i=1}^{\mathcal{N}_r} \omega_{n_i} \varphi_i = \lfloor \omega_{\mathcal{N}_r} \rfloor^t \lfloor \varphi_{\mathcal{N}_r} \rfloor \qquad [4.330]$$

Concerning the vector potential **T** (see equation [3.347]), in the continuous domain, it belongs to the function space H(**curl**, Ω). It is therefore decomposed in the space of edge elements. Moreover, this potential is zero on $\Omega - \Omega_c$. Finally, the focus is on the restriction of **T** in Ω_c (see equation [3.348]). In conclusion, the vector potential $\mathbf{T}_d(x) \in W^1{}_{\Gamma j}(\Omega_{cd})$. Boundary conditions should therefore be imposed on the boundary Γ_j, which represents the interface between the conductor and the domain Ω_0.

To gauge the discretized form $\mathbf{T}_d(x)$ and impose the boundary condition on Γ_j, we proceed similarly to section 4.4.4.2.1. We build a tree by considering first the edges located on the boundary Γ_j, and then those located inside the domain Ω_{cd}. Then, we denote by $\mathcal{A}_j$ the edges of the boundary Γ_j and by $\mathcal{A}_{ja}$ those of Γ_j belonging to the edge tree. Finally, the circulation of $\mathbf{T}_d(x)$ is set equal to zero on the edges belonging to the tree and also on the edges of boundary Γ_j that are not included in the tree. Under these conditions, the number of unknown edges $\mathcal{A}_{rc}$ of the electric vector potential has the following expression:

$$\mathcal{A}_{rc} = \mathcal{A}_c - (\mathcal{N}_c - 1) - (\mathcal{A}_j - \mathcal{A}_{ja}) \qquad [4.331]$$

The arrangement of the edges of the mesh is such that those belonging to the edge tree or to Γ_j have an index that is higher than $\mathcal{A}_{rc}$. Under these conditions, the expression of the discretized electric vector potential $\mathbf{T}_d(x)$ is:

$$\mathbf{T}_d(x) = \sum_{i=1}^{\mathcal{A}_{rc}} \boldsymbol{\omega}_{a_i} T_i = \left[\boldsymbol{\omega}_{\mathcal{A}_{rc}}\right]^t \left[T_{\mathcal{A}_{rc}}\right] \qquad [4.332]$$

NOTE.– It can be noted that the functions $\boldsymbol{\omega}_{ai}$ appearing in the expression [4.332] are all associated with the edges located inside Ω_{cd} and are therefore all zero on $\Omega_d - \Omega_{cd}$. Since $\mathbf{T}_d(x)$ can be extended to $\Omega_d - \Omega_{cd}$ keeping it equal to zero, the expression is still valid on $\Omega_d - \Omega_{cd}$ [4.332].

4.4.6.2.1. Imposed current intensity

For this formulation, when the source term is the current intensity I, the weak form of the equations to be solved is given by relations [4.59] and [4.62], which are recalled as follows:

$$\iiint_{\Omega_c} (\sigma^{-1}(\mathbf{curl T} + \mathrm{I}\,\mathbf{curl}\boldsymbol{\chi}_I).\mathbf{curl}\boldsymbol{\Psi} \\ + \frac{\partial}{\partial t}\mu(\mathbf{T} + \mathrm{I}\boldsymbol{\chi}_I - \mathbf{grad}\varphi).\boldsymbol{\Psi})d\tau = 0 \qquad [4.333]$$

$$\iiint_{\Omega} \mu(\mathbf{T} + \mathrm{I}\boldsymbol{\chi}_I - \mathbf{grad}\varphi).\mathbf{grad}\psi d\tau = 0 \qquad [4.334]$$

with $\boldsymbol{\Psi} \in H_{\Gamma j}(\mathbf{curl}, \Omega_c)$ and $\psi \in H(\mathbf{grad}, \Omega)$. As for the associated potential $\boldsymbol{\chi}_I$, as shown by equation [3.346], it belongs to $H(\mathbf{curl}, \Omega)$ (see Table 4.8).

Applying the finite element method requires the prior discretization of the potential $\boldsymbol{\chi}_I$, besides the potentials $\mathbf{T}$ and φ.

It should be recalled that the associated vector potential $\boldsymbol{\chi}_I$ is defined, as shown by equations [3.345] and [3.346], based on the support field $\boldsymbol{\lambda}_I \in H(div0, \Omega)$ which makes it possible to impose a current in the domain Ω_c. Referring to Table 4.18, the term $\boldsymbol{\lambda}_I$ is decomposed in the space of facet elements $W^2(\Omega_d)$. Moreover (see relation [3.309]), it must verify the boundary conditions on the boundaries Γ_{e1}, Γ_{e2} and Γ_j. To calculate this source term, namely the flux through the facets λ_{Ik} (with $1 \leq k \leq \mathcal{F}$), we apply only on the domain Ω_{cd} the method introduced in section 4.3.6.3 (see case II). The domain of definition is then extended to Ω_d by imposing a zero flux λ_{Ik} on the facets belonging to Ω_d-Ω_{cd}. It is then possible to write:

$$\boldsymbol{\lambda}_{Id}(x) = \sum_{k=1}^{\mathcal{F}} \boldsymbol{\omega}_{f_k}(x)\lambda_{I_k} = [\boldsymbol{\omega}_{\mathcal{F}}]^t[\lambda_{I_{\mathcal{F}}}] \qquad [4.335]$$

The source term $\boldsymbol{\chi}_{Id}(x)$ can then be calculated on Ω_d with the method proposed in section 4.3.6.2. Under these conditions, $\boldsymbol{\chi}_{Id}(x)$ is written as:

$$\boldsymbol{\chi}_{Id}(x) = \sum_{i=1}^{\mathcal{A}} \boldsymbol{\omega}_{a_i}(x)\chi_{Ia_i} = [\boldsymbol{\omega}_{\mathcal{A}}]^t[\chi_{I_{\mathcal{A}}}] \qquad [4.336]$$

To build the matrix equation to be solved, we first consider equation [4.333]. The potentials $\mathbf{T}$,φ and the associated potential $\boldsymbol{\chi}_I$ are replaced by their discrete form.

Applying the Ritz–Galerkin method, we obtain:

$$\begin{aligned}
&\sum_{i=1}^{\mathcal{A}_{rc}} \iiint_{\Omega_{cd}} \sigma^{-1}(\mathbf{curl}\boldsymbol{\omega}_{a_j}.\mathbf{curl}\boldsymbol{\omega}_{a_i})T_i d\tau + \\
&I\sum_{k=1}^{\mathcal{A}} \iiint_{\Omega_{dc}} \sigma^{-1}(\mathbf{curl}\boldsymbol{\omega}_{a_j}.\mathbf{curl}\boldsymbol{\omega}_{a_k})\chi_{I_k} d\tau + \\
&\sum_{l=1}^{\mathcal{A}_{rc}} \iiint_{\Omega_{cd}} \mu\boldsymbol{\omega}_{a_j}.\boldsymbol{\omega}_{a_l}\frac{dT_l}{dt}d\tau + \sum_{m=1}^{\mathcal{A}} \iiint_{\Omega_{dc}} \mu\boldsymbol{\omega}_{a_j}.\boldsymbol{\omega}_{a_m}\frac{dI\chi_{I_m}}{dt}d\tau - \\
&\sum_{n=1}^{\mathcal{N}_r} \iiint_{\Omega_{cd}} \mu\boldsymbol{\omega}_{a_j}.\mathbf{grad}\omega_{n_n}\frac{d\varphi_{n_n}}{dt}d\tau = 0 \quad \text{for} \quad 1 \leq j \leq \mathcal{A}_{rc}
\end{aligned} \qquad [4.337]$$

Expression [4.337] must be verified for the $\mathcal{A}_{rc}$ weighting functions $\boldsymbol{\omega}_{aj}$; therefore, we can write the following system of equations:

$$\begin{aligned}
&\iiint_{\Omega_{cd}} \sigma^{-1}(\mathbf{curl}\left[\boldsymbol{\omega}_{\mathcal{A}_{rc}}\right]^{t})^{t}(\mathbf{curl}\left[\boldsymbol{\omega}_{\mathcal{A}_{rc}}\right]^{t})\left[T_{\mathcal{A}_{rc}}\right])d\tau+\\
&I\iiint_{\Omega_{cd}} \sigma^{-1}(\mathbf{curl}\left[\boldsymbol{\omega}_{\mathcal{A}_{rc}}\right]^{t})^{t}(\mathbf{curl}\left[\boldsymbol{\omega}_{\mathcal{A}}\right]^{t})\left[\chi_{I\mathcal{A}}\right]d\tau+\\
&\iiint_{\Omega_{dc}} \mu\left[\boldsymbol{\omega}_{\mathcal{A}_{rc}}\right]\left[\boldsymbol{\omega}_{\mathcal{A}_{rc}}\right]^{t})\frac{d}{dt}\left[T_{\mathcal{A}_{rc}}\right]d\tau+\\
&\iiint_{\Omega_{cd}} \mu\left[\boldsymbol{\omega}_{\mathcal{A}_{rc}}\right]\left[\boldsymbol{\omega}_{\mathcal{A}}\right]^{t}\left[\chi_{I\mathcal{A}}\right]\frac{dI}{dt}d\tau-\\
&\iiint_{\Omega_{cd}} \mu\left[\boldsymbol{\omega}_{\mathcal{A}_{rc}}\right](\mathbf{grad}\left[\omega_{\mathcal{N}_{r}}\right]^{t})\frac{d}{dt}\left[\varphi_{\mathcal{N}_{r}}\right]d\tau=0
\end{aligned} \qquad [4.338]$$

Let us now use the properties of discrete operators: curl (see equation [4.124]) and gradient (see equation [4.110]).

Then, we can write:

$$\begin{aligned}
&\iiint_{\Omega_{cd}} \sigma^{-1}\left[\mathbf{R}_{\mathcal{F}\mathcal{A}_{rc}}\right]^{t}\left[\boldsymbol{\omega}_{\mathcal{F}}\right]\left[\boldsymbol{\omega}_{\mathcal{F}}\right]^{t}\left[\mathbf{R}_{\mathcal{F}\mathcal{A}_{rc}}\right]\left[T_{\mathcal{A}_{rc}}\right]d\tau+\\
&I\iiint_{\Omega_{cd}} \sigma^{-1}\left[\mathbf{R}_{\mathcal{F}\mathcal{A}_{rc}}\right]^{t}\left[\boldsymbol{\omega}_{\mathcal{F}}\right]\left[\boldsymbol{\omega}_{\mathcal{F}}\right]^{t}\left[\mathbf{R}_{\mathcal{F}\mathcal{A}}\right]\left[\chi_{I\mathcal{A}}\right]d\tau+\\
&\iiint_{\Omega_{dc}} \mu\left[\boldsymbol{\omega}_{\mathcal{A}_{rc}}\right]\left[\boldsymbol{\omega}_{\mathcal{A}_{rc}}\right]^{t})\frac{d}{dt}\left[T_{\mathcal{A}_{rc}}\right]d\tau\\
&+\iiint_{\Omega_{cd}} \mu\left[\boldsymbol{\omega}_{\mathcal{A}_{rc}}\right]\left[\boldsymbol{\omega}_{\mathcal{A}}\right]^{t}\left[\chi_{I\mathcal{A}}\right]\frac{dI}{dt}d\tau-\\
&\iiint_{\Omega_{cd}} \mu\left[\boldsymbol{\omega}_{\mathcal{A}_{rc}}\right]\left[\boldsymbol{\omega}_{\mathcal{A}_{c}}\right]^{t}\left[\mathbf{G}_{\mathcal{A}\mathcal{N}_{r}}\right]\frac{d}{dt}\left[\varphi_{\mathcal{N}_{r}}\right]d\tau=0
\end{aligned} \qquad [4.339]$$

NOTE.– In the nonlinear case, which is not applicable for this example, magnetic permeability depends on the space (see equation [1.25]), as well as on the field magnitude. Therefore, the expression cannot be decomposed in the form: $\partial\mathbf{B}/\partial t = \mu\partial\mathbf{H}/\partial t = \mu\partial(\mathbf{T} + I\,\boldsymbol{\chi} - \mathbf{grad}\varphi)/\partial t$ since μ is no longer a constant, but it depends explicitly on the field $\mathbf{H}$ and so implicitly on time.

As the incidence matrices $\left[\mathbf{R}_{\mathcal{F}\mathcal{A}_{rc}}\right]$ and $\left[\mathbf{G}_{\mathcal{A}\mathcal{N}_{r}}\right]$, the vectors $\left[T_{\mathcal{A}_{rc}}\right]$, $\left[\varphi_{\mathcal{N}_{r}}\right]$ and $\left[\chi_{I\mathcal{A}}\right]$ are independent of space, this equation can be written in the following form:

$$\left[\mathbf{R}_{\mathcal{FA}_{rc}}\right]^t\left[M^{\sigma-1}_{\mathcal{FF}}\right]\left[\mathbf{R}_{\mathcal{FA}_{rc}}\right]\left[T_{\mathcal{A}_{rc}}\right]+I\left[\mathbf{R}_{\mathcal{FA}_{rc}}\right]^t\left[M^{\sigma-1}_{\mathcal{FF}}\right]\left[\mathbf{R}_{\mathcal{FA}}\right]\left[\chi_{I\mathcal{A}}\right]+ \\ \left[M^{\mu}_{\mathcal{A}_{rc}\mathcal{A}_{rc}}\right]\frac{d}{dt}\left[T_{\mathcal{A}_{rc}}\right]+\left[M^{\mu}_{\mathcal{A}_{rc}\mathcal{A}}\right]\left[\chi_{I\mathcal{A}}\right]\frac{dI}{dt}- \\ \left[M^{\mu}_{\mathcal{A}_{rc}\mathcal{A}}\right]\left[\mathbf{G}_{\mathcal{AN}_r}\right]\frac{d}{dt}\left[\varphi_{\mathcal{N}_r}\right]=0 \qquad [4.340]$$

In equation [4.340], the elementary term of the matrix $\left[M^{\sigma^{-1}}_{\mathcal{FF}}\right]$ is written as:

$$M^{\sigma^{-1}}_{f_i f_j}=\iiint_{\Omega_{cd}}\sigma^{-1}\boldsymbol{\omega}_{f_i}.\boldsymbol{\omega}_{f_j}d\tau \qquad [4.341]$$

For the matrices $\left[M^{\sigma^{-1}}_{\mathcal{FF}}\right]$ and $\left[M^{\mu}_{\mathcal{A}_{rc}\mathcal{A}}\right]$, the elementary terms have the following form:

$$M^{\mu}_{a_i a_j}=\iiint_{\Omega_{cd}}\mu\boldsymbol{\omega}_{a_i}.\boldsymbol{\omega}_{a_j}d\tau \qquad [4.342]$$

Let us now consider equation [4.333], which must also be discretized. For this, we introduce the discrete forms of potentials **T** and φ and of the associated potential $\boldsymbol{\chi}_I$. The Ritz–Galerkin method can be used to write:

$$\sum_{i=1}^{\mathcal{A}_{rc}}\iiint_{\Omega_d}\mu(\mathbf{grad}\omega_{n_j}.\boldsymbol{\omega}_{a_i}T_i)d\tau+ \\ I\sum_{k=1}^{\mathcal{A}}\iiint_{\Omega_d}\mu(\mathbf{grad}\omega_{n_j}.\boldsymbol{\omega}_{a_k}\chi_{I_k})d\tau- \\ \sum_{l=1}^{\mathcal{N}_r}\iiint_{\Omega_d}\mu(\mathbf{grad}\omega_{n_j}.\mathbf{grad}\omega_{n_l})\varphi_l d\tau=0 \text{ avec } 1\leq j\leq\mathcal{N}_r \qquad [4.343]$$

It should be noted that the first term of this expression concerns the electric vector potential, defined on Ω_{cd} and the $\mathcal{N}_r$ test functions defined on the entire domain.

The same steps can be taken for the development of equation [4.337]. Since equation [4.343] must be verified for the $\mathcal{N}_r$ test functions, the first step is to obtain a system of $\mathcal{N}_r$ equations. Then, we introduce the properties of discrete vector operators. Finally, given that the incidence matrices and the vectors of discrete quantities are independent of space, we obtain after development the following system of equations:

$$\begin{aligned}&\left[\mathbf{G}_{\mathcal{AN}_r}\right]^t\left[M^{\mu}_{\mathcal{AA}_{rc}}\right]\left[T_{\mathcal{A}_{rc}}\right]+I\left[\mathbf{G}_{\mathcal{AN}_r}\right]^t\left[M^{\mu}_{\mathcal{AA}}\right]\left[\chi_{I\mathcal{A}}\right]\\&-\left[\mathbf{G}_{\mathcal{AN}_r}\right]^t\left[M^{\mu}_{\mathcal{AA}}\right]\left[\mathbf{G}_{\mathcal{AN}_r}\right]\left[\varphi_{\mathcal{N}_r}\right]=0\end{aligned} \quad [4.344]$$

In this expression, the elementary terms of the matrices $\left[M^{\mu}_{\mathcal{AA}_{rc}}\right]$ and $\left[M^{\mu}_{\mathcal{AA}}\right]$ are given by equation [4.342], but by integration over the entire domain Ω_d. Gathering the equations [4.340] and [4.344] then leads to the matrix equation below:

$$\begin{aligned}&\begin{bmatrix}\left[\mathbf{R}_{\mathcal{FA}_{rc}}\right]^t\left[M^{\sigma-1}_{\mathcal{FF}}\right]\left[\mathbf{R}_{\mathcal{FA}_{rc}}\right] & [0]\\ \left[\mathbf{G}_{\mathcal{AN}_r}\right]^t\left[M^{\mu}_{\mathcal{AA}_{rc}}\right] & -\left[\mathbf{G}_{\mathcal{AN}_r}\right]^t\left[M^{\mu}_{\mathcal{AA}}\right]\left[\mathbf{G}_{\mathcal{AN}_r}\right]\end{bmatrix}\begin{bmatrix}\left[T_{\mathcal{A}_{rc}}\right]\\ \left[\varphi_{\mathcal{N}_r}\right]\end{bmatrix}\\&+\begin{bmatrix}\left[M^{\mu}_{\mathcal{A}_{rc}\mathcal{A}_{rc}}\right] & -\left[M^{\mu}_{\mathcal{A}_{rc}\mathcal{A}}\right]\left[\mathbf{G}_{\mathcal{AN}_r}\right]\\ [0] & [0]\end{bmatrix}\frac{d}{dt}\begin{bmatrix}\left[T_{\mathcal{A}_{rc}}\right]\\ \left[\varphi_{\mathcal{N}_r}\right]\end{bmatrix}\\&=-I\begin{bmatrix}\left[M^{\sigma-1}_{\mathcal{A}_{rc}\mathcal{A}}\right]\left[\chi_{I\mathcal{A}}\right]\\ \left[\mathbf{G}_{\mathcal{AN}_r}\right]^t\left[M^{\mu}_{\mathcal{AA}}\right]\left[\chi_{I\mathcal{A}}\right]\end{bmatrix}-\frac{dI}{dt}\begin{bmatrix}\left[M^{\mu}_{\mathcal{A}_{rc}\mathcal{A}}\right]\left[\chi_{I\mathcal{A}}\right]\\ [0]\end{bmatrix}\end{aligned} \quad [4.345]$$

Here, the unknowns are the circulations of the electric vector potential, the nodal values of the magnetic scalar potential and also their time derivatives.

4.4.6.2.2. Imposed electromotive force

If an electromotive force is imposed instead of the current intensity, the matrix system [4.345] is unchanged, but we must add equation [3.373], which is recalled as follows:

$$\begin{aligned}e=&\iiint_{\Omega}\sigma^{-1}(\mathbf{curlT}+I\mathbf{curl}\chi_I).\lambda_I)d\tau\\&+\iiint_{\Omega}\frac{\partial}{\partial t}\mu(\mathbf{T}-\mathbf{grad}\varphi+I\chi_I).\chi_I d\tau\end{aligned} \quad [4.346]$$

In order to associate this equation with the matrix equation [4.345], we must introduce the various discrete forms of the terms, i.e. $\mathbf{T}_d(x)$, $\varphi_d(x)$, $\boldsymbol{\lambda}_{Id}(x)$ and $\boldsymbol{\chi}_{Id}(x)$). For this purpose, we use the same process as that used in section 4.4.6.1.2. We then obtain, if the magnetic permeability is linear (see note after equation [4.339]), the following system of equations:

$$
\begin{aligned}
&[\lambda_{\mathrm{I}\mathcal{F}}]^{t}\left[\mathrm{M}_{\mathcal{FF}}^{\sigma-1}\right]\left[\mathbf{R}_{\mathcal{FA}_{rc}}\right]\left[\mathrm{T}_{\mathcal{A}_{rc}}\right]+\mathrm{I}[\lambda_{\mathrm{I}\mathcal{F}}]^{t}\left[\mathrm{M}_{\mathcal{FF}}^{\sigma-1}\right]\left[\mathbf{R}_{\mathcal{FA}}\right][\chi_{\mathrm{I}\mathcal{A}}]\\
&+[\chi_{\mathrm{I}\mathcal{A}}]^{t}\left[\mathrm{M}_{\mathcal{AA}_{rc}}^{\mu}\right]\frac{d}{dt}\left[\mathrm{T}_{\mathcal{A}_{rc}}\right]-[\chi_{\mathrm{I}\mathcal{A}}]^{t}\left[\mathrm{M}_{\mathcal{AA}}^{\mu}\right]\left[\mathbf{G}_{\mathcal{AN}_{r}}\right]\frac{d}{dt}\left[\varphi_{\mathcal{N}_{r}}\right]\\
&+[\chi_{\mathrm{I}\mathcal{A}}]^{t}\left[\mathrm{M}_{\mathcal{AA}}^{\mu}\right][\chi_{\mathrm{I}\mathcal{A}}]\frac{d\mathrm{I}}{dt}=e
\end{aligned}
\qquad [4.347]
$$

In this expression, the elementary term of the matrix $\left[\mathrm{M}_{\mathcal{FF}}^{\sigma^{-1}}\right]$ is given by relation [4.341], with integration on Ω_{cd} (see the domain of definition of σ, equation [3.307]). For the matrices $\left[\mathrm{M}_{\mathcal{AA}_{rc}}^{\mu}\right]$ and $\left[\mathrm{M}_{\mathcal{AA}}^{\mu}\right]$, we refer to equation [4.342], but by an integration over the domain Ω_d.

Under these conditions, when the electromotive force is imposed between the gates Γ_{e1} and Γ_{e2}, the equations to be solved are given by the systems of equations [4.345] and [4.347]. The unknowns are then the circulations of **T**, the nodal values of φ and the current intensity I.

References

Albanese, A. and Rubinacci, G. (1990). Magnetostatic field computations in terms of two components vector potentials. *Int. J. Numer. Meth. Eng.*, 29, 515–532.

Alonso Rodriguez, A. and Valli, A. (2010). *Eddy Current Approximation of Maxwell Equations*. Springer, Berlin.

Bastos, J.P.A. and Sadowski, N. (2014). *Magnetic Materials and 3D Finite Element Modeling*. CRC Press, Taylor & Francis, New York.

Benabou, A. (2002). Contribution à la caractérisation et à la modélisation de matériaux magnétiques en vue d'une implantation dans un code de calcul de champ. PhD Thesis, Université de Lille, Lille.

Bossavit, A. (1988). Magnetostatic problems in multiply connected regions: Some properties of the curl operator. *IEE Proceedings*, 135(3), 179–187.

Bossavit, A. (1991). *Électromagnétisme en vue de la modélisation*. Springer, Berlin.

Bossavit, A. (1997). *Computational Electromagnetism: Variational Formulations, Complementary, Edge Elements*. Academic Press, Elsevier, Oxford.

Bouillault, F. and Ren, Z. (2008). Magnetodynamic formulations. In *The Finite Element Method for Electromagnetic Modeling*, Meunier, G. (ed.). ISTE Ltd, London, and John Wiley & Sons, New York.

Bozorth, R.M. (1993). *Ferromagnetism*. Wiley-IEEE Press, Piscataway.

Cardoso, J.R. (2016). *Electromagnetics through the Finite Element Method: A Simplified Approach Using Maxwell's Equations*. CRC Press, Taylor & Francis, New York.

Dular, P. and Piriou, F. (2008). Static formulations: Electrostatic, electrokinetic, magnetostatics. In *The Finite Element Method for Electromagnetic Modeling*, Meunier, G. (ed.). ISTE Ltd, London, and John Wiley & Sons, New York.

Geuzaine, C. (2001). High order hybrid finite element schemes for Maxwell's equations taking thin structures and global quantities into account. PhD Thesis, Université de Liège, Liège.

Haymer, K. and Belmans, R. (1999). *Numerical Modelling and Design of Electrical Machines and Devices*. WIT Press, Southampton.

Henneron, T., Clénet, S., Piriou, F. (2004). Comparison of 3D magnetodynamic formulations in terms of potential with imposed electric global quantities. *COMPEL*, 23(4), 885–895.

Ida, N. (2020). *Engineering Electromagnetics*. Springer, Berlin.

Ida, N. and Bastos, J.P.A. (1997). *Electromagnetics and Calculation of Fields*. Springer, Berlin.

Le Menach, Y., Clenet, S., Piriou, F. (1998). Determination and utilization of the source field in 3D magnetostatic problems. *IEEE Trans. Mag.*, 34, 2509–2512.

Ren, Z. (1996). Influence of the R.H.S. on the convergence behaviour of the curl-curl equation. *IEEE Trans. Mag.*, 32, 655–658.

Russenschuck, S. (2010). *Field Computation for Accelerator Magnets: Analytical and Numerical Methods for Electromagnetic Design and Optimization*. John Wiley & Sons, New York.

Stratton, J.A. (1941). *Electromagnetic Theory*. McGraw-Hill Book Company, Inc., New York and London.

Index

G, I

K, L

M, N, O

P, R

S, T, W

Other titles from

ISTE

in

Numerical Methods in Engineering

2023

BOUCLIER Robin, PASSIEUX Jean-Charles
IGA: Non-Invasive Coupling with FEM and Regularization of Digital Image Correlation Problems
(Isogeometric Analysis Tools for Optimization Applications in Structural Mechanics Set – Volume 2)

CIARLET Patrick, LUNÉVILLE Eric
The Finite Element Method: From Theory to Practice

2022

BOUCLIER Robin, HIRSCHLER Thibault
IGA: Non-conforming Coupling and Shape Optimization of Complex Multipatch Structures
(Isogeometric Analysis Tools for Optimization Applications in Structural Mechanics Set – Volume 1)

DERVIEUX Alain, ALAUZET Frédéric, LOSEILLE Adrien, KOOBUS Bruno
Mesh Adaptation for Computational Fluid Dynamics 1: Continuous Riemannian Metrics and Feature-based Adaptation
Mesh Adaptation for Computational Fluid Dynamics 2: Unsteady and Goal-oriented Adaptation

GRANGE Stéphane, SALCIARINI Diana
Deterministic Numerical Modeling of Soil–Structure Interaction

2021

GENTIL Christian, GOUATY Gilles, SOKOLOV Dmitry
Geometric Modeling of Fractal Forms for CAD
(Geometric Modeling and Applications Set – Volume 5)

2020

GEORGE Paul Louis, ALAUZET Frédéric, LOSEILLE Adrien, MARÉCHAL Loïc
Meshing, Geometric Modeling and Numerical Simulation 3: Storage, Visualization and In Memory Strategies
(Geometric Modeling and Applications Set – Volume 4)

SIGRIST Jean-François
Numerical Simulation, An Art of Prediction 2: Examples

2019

DA Daicong
Topology Optimization Design of Heterogeneous Materials and Structures

GEORGE Paul Louis, BOROUCHAKI Houman, ALAUZET Frédéric, LAUG Patrick, LOSEILLE Adrien, MARÉCHAL Loïc
Meshing, Geometric Modeling and Numerical Simulation 2: Metrics, Meshes and Mesh Adaptation
(Geometric Modeling and Applications Set – Volume 2)

MARI Jean-Luc, HÉTROY-WHEELER Franck, SUBSOL Gérard
Geometric and Topological Mesh Feature Extraction for 3D Shape Analysis
(Geometric Modeling and Applications Set – Volume 3)

SIGRIST Jean-François
Numerical Simulation, An Art of Prediction 1: Theory

2017

BOROUCHAKI Houman, GEORGE Paul Louis
Meshing, Geometric Modeling and Numerical Simulation 1: Form Functions, Triangulations and Geometric Modeling
(Geometric Modeling and Applications Set – Volume 1)

2016

KERN Michel
Numerical Methods for Inverse Problems

ZHANG Weihong, WAN Min
Milling Simulation: Metal Milling Mechanics, Dynamics and Clamping Principles

2015

ANDRÉ Damien, CHARLES Jean-Luc, IORDANOFF Ivan
3D Discrete Element Workbench for Highly Dynamic Thermo-mechanical Analysis
(Discrete Element Model and Simulation of Continuous Materials Behavior Set – Volume 3)

JEBAHI Mohamed, ANDRÉ Damien, TERREROS Inigo, IORDANOFF Ivan
Discrete Element Method to Model 3D Continuous Materials
(Discrete Element Model and Simulation of Continuous Materials Behavior Set – Volume 1)

JEBAHI Mohamed, DAU Frédéric, CHARLES Jean-Luc, IORDANOFF Ivan
Discrete-continuum Coupling Method to Simulate Highly Dynamic Multi-scale Problems: Simulation of Laser-induced Damage in Silica Glass
(Discrete Element Model and Simulation of Continuous Materials Behavior Set – Volume 2)

SOUZA DE CURSI Eduardo
Variational Methods for Engineers with Matlab®

2014

BECKERS Benoit, BECKERS Pierre
Reconciliation of Geometry and Perception in Radiation Physics

BERGHEAU Jean-Michel
Thermomechanical Industrial Processes: Modeling and Numerical Simulation

BONNEAU Dominique, FATU Aurelian, SOUCHET Dominique
Hydrodynamic Bearings – Volume 1
Mixed Lubrication in Hydrodynamic Bearings – Volume 2
Thermo-hydrodynamic Lubrication in Hydrodynamic Bearings – Volume 3
Internal Combustion Engine Bearings Lubrication in Hydrodynamic Bearings – Volume 4

DESCAMPS Benoît
Computational Design of Lightweight Structures: Form Finding and Optimization

2013

YASTREBOV Vladislav A.
Numerical Methods in Contact Mechanics

2012

DHATT Gouri, LEFRANÇOIS Emmanuel, TOUZOT Gilbert
Finite Element Method

SAGUET Pierre
Numerical Analysis in Electromagnetics

SAANOUNI Khemais
Damage Mechanics in Metal Forming: Advanced Modeling and Numerical Simulation

2011

CHINESTA Francisco, CESCOTTO Serge, CUETO Elias, LORONG Philippe
Natural Element Method for the Simulation of Structures and Processes

DAVIM Paulo J.
Finite Element Method in Manufacturing Processes

POMMIER Sylvie, GRAVOUIL Anthony, MOËS Nicolas, COMBESCURE Alain
Extended Finite Element Method for Crack Propagation

2010

SOUZA DE CURSI Eduardo, SAMPAIO Rubens
Modeling and Convexity

2008

BERGHEAU Jean-Michel, FORTUNIER Roland
Finite Element Simulation of Heat Transfer

EYMARD Robert
Finite Volumes for Complex Applications V: Problems and Perspectives

FREY Pascal, GEORGE Paul Louis
Mesh Generation: Application to finite elements – 2nd edition

GAY Daniel, GAMBELIN Jacques
Modeling and Dimensioning of Structures

MEUNIER Gérard
The Finite Element Method for Electromagnetic Modeling

2005

BENKHALDOUN Fayssal, OUAZAR Driss, RAGHAY Said
Finite Volumes for Complex Applications IV: Problems and Perspectives

Printed and bound by CPI Group (UK) Ltd, Croydon, CR0 4YY

24/03/2024

14475064-0002